DOE/EIA-0484(2000)

International Energy Outlook

2000

March 2000

Energy Information Administration
Office of Integrated Analysis and Forecasting
U.S. Department of Energy
Washington, DC 20585

This publication is on the WEB at:
www.eia.doe.gov/oiaf/ieo/index.html.

Contacts

The International Energy Outlook is prepared by the Energy Information Administration (EIA). General questions concerning the contents of the report should be referred to Mary J. Hutzler (202/586-2222), Director, Office of Integrated Analysis and Forecasting. Specific questions about the report should be referred to Linda E. Doman (202/586-1041) or the following analysts:

World Energy Consumption	Linda E. Doman	(linda.doman@eia.doe.gov,	202/586-1041)
World Oil Markets	G. Daniel Butler	(george.butler@eia.doe.gov,	202/586-9503)
	Bruce Bawks	(bruce.bawks@eia.doe.gov,	202/586-6579)
Natural Gas	Sara Banaszak	(sara.banaszak@eia.doe.gov,	202/586-2066)
	Phyllis Martin	(phyllis.martin@eia.doe.gov,	202/586-9592)
Gas-to-Liquids Technology	William Trapmann	(william.trapmann@eia.doe.gov,	202/586-6408)
Coal	Michael Mellish	(michael.mellish@eia.doe.gov,	202/586-2136)
Nuclear	Laura Church	(laura.church@eia.doe.gov,	202/586-1494)
Renewable Energy	Linda E. Doman	(linda.doman@eia.doe.gov,	202/586-1041)
Electricity	Kevin Lillis	(kevin.lillis@eia.doe.gov,	202/586-1395)
Transportation	Linda E. Doman	(linda.doman@eia.doe.gov,	202/586-1041)
Environmental Issues	Perry Lindstrom	(perry.lindstrom@eia.doe.gov,	202/586-0934)
	Pia Hartman	(pia.hartman@eia.doe.gov,	202/586-2873)
Economic Growth	Kay A. Smith	(kay.smith@eia.doe.gov,	202/586-1455)

Electronic Access and Related Reports

The *IEO2000* will be available on CD-ROM and the EIA Home Page (*http://www.eia.doe.gov/oiaf/ieo/index.html*) by May 2000, including text, forecast tables, and graphics. To download the entire publication in Portable Document Format (PDF), go to *ftp://ftp.eia.doe.gov/pub/pdf/international/0484(2000).pdf*.

For ordering information and questions on other energy statistics available from EIA, please contact EIA's National Energy Information Center. Addresses, telephone numbers, and hours are as follows:

> National Energy Information Center, EI-231
> Energy Information Administration
> Forrestal Building, Room 1F-048
> Washington, DC 20585

Telephone: 202/586-8800
TTY: For people who are deaf or hard of hearing: 202/586-1181

E-mail: *infoctr@eia.doe.gov*
World Wide Web Site: *http://www.eia.doe.gov*
Gopher Site: *gopher://gopher.eia.doe.gov*

Contents

Contents (Continued)

Figures

Preface

*This report presents international energy projections through 2020,
prepared by the Energy Information Administration, including outlooks for
major energy fuels and issues related to electricity, transportation, and the environment.*

The *International Energy Outlook 2000 (IEO2000)* presents an assessment by the Energy Information Administration (EIA) of the outlook for international energy markets through 2020. The report is an extension of the EIA's *Annual Energy Outlook 2000 (AEO2000)*, which was prepared using the National Energy Modeling System (NEMS). U.S. projections appearing in the *IEO2000* are consistent with those published in the *AEO2000*. *IEO2000* is provided as a statistical service to energy managers and analysts, both in government and in the private sector. The projections are used by international agencies, Federal and State governments, trade associations, and other planners and decisionmakers. They are published pursuant to the Department of Energy Organization Act of 1977 (Public Law 95-91), Section 205(c). The *IEO2000* projections are based on U.S. and foreign government policies in effect on October 1, 1999.

Projections in *IEO2000* are displayed according to six basic country groupings (Figure 1). The industrialized region includes projections for nine individual countries—the United States, Canada, Mexico, Japan, France, Germany, Italy, the Netherlands, and the United Kingdom—plus the subgroups Other Europe and Australasia (the latter defined as Australia, New Zealand, and the U.S. Territories). The developing countries are represented by four separate regional subgroups: developing Asia, Africa, Middle East, and Central and South America. China, India, and South Korea are represented in developing Asia; Brazil is represented in Central and South America; and Turkey is represented in the Middle East.

The nations of Eastern Europe and the former Soviet Union (EE/FSU) are considered as a separate country grouping. In addition, in this year's report, the EE/FSU nations are further separated into Annex I and non-Annex I member countries participating in the Kyoto Climate Change Protocol on Greenhouse Gas Emissions. The new groupings are used to assess the potential role of Annex I EE/FSU countries in reaching the Annex I emissions targets of the Kyoto Climate Change Protocol.

The report begins with a review of world trends in energy demand. The historical time frame begins with data from 1970 and extends to 1997, providing readers with a 27-year historical view of energy demand. The *IEO2000* projections cover a 23-year period.

High economic growth and low economic growth cases were developed to depict a set of alternative growth paths for the energy forecast. The two cases consider alternative growth paths for regional gross domestic product (GDP). The resulting projections and the uncertainty associated with making international energy projections in general are discussed in the first chapter of the report. The status of environmental issues, including global carbon emissions, is reviewed. Comparisons of the *IEO2000* projections with other available international energy forecasts are included in the first chapter.

The next part of the report is organized by energy source. Regional consumption projections for oil, natural gas, coal, nuclear power, and renewable energy (hydroelectricity, geothermal, wind, solar, and other renewables) are presented in the five fuel chapters, along with a review of the current status of each fuel on a worldwide basis. Chapters on energy consumed by electricity producers and energy use in the transportation sector follow. The report ends with a discussion of energy and environmental issues, with particular attention to the outlook for global carbon emissions and the Kyoto Protocol.

Appendix A contains summary tables of the *IEO2000* reference case projections for world energy consumption, gross domestic product (GDP), energy consumption by fuel, electricity consumption, carbon emissions, nuclear generating capacity, energy consumption measured in oil-equivalent units, and regional population growth. The reference case projections of total foreign energy consumption and consumption of oil, natural gas, coal, and renewable energy were prepared using EIA's World Energy Projection System (WEPS) model, as were projections of net electricity consumption, energy consumed by fuel for the purpose of electricity generation, and carbon emissions. In addition, the National Energy Modeling System's (NEMS) Coal Export Submodule (CES) was used to derive flows in international coal trade, presented in the coal chapter. Nuclear *consumption* projections for the reference case were derived from the International Nuclear Model, PC Version (PC-INM). Nuclear *capacity* projections for the reference case were based on analysts' knowledge of the nuclear programs in different countries.

Appendix B and C present projections for the high and low economic growth cases, respectively. Nuclear *capacity* projections for the high and low growth cases were based on analysts' knowledge of nuclear programs. Nuclear *consumption* projections for both cases were derived from WEPS. Appendix D contains summary tables of projections for world oil production capacity and oil production in the reference case and four alternative cases: high oil price, low oil price, high non-OPEC supply, and low non-OPEC supply. The projections were derived from WEPS and from the "DESTINY" International Energy Forecast Software. Appendix E presents regional forecasts of transportation energy use in the reference case, derived from the WEPS model. Appendix F describes the WEPS model.

The six basic country groupings used in this report (Figure 1) are defined as follows:

- **Industrialized Countries** (the industrialized countries contain 18 percent of the 1999 world population): Australia, Austria, Belgium, Canada, Denmark, Finland, France, Germany, Greece, Iceland, Ireland, Italy, Japan, Luxembourg, Mexico, the Netherlands, New Zealand, Norway, Portugal, Spain, Sweden, Switzerland, the United Kingdom, and the United States.

- **Eastern Europe and the Former Soviet Union (EE/FSU)** (7 percent of the 1999 world population):

  - **Eastern Europe:** Albania, Bosnia and Herzegovina, Bulgaria, Croatia, Czech Republic, Hungary, Macedonia, Poland, Romania, Serbia and Montenegro, Slovakia, and Slovenia.

  - **Former Soviet Union:** Armenia, Azerbaijan, Belarus, Estonia, Georgia, Kazakhstan, Kyrgyzstan, Latvia, Lithuania, Moldova, Russia, Tajikistan, Turkmenistan, Moldova, Russia, Tajikistan, Turkmenistan, Ukraine, and Uzbekistan.

- **Developing Asia** (54 percent of the 1999 world population): Afghanistan, Bangladesh, Bhutan, Brunei, Cambodia (Kampuchea), China, Fiji, French Polynesia, Hong Kong, India, Indonesia, Kiribati, Laos, Malaysia, Macau, Maldives, Mongolia, Myanmar (Burma), Nauru, Nepal, New Caledonia, Niue, North Korea, Pakistan, Papua New Guinea, Philippines, Samoa, Singapore, Solomon Islands, South Korea, Sri Lanka, Taiwan, Thailand, Tonga, Tuvalu, Vanuatu, and Vietnam.

- **Middle East** (4 percent of the 1999 world population): Bahrain, Cyprus, Iran, Iraq, Israel, Jordan, Kuwait, Lebanon, Oman, Qatar, Saudi Arabia, Syria, Turkey, the United Arab Emirates, and Yemen.

Figure 1. Map of the Six Basic Country Groupings

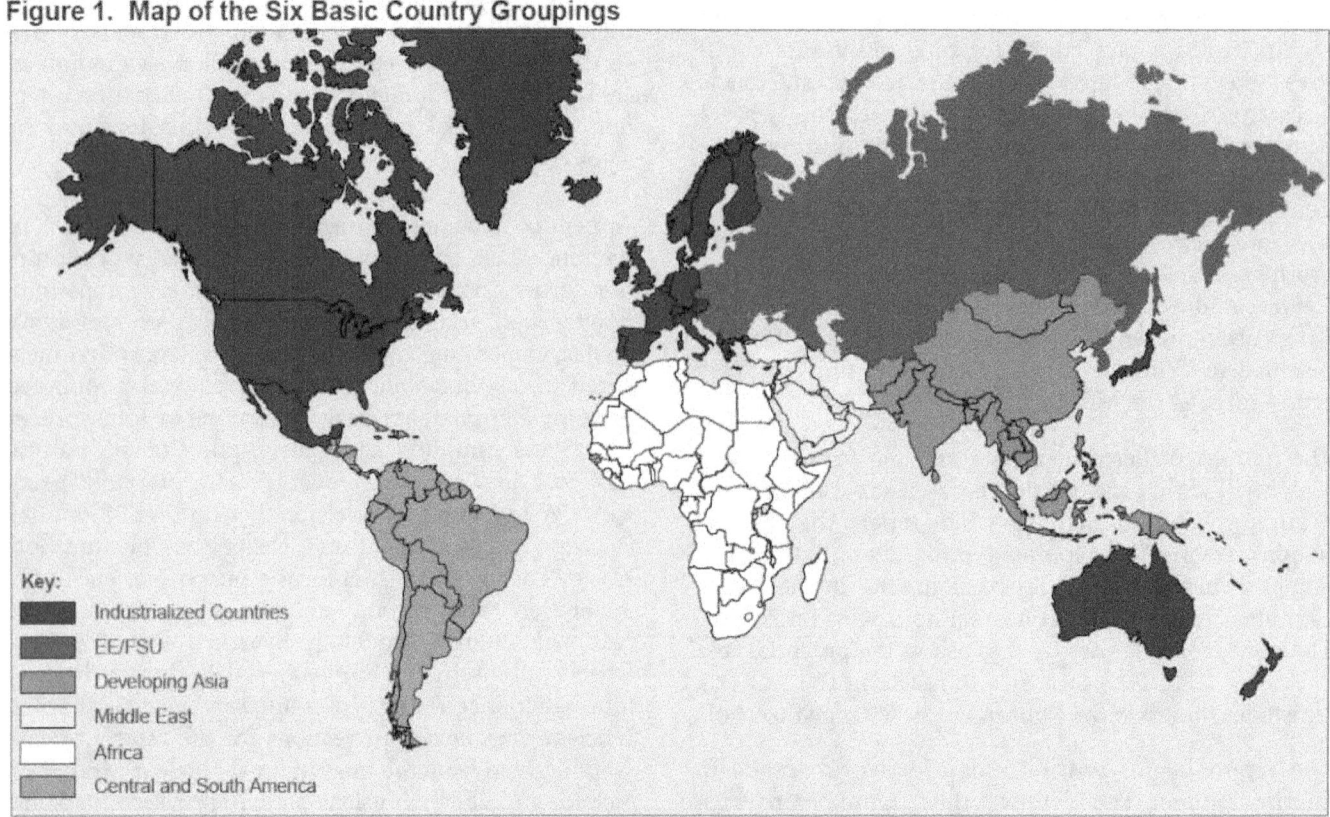

Key:
- Industrialized Countries
- EE/FSU
- Developing Asia
- Middle East
- Africa
- Central and South America

Source: Energy Information Administration, Office of Integrated Analysis and Forecasting.

- **Africa** (10 percent of the 1999 world population): Algeria, Angola, Benin, Botswana, Burkina Faso, Burundi, Cameroon, Cape Verde, Central African Republic, Chad, Comoros, Congo (Brazzaville), Congo (Kinshasa), Djibouti, Egypt, Equatorial Guinea, Eritrea, Ethiopia, Gabon, Gambia, Ghana, Guinea, Guinea-Bissau, Ivory Coast, Kenya, Lesotho, Liberia, Libya, Madagascar, Malawi, Mali, Mauritania, Mauritius, Morocco, Mozambique, Namibia, Niger, Nigeria, Reunion, Rwanda, Sao Tome and Principe, Senegal, Seychelles, Sierra Leone, Somalia, South Africa, St. Helena, Sudan, Swaziland, Tanzania, Togo, Tunisia, Uganda, Western Sahara, Zambia, and Zimbabwe.

- **Central and South America** (6 percent of the 1999 world population): Antarctica, Antigua and Barbuda, Argentina, Aruba, Bahama Islands, Barbados, Belize, Bolivia, Brazil, British Virgin Islands, Cayman Islands, Chile, Colombia, Costa Rica, Cuba, Dominica, Dominican Republic, Ecuador, El Salvador, Falkland Islands, French Guiana, Grenada, Guadeloupe, Guatemala, Guyana, Haiti, Honduras, Jamaica, Martinique, Montserrat, Netherlands Antilles, Nicaragua, Panama Republic, Paraguay, Peru, St. Kitts-Nevis, St. Lucia, St. Vincent/Grenadines, Suriname, Trinidad and Tobago, Uruguay, and Venezuela.

In addition, the following commonly used country groupings are referenced in this report:

- **Annex I Countries** (countries participating in the Kyoto Climate Change Protocol on Greenhouse Gas Emissions): Australia, Austria, Belgium, Bulgaria, Canada, Croatia, Czech Republic, Denmark, Estonia, European Community, Finland, France, Germany, Greece, Hungary, Iceland, Ireland, Italy, Japan, Latvia, Liechtenstein, Lithuania, Luxembourg, Monaco, the Netherlands, New Zealand, Norway, Poland, Portugal, Romania, Russia, Slovakia, Slovenia, Spain, Sweden, Switzerland, Ukraine, the United Kingdom, and the United States.[1]

- **European Union (EU):** Austria, Belgium, Denmark, Finland, France, Germany, Greece, Ireland, Italy, Luxembourg, the Netherlands, Portugal, Spain, Sweden, and the United Kingdom.

- **Mercosur Trading Block:** Argentina, Brazil, Paraguay, and Uruguay. Chile, and Bolivia are Associate Members.

- **North American Free Trade Agreement (NAFTA) Member Countries:** Canada, Mexico, and the United States.

- **Organization for Economic Cooperation and Development (OECD):** Australia, Austria, Belgium, Canada, Czech Republic, Denmark, Finland, France, Germany, Greece, Hungary, Iceland, Ireland, Italy, Japan, Luxembourg, Mexico, the Netherlands, New Zealand, Norway, Poland, Portugal, South Korea, Spain, Sweden, Switzerland, Turkey, the United Kingdom, and the United States.

- **Organization of Petroleum Exporting Countries (OPEC):** Algeria, Indonesia, Iran, Iraq, Kuwait, Libya, Nigeria, Qatar, Saudi Arabia, the United Arab Emirates, and Venezuela.

- **Pacific Rim Developing Countries:** Hong Kong, Indonesia, Malaysia, Philippines, Singapore, South Korea, Taiwan, and Thailand.

- **Persian Gulf:** Bahrain, Iran, Iraq, Kuwait, Qatar, Saudi Arabia, and the United Arab Emirates.

Objectives of the *IEO2000* Projections

The projections in *IEO2000* are not statements of what will happen, but what might happen given the specific assumptions and methodologies used. These projections provide an objective, policy-neutral reference case that can be used to analyze international energy markets. As a policy-neutral data and analysis organization, EIA does not propose, advocate, or speculate on future legislative and regulatory changes. The projections are based on current U.S. and foreign government policies. Assuming current policies, even knowing that changes will occur, will naturally result in projections that differ from the final data.

Models are abstractions of energy production and consumption activities, regulatory activities, and producer and consumer behavior. The forecasts are highly dependent on the data, analytical methodologies, model structures, and specific assumptions used in their development. Trends depicted in the analysis are indicative of tendencies in the real world rather than representations of specific real-world outcomes. Even where trends are stable and well understood, the projections are subject to uncertainty. Many events that shape energy markets are random and cannot be anticipated, and assumptions concerning future technology characteristics, demographics, and resource availability cannot be known with any degree of certainty.

---

[1]Turkey and Belarus are Annex I nations that have not ratified the Framework Convention on Climate Change and did not commit to quantifiable emissions targets under the Kyoto Protocol.

Oil Market Volatility: The Long-Term Perspective

The recent escalation of world oil prices leads to questions about how short-term events influence projections in the *International Energy Outlook 2000* (*IEO2000*). *IEO2000* provides an assessment of intermediate- and long-term trends in world energy markets. It is not intended as an analysis of short-term market fluctuations.

The turbulence of world oil prices has strongly influenced short-term markets, as documented in monthly issues of the Energy Information Administration's *Short-Term Energy Outlook* (*STEO*). *IEO2000* represents several months of effort, and although the graphics and discussion of world oil markets are consistent with projections from the February 2000 *STEO*, it is impossible to incorporate the volatility of oil prices into the construction of a mid-term oil market outlook. Oil prices have been quite volatile in the past, and volatile price behavior can be expected in the future, principally as the result of unforeseen political and social circumstances. The *IEO2000* mid-term projections do not attempt to predict volatility. Because of these assumptions, short-term price movements do not affect the long-term price projections in this report, and the *IEO2000* price path largely converges with last year's projections by 2005.

Last year, the *International Energy Outlook 1999* (*IEO99*) was prepared in a period during which world oil prices were at their lowest point (in real terms) of the past 50 years. An oil supply glut in 1998 resulted from lower growth in worldwide oil demand than had been expected, and from the failure of production management efforts by the Organization of Petroleum Exporting Countries (OPEC) to provide a significant reduction of oil supplies. *IEO2000*, in contrast, has been prepared during a period when oil prices (in nominal terms) have risen by more than 150 percent over a 12-month period (December 1998 to December 1999). The dramatic escalation of world oil prices was brought on by disciplined adherence to oil production cutback strategies by OPEC members and several non-OPEC producers as well and by stronger-than-anticipated growth in oil demand in recovering Asian economies. The figure opposite illustrates the oil-price roller coaster that has defined the world oil market over the past 3 years.

Some readers may expect the mid-term oil price forecast in *IEO2000* to be influenced significantly by the near-term behavior of oil markets. The following question has been asked over the past several months: With current oil prices higher than any of last year's oil price paths, what are the *IEO2000* price paths going to look like? The answer is that the long-term oil price outlook has not changed significantly.

Historically, only disruptions in oil supply brought about by politically motivated actions (e.g., the oil embargo of 1974) or conflicts involving major oil producers (e.g., the Iranian Revolution and the Iran-Iraq War) have had lingering, long-term impacts on oil prices. The oil market volatility over the past 2 years has been the result of oil market fundamentals that are reasonably well understood but nearly impossible to predict. Traditionally, such near-term oil market gyrations are considered unlikely to have much significant impact on long-term markets.

Current high prices are expected to fall for several reasons:

- OPEC is expected to ease production restraints at some point in the second or third quarter of 2000, based on recent pronouncements by Saudi Arabia, Venezuela, and Mexico. While OPEC and several non-OPEC producers have successfully managed the market in order to boost prices and increase revenues, they are clearly aware of the connected nature of the global economy. Sustained high oil prices would have the potential to damage the economic strength of consuming nations and to delay the economic recovery of Asia's developing nations. Neither would be commensurate with OPEC's long-term objective of vigorous growth in world oil demand.

- Continued high prices would begin to show a significant impact on oil demand, slowing the growth of worldwide demand by the second half of 2000.

(continued on page xiii)

Refiners' Acquisition Cost of Imported Crude Oil, 1997-1999

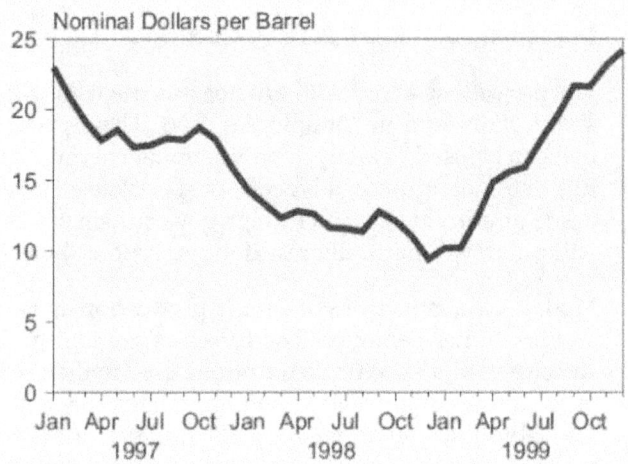

Sources: **1997-1998:** Energy Information Administration (EIA), *Monthly Energy Review,* DOE/EIA-0035(2000/01) (Washington, DC, January 2000). **1999:** EIA, *Weekly Petroleum Status Report,* DOE/EIA-0208(2000-06) (Washington, DC, February 11, 2000).

Oil Market Volatility: The Long-Term Perspective (Continued)

High prices would fuel economic inflation and rising interest rates, which in turn would adversely affect global stock markets, consumer confidence, and oil consumption.

• As has been seen in previous periods of high oil prices, non-OPEC producers would be stimulated to increase their output levels. The upward movement of prices in 1999 already has enabled oil companies to increase their spending on field rehabilitation and maintenance projects. In addi-

tion, many exploration projects that were delayed in the low price environment of 1998 have been given the green light because they now meet profitability standards. In fact, non-OPEC oil production (including some of the countries that agreed to cut production along with OPEC, such as Mexico and Norway) was 280,000 barrels per day higher in the fourth quarter of 1999 than had been estimated in the February 2000 edition of the *STEO*, an indication that non-OPEC production may be increasing faster as a result of higher oil prices.

Highlights

World energy consumption is projected to increase by 60 percent from 1997 to 2020.
Recent price developments in world oil markets and economic recovery in
Southeast Asia have altered projections relative to last year's report.

In the reference case projections for the *International Energy Outlook 2000 (IEO2000)*, world energy consumption increases by 60 percent over a 23-year forecast period, from 1997 to 2020. Energy use worldwide increases from 380 quadrillion British thermal units (Btu) in 1997 to 608 quadrillion Btu in 2020 (Figure 2 and Table 1). Many developments in 1999 are reflected in this year's outlook. Shifting short-term world oil markets, the beginnings of strong recovery for the economies of southeast Asia, and a faster than expected economic recovery in the former Soviet Union (FSU) have all influenced the mid-term forecast for the world's energy markets.

World oil prices recovered substantially in 1999 from their record lows of 1998, mostly because members of the Organization of Petroleum Exporting Countries (OPEC) and non-OPEC producers—notably, Mexico and Norway—were able to sustain the oil production cuts set by the cartel in March 1999, and because oil demand began to recover among the Southeast Asian nations that had been stuck in economic recession since mid-1997. Many Asian countries had strong economic growth and a corresponding rise in energy demand in 1999 that seemed to mark the ending of the recession. In South Korea, for example, energy consumption in 1999 exceeded pre-recession levels.

High oil prices also helped the Russian economy post positive economic growth in 1999. In fact, Russia's gross domestic product (GDP) grew more in 1999 than it had

since the late 1980s. There are, however, many current events in Russia that add considerable uncertainty to the projections for the FSU, including the upcoming presidential elections following the resignation of Russian President Boris Yeltsin, the banking scandal uncovered in 1999 at the Bank of New York and the resulting delay of International Monetary Fund credits, and the ongoing Russian war in Chechnya. Nevertheless, stronger economic performance in Russia and Ukraine—the region's

Figure 2. World Energy Consumption, 1970-2020

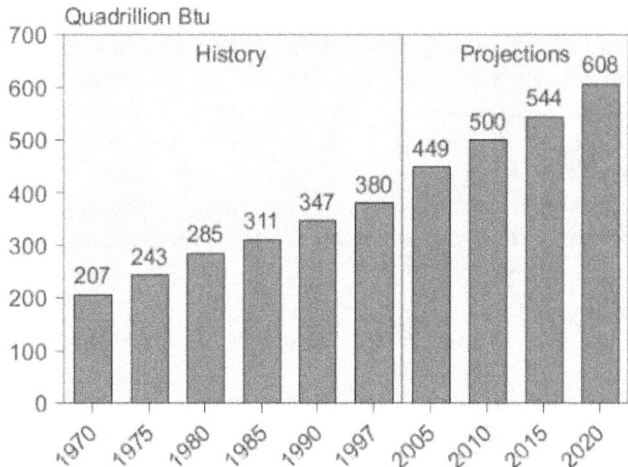

Sources: **History:** Energy Information Administration (EIA), Office of Energy Markets and End Use, International Statistics Database and *International Energy Annual 1997*, DOE/EIA-0219(97) (Washington, DC, April 1999). **Projections:** EIA, World Energy Projection System (2000).

Table 1. Energy Consumption and Carbon Emissions by Region, 1990-2020

Region	Energy Consumption (Quadrillion Btu)				Carbon Emissions (Million Metric Tons)			
	1990	1997	2010	2020	1990	1997	2010	2020
Industrialized	182.8	203.7	238.7	259.9	2,850	3,039	3,563	3,928
EE/FSU	76.4	53.3	63.0	75.7	1,337	878	992	1,151
Developing								
Asia	51.4	75.3	126.4	172.6	1,067	1,522	2,479	3,380
Middle East	13.1	17.9	26.2	34.3	229	297	422	552
Africa.	9.3	11.4	15.8	20.6	180	214	292	380
Central and South America . .	13.7	18.3	30.1	44.7	174	225	399	617
Total	87.6	122.9	198.5	272.1	1,649	2,258	3,591	4,930
Total World	346.7	379.9	500.2	607.7	5,836	6,175	8,146	10,009

Sources: **History:** Energy Information Administration (EIA), *International Energy Annual 1997*, DOE/EIA-0219(97) (Washington, DC, April 1999). **Projections:** EIA, World Energy Projection System (2000).

largest economies—has led to a 12-percent upward revision in the *IEO2000* projection for FSU energy demand in 2020, as compared with last year's *Outlook*. The projections for the transitional economies of Eastern Europe—where economic recovery has been sustained for the most part since 1993—remain largely unchanged from those in last year's report.

In the *IEO2000* reference case, much of the growth in worldwide energy use is projected for the developing world (Figure 3). In particular, energy demand in developing Asia and Central and South America is projected to more than double between 1997 and 2020. Both regions are expected to sustain energy demand growth of more than 3 percent annually throughout the forecast, accounting for more than one-half of the total projected increment in world energy consumption and 83 percent of the increment for the developing world alone.

In the industrialized countries, one of the primary sources of uncertainty in the forecast is the potential impact of the Kyoto Climate Change Protocol, which would require reductions or limits to the growth of carbon emissions within the Annex I countries[2] between 2008 and 2012, resulting in a combined 4 percent reduction in emissions relative to the 1990 levels. As of January 2000, 83 countries and the European Commission had signed the treaty; however, none of the Annex I countries had ratified it by the time the *IEO2000* was prepared for publication. Should the Kyoto Protocol

enter into force, it could have profound effects on the use of energy in the industrialized world.

The industrialized countries are expected to account for about 30 percent of the increment in worldwide energy use over the 1997-2010 time period in the reference case. Achieving the Protocol's targets solely by reducing fossil fuel use in the industrialized world might mean a reduction of between 30 and 60 quadrillion Btu—equivalent to between 15 and 30 million barrels of oil per day—depending on the mix of fossil fuels used to achieve such a reduction. On the other hand, it is more likely that fuel-switching opportunities will be used and that a more modest reduction in total fossil fuel use will be required. Emissions trading and other strategies—such as conservation measures, reforestation, and joint implementation programs, among others allowed under the Protocol—could further lower the need for fossil fuel reductions, although the specific mechanisms for such offsets have not yet been established.

World carbon emissions are projected to rise from 6.2 billion metric tons in 1997 to 8.1 billion metric tons in 2010 and 10.0 billion metric tons in 2020 in the reference case projections, which do not take into account the potential impact of the Kyoto Protocol. In this forecast, world carbon emissions exceed their 1990 levels by 40 percent in 2010 and by 72 percent in 2020 (Figure 4). Emissions in the industrialized world grow by 1.1 billion metric tons between 1990 and 2020, with nearly one-half

Figure 3. World Energy Consumption by Region, 1970-2020

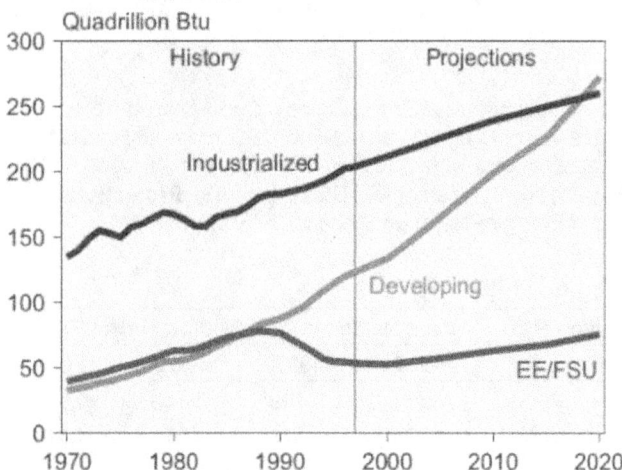

Sources: **History:** Energy Information Administration (EIA), Office of Energy Markets and End Use, International Statistics Database and *International Energy Annual 1997*, DOE/EIA-0219(97) (Washington, DC, April 1999). **Projections:** EIA, World Energy Projection System (2000).

Figure 4. World Carbon Emissions by Region, 1990-2020

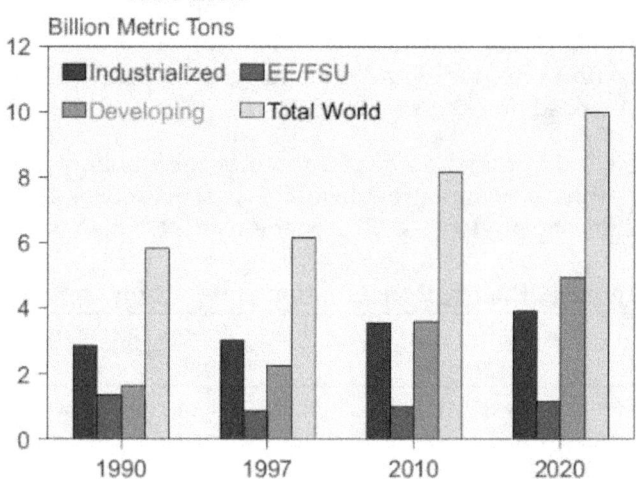

Sources: **History:** Energy Information Administration (EIA), Office of Energy Markets and End Use, International Statistics Database and *International Energy Annual 1997*, DOE/EIA-0219(97) (Washington, DC, April 1999). **Projections:** EIA, World Energy Projection System (2000).

[2]The Annex I countries under the Framework Convention on Climate Change are Australia, Austria, Belgium, Bulgaria, Canada, Croatia, Czech Republic, Denmark, Estonia, Finland, France, Germany, Greece, Hungary, Iceland, Ireland, Italy, Japan, Latvia, Lithuania, Luxembourg, the Netherlands, New Zealand, Norway, Poland, Portugal, Romania, Russia, Slovakia, Slovenia, Spain, Sweden, Switzerland, the Ukraine, the United Kingdom, and the United States. Turkey and Belarus are also considered Annex I countries, but neither has agreed to any limits on greenhouse gas emissions.

of the increment attributed to an increase in natural gas use. Natural gas is increasingly seen as a fuel of choice among the industrialized countries for new electric power generation. Gas-fired power plants run more efficiently than other fossil fuel generators, and natural gas, as the least carbon intensive of the fossil fuels, is an attractive alternative to coal or oil for electricity generation and industrial uses.

Much of the increase in carbon emissions is expected to occur in the developing world, where emerging economies produce the highest growth rates for energy use in the forecast. Emissions in the developing countries accounted for about 28 percent of the world total in 1990, but they are projected to make up 44 percent of the total by 2010 and nearly 50 percent by 2020. As a result, even if the Annex I countries were able to meet the emissions limits or reductions prescribed in the Kyoto Protocol, worldwide carbon emissions still would grow substantially (Figure 5). The increase is expected to be caused both by rapid economic expansion, accompanied by growing demand for energy, and by continued heavy reliance on coal (the most carbon intensive of the fossil fuels), particularly in developing Asia. Coal accounts for 41 percent of the projected increment in carbon emissions in the developing world between 1990 and 2020, followed closely by oil's contribution of 36 percent. Gas accounts for 22 percent of the developing world's increase in emissions.

The crude oil market rebounded dramatically in 1999, with prices rising from the low monthly average of $9.39 per barrel (nominal U.S. dollars) in December 1998 to $25 per barrel in January 2000. Prices were influenced by the successful adherence to announced cutbacks in production by OPEC and key non-OPEC members, notably, Mexico and Norway, along with strong growth in oil consumption in the industrialized countries (which accounted for 60 percent of the growth in demand in 1999) and the recovery of demand in Southeast Asia as the economies began to recover from the recession of 1997-1998. World oil prices are expected to reach $22 per barrel in constant 1998 U.S. dollars ($36 per barrel in nominal dollars) at the end of the projection period—about the same as in last year's forecast (Figure 6).

For the near term, the *IEO2000* projections are substantially altered by the strong recovery of world oil prices in 1999. Incorporating the recent price turbulence into the construction of an intermediate- and long-term oil market outlook is, however, impossible. Oil prices have been quite volatile in the past, and volatile price behavior can be expected in the future, principally as the result of unforeseen political and social circumstances. The *IEO2000* projections do not attempt to predict volatility. Because of these assumptions, short-term price movements do not affect the long-term price projections in this report, and the price path largely converges with last year's projections by 2005.

Figure 5. World Carbon Emissions in the *IEO2000* Reference Case and Under the Kyoto Protocol

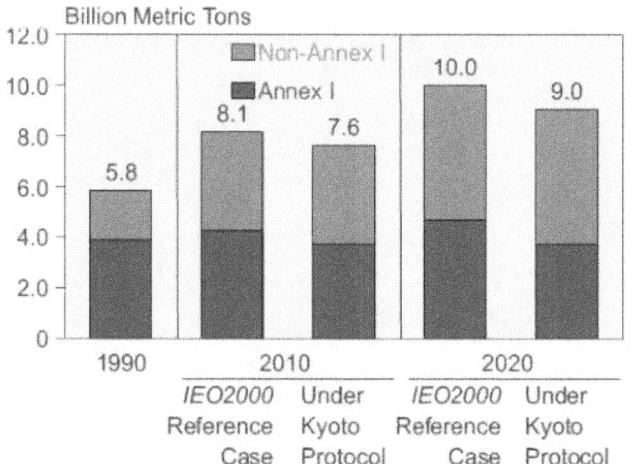

Sources: **1990:** Energy Information Administration (EIA), Office of Energy Markets and End Use, International Statistics Database and *International Energy Annual 1997*, DOE/EIA-0219(97) (Washington, DC, April 1999). **Projections:** EIA, World Energy Projection System (2000).

Figure 6. Comparison of 1999 and 2000 World Oil Price Projections

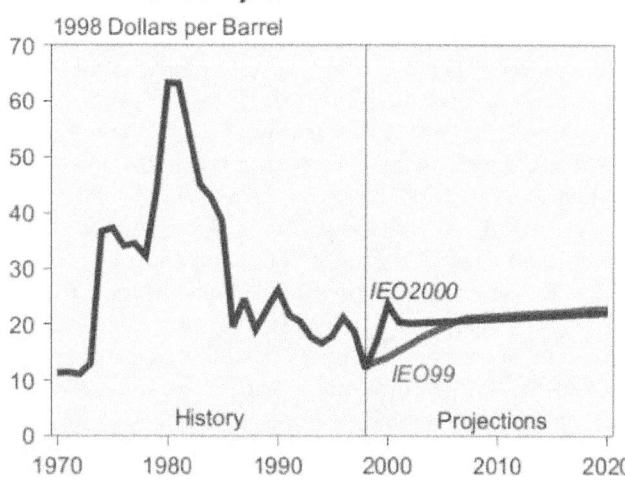

Sources: **History:** Energy Information Administration (EIA), *Annual Energy Review 1998*, DOE/EIA-0384(98) (Washington, DC, July 1999). ***IEO99:*** EIA, *International Energy Outlook 1999*, DOE/EIA-0484(99) (Washington, DC, March 1999). ***IEO2000:*** 1999-2001—EIA, *Short-Term Energy Outlook*, on-line version (February 7, 2000), web site www.eia.doe.gov/emeu/steo/pub/contents.html. 2002-2020—EIA, *Annual Energy Outlook 2000*, DOE/EIA-0383(2000) (Washington, DC, December 1999).

Worldwide oil demand reaches almost 113 million barrels per day by 2020 in the reference case—about 2 percent higher than in last year's forecast, based primarily on more optimistic expectations for economic recovery in the FSU—requiring an increment of almost 40 million barrels per day relative to current capacity. OPEC producers are expected to be the major beneficiaries of increased production requirements, but non-OPEC supply is expected to remain competitive, with major increments of supply coming from offshore resources, especially in the Caspian Basin and deepwater West Africa. Deepwater exploration and development initiatives are generally expected to be sustained worldwide, with offshore West Africa emerging as a major future source of oil production. Technology and resource availability can sustain large increments in oil production capability at the reference case prices. The low price environment of 1998 and early 1999 did slow the pace of development in some prospective production areas, and especially in the Caspian Basin region.

Oil currently provides a larger share of world energy consumption than any other energy source and is expected to remain in that position throughout the forecast period. Its share of total energy consumption declines slightly, however, from 39 percent in 1997 to 38 percent in 2020, as countries in many parts of the world switch to natural gas and other fuels, particularly for electricity generation. World oil consumption is projected to increase by 1.9 percent annually over the 23-year projection period, from 73 million barrels per day in 1997 to 113 million barrels per day in 2020. Petroleum is used heavily in the transportation sector and also to provide heat, power, and feedstocks for industry.

In the industrialized countries, most of the growth in oil use is projected for the transportation sector, where few alternatives are currently economical. In the developing countries, the transportation sector also shows the fastest projected growth in oil use; however, in contrast to the industrialized countries, oil use for purposes other than transportation is projected to contribute 42 percent of the total increase in petroleum consumption in the developing countries. The growth in nontransportation oil use in the developing countries is caused in part by the substitution of petroleum products for noncommercial fuels (such as wood burning for home heating and cooking) as incomes rise and the energy infrastructure matures.

Natural gas remains the fastest growing component of primary world energy consumption. Over the *IEO2000* forecast period, gas use is projected to more than double in the reference case, reaching 167 trillion cubic feet (Figure 7). The gas share of total energy consumption increases from 22 percent in 1997 to 29 percent in 2020. Moreover, natural gas accounts for the largest increment

in electricity generation (increasing by 33 quadrillion Btu or 41 percent of the total increment in energy used for electricity generation). Combined-cycle gas turbine power plants offer some of the highest commercially available plant efficiencies, and natural gas is environmentally attractive because it emits less sulfur dioxide, carbon dioxide, and particulate matter than does oil or coal.

In the industrialized world, natural gas consumption has the largest projected increase among the major fuels, increasingly becoming the choice for new power generation because of its environmental and economic advantages. Increments in gas use in the developing countries are expected to supply both power generation and other uses, such as town gas and fuel for industry. In China, for example, natural gas use is projected to grow at a robust rate of 11.2 percent per year over the forecast period.

In the *IEO2000* reference case, coal's share of total energy consumption falls only slightly, from 24 percent in 1997 to 22 percent in 2020 (Figure 8). Its historical share is nearly maintained, because large increases in energy use are projected for the developing countries of Asia, where coal continues to dominate many national fuel markets. Together, two of the key countries in the region—China and India—are projected to account for 97 percent of the world's total increase in coal use (on a Btu basis). Coal continues to be a major fuel source for electricity generation worldwide, and virtually all of the projected growth in the world's consumption of coal is for electricity. The exception is China, where coal continues to be the primary energy source in a rapidly expanding industrial

Figure 7. World Energy Consumption by Fuel Type, 1970-2020

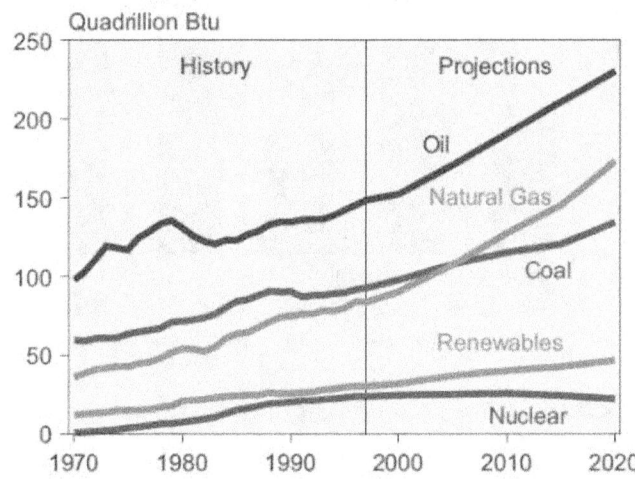

Sources: **History:** Energy Information Administration (EIA), Office of Energy Markets and End Use, International Statistics Database and *International Energy Annual 1997*, DOE/EIA-0219(97) (Washington, DC, April 1999). **Projections:** EIA, World Energy Projection System (2000).

sector because of the nation's abundant coal reserves and limited access to alternative sources of energy.

The prospects for nuclear power to continue its role of meeting a significant share of worldwide electricity consumption are uncertain, despite projected growth of 2.5 percent per year in total electricity demand through 2020. In the *IEO2000* reference case, worldwide nuclear capacity is projected to increase to 368 gigawatts in 2010, then begin to decline, falling to 303 gigawatts in 2020. Aggressive plans to expand nuclear capacity, mainly in the Far East, lead to the near-term increase, but plant retirements in the United States and other countries exceed total new additions worldwide and produce a decline later in the forecast. Developing Asian countries are projected to add 30 gigawatts of new nuclear capacity by 2020, while the industrialized countries overall lose 64 gigawatts. Nuclear safety issues moved to the forefront in Asia in 1999 after several leaks at nuclear power plants in South Korea and China, as well as the serious accident in a reprocessing facility in Tokaimura, Japan. These incidents are likely to cause further public concern about the aggressive plans for nuclear capacity expansion in the Far East.

The development of renewable resources is constrained in the *IEO2000* reference case projections by expectations that fossil fuel prices will remain relatively low over the forecast horizon and that, as a result, renewables will have a difficult time competing. Failing a strong worldwide commitment to environmental programs, such as the limitations and reductions of greenhouse gases outlined in the Kyoto Protocol, it is difficult to foresee significant widespread increases in renewable

energy use. Modest growth in renewable energy is projected to continue, maintaining an 8-percent share of total energy consumption over the forecast horizon. Most of the increase is expected from large-scale hydroelectric projects that are under construction or planned, particularly in developing Asia. For environmental reasons, higher growth rates are expected for alternative renewable energy sources—notably, wind—in the industrialized countries. In addition, in developing countries such as China and Brazil renewables are expected to be used to reach rural populations that do not have access to national electricity grids.

Electricity consumption worldwide increases by 76 percent in the reference case, from 12 trillion kilowatthours in 1997 to 22 trillion kilowatthours in 2020. Long-term growth in electricity consumption is expected to be strongest in the developing countries of Asia, followed by those of Central and South America. Those two regions alone account for 52 percent of the world's net electricity consumption increment in the *IEO2000* reference case (Figure 9). Rapid growth in population and income, along with greater industrialization and more widespread household electrification are responsible for the increase.

To a large degree, future growth in the world's electricity generation will depend on progress made in connecting more of the world's population to national electricity grids. Electricity demand and investment in the electric power sector infrastructure have responded positively to the recent net improvement in global economic conditions, and to the movement toward privatization in many parts of the world. Many developing countries

Figure 8. World Energy Consumption Shares by Fuel Type, 1970-2020

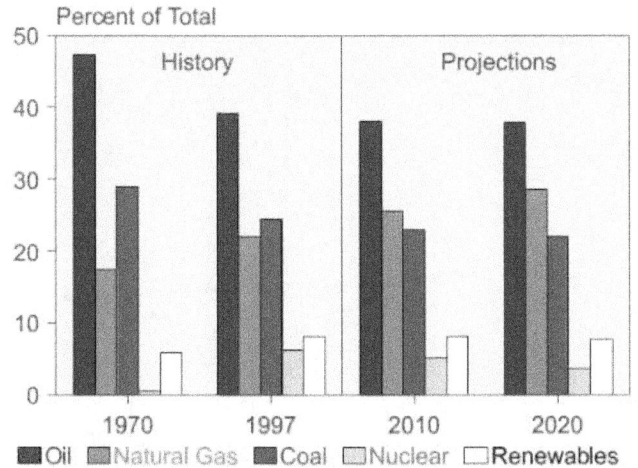

Sources: **History:** Energy Information Administration (EIA), Office of Energy Markets and End Use, International Statistics Database and *International Energy Annual 1997*, DOE/EIA-0219(97) (Washington, DC, April 1999). **Projections:** EIA, World Energy Projection System (2000).

Figure 9. World Net Electricity Consumption by Region, 1990-2020

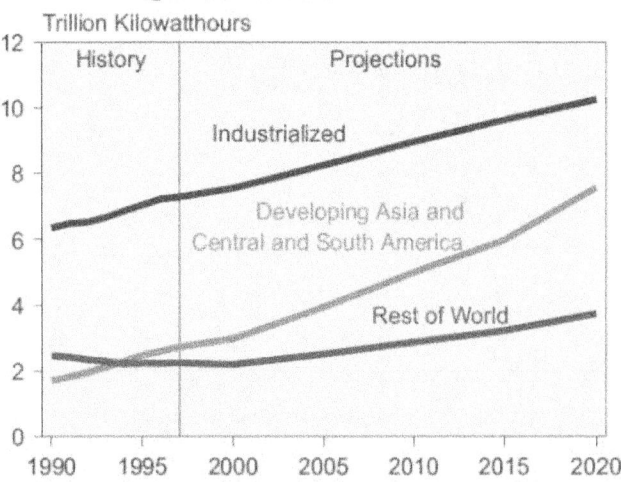

Sources: **History:** Energy Information Administration (EIA), Office of Energy Markets and End Use, International Statistics Database and *International Energy Annual 1997*, DOE/EIA-0219(97) (Washington, DC, April 1999). **Projections:** EIA, World Energy Projection System (2000).

have been motivated to encourage various forms of private investment to raise the capital necessary to meet rapidly growing demand for electricity. In the developing world, $142 billion in private capital has flowed into electricity projects since 1990.

Transportation remains the most important oil-consuming sector throughout the projection period. With little present competition from alternative fuels, oil is projected to be the primary energy source fueling transportation around the globe. Road transport retains the largest share of energy use in the transportation sector, and the projections indicate strong growth in demand for personal motor vehicles over the next two decades, particularly in the developing world (Figure 10). As per capita income expands in the emerging economies and standards of living rise, fast-paced growth in the demand for personal transportation is expected.

In urban centers of the developing world, car ownership is often seen as one of the first symbols of emerging prosperity. Per capita motorization in much of the developing world is projected to more than double between 1997 and 2020, although population growth is expected to keep motorization levels low relative to those in the industrialized world. For example, the U.S. per capita motorization level in 2020 is projected at 797 vehicles per thousand persons, but in China—where motorization is expected to grow fivefold over the forecast horizon—the projected motorization level in 2020 is only 54 vehicles per thousand persons.

The *IEO2000* projections, like all forecasts, are accompanied by a measure of uncertainty. One way to quantify the uncertainty is to consider the relationship between energy consumption and GDP growth (that is, energy intensity) over time. In the industrialized countries, history shows the link between energy consumption and economic growth to be a relatively weak one, with growth in energy demand lagging behind economic growth. In the developing countries, the two have been more closely correlated in the past, with energy demand growing in parallel with economic expansion.

In the *IEO2000* forecast, energy intensity in the industrialized countries is expected to improve (decrease) by 1.1 percent per year between 1997 and 2020, slightly slower than the 1.3-percent annual improvement for the region from 1970 to 1997. Energy intensity is also projected to improve in the developing countries—by 1.0 percent per year—as their economies begin to behave more like those of the industrialized countries as a result of improving standards of living that accompany the projected economic expansion (Figure 11). The EE/FSU has always maintained a much higher level of energy intensity than either the industrialized or the developing countries. Over the forecast horizon, energy intensity is projected to improve in the EE/FSU region in concert with expected recovery from the economic and social declines of the early 1990s; however, it still is expected to be twice as high as in the developing world and five times as high as in the industrialized world.

Figure 10. Motorization Levels in Selected Countries, 1997 and 2020

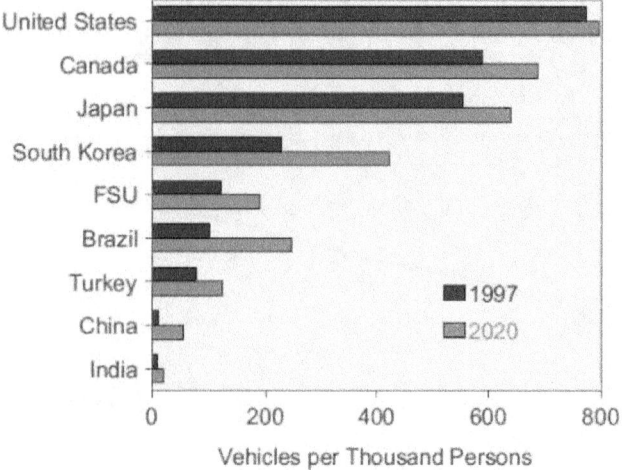

Sources: **1997:** American Automobile Manufacturers Association, *World Motor Vehicle Data* (Detroit, MI, 1997). **Projections:** EIA, World Energy Projection System (2000).

Figure 11. World Energy Intensity by Region, 1970-2020

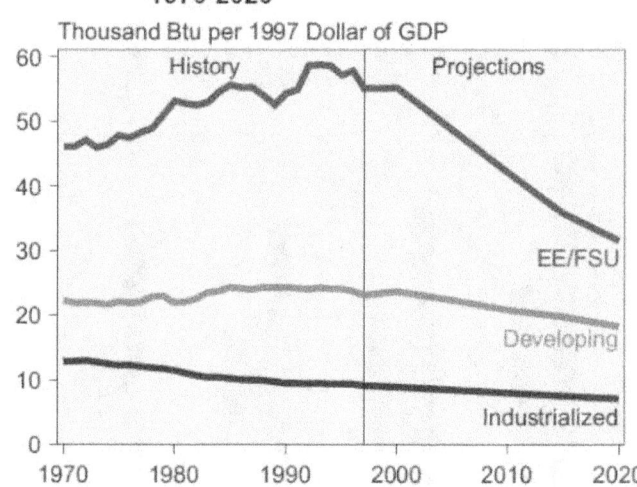

Sources: **History:** Energy Information Administration (EIA), Office of Energy Markets and End Use, International Statistics Database and *International Energy Annual 1997*, DOE/EIA-0219(97) (Washington, DC, April 1999). **Projections:** EIA, World Energy Projection System (2000).

World Energy Consumption

The IEO2000 projections indicate continued growth in world energy use, including large increases for the developing economies of Asia and South America. Energy resources are thought to be adequate to support the growth expected through 2020.

Current Trends Influencing World Energy Demand

Changing world events and their effects on world energy markets shape the long-term view of trends in energy demand. Several developments in 1999—shifting short-term world oil markets, the recovery of developing Asian markets, and a faster than expected recovery in the economies of the former Soviet Union—are reflected in the projections presented in this year's *International Energy Outlook 2000 (IEO2000)*.

In 1998, oil prices reached 20-year lows as a result of oil surpluses caused by a combination of the economic recession in Southeast Asia and the milder than expected winters of 1997 and 1998 in North America and Western Europe. In 1999, the start of economic and energy demand recovery among many of the Asian economies, along with a tightening of the world's oil supply orchestrated by production cuts by the Organization of Petroleum Exporting Countries (OPEC) and by Mexico, a key non-OPEC producer, have strengthened oil prices to above their pre-1997 levels. In late November 1999, Iraq decided to shut down oil exports of more than 2 million barrels per day, protesting its treatment by the United Nations over the terms for renewal of the "oil for food and medicine" agreement. The Iraqi move tightened the oil supply market even further.

World oil prices recovered substantially from their record 1998 lows largely because OPEC member countries and Mexico were able to sustain the production cuts set by the cartel in March 1999 (and reaffirmed in September 1999), and because the demand for oil in Asian economies began to recover from the economic recession that had plagued the region since mid-1997. Compliance rates among the OPEC member countries were estimated to have been as high as 80 to 90 percent in 1999, and most analysts believe the production cut agreements will be maintained, by and large, at least until the end of March 2000, when they expire.

In Asia, oil demand began to strengthen again as economic recovery appeared to take hold. Although oil consumption in Japan remained level in 1999 (after falling by 200,000 barrels per day in 1998), oil use in China and the other parts of Asia increased by 400,000 barrels per day between 1998 and 1999, more than compensating for the flat performance in Japan [1]. Oil use in developing Asia (including China, India, and South Korea) is expected to rise to 13.6 million barrels per day in 2000 (an increase of 400,000 barrels per day) with expectations for continued economic recovery in the region.

In *IEO2000*, world energy consumption is projected to increase by 60 percent between 1997 and 2020, reaching 608 quadrillion British thermal units (Btu), about the same as in last year's forecast (Figure 12). As in other recent *IEOs*, much of the growth is projected for regions of the developing world, particularly developing Asia and Central and South America, where energy consumption is expected to more than double over the 23-year projection period. Both of these regions are expected to sustain energy demand growth of more than 3 percent annually through 2020, accounting for more than one-half of the world's total increment in energy consumption and 83 percent of the increment of the developing world alone (Figure 13).

Many of the developing Asian countries that had been mired in economic recession appeared to begin strong

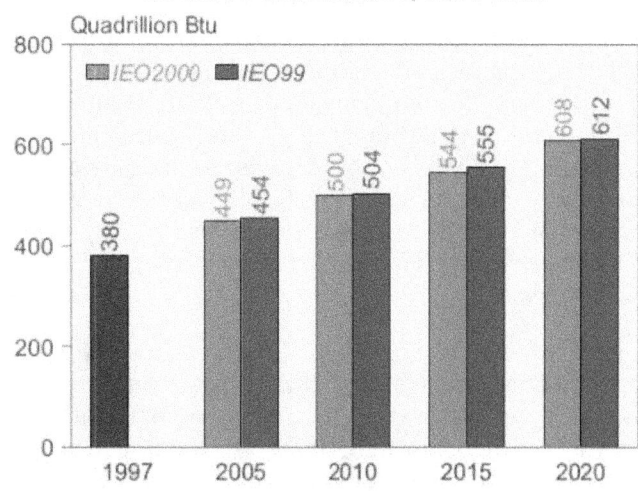

Figure 12. World Energy Consumption Projections in *IEO99* and *IEO2000*, 2005-2020

Sources: **1997**: Energy Information Administration (EIA), *International Energy Annual 1997*, DOE/EIA-0219(97) (Washington, DC, April 1999). *IEO99*: EIA, *International Energy Outlook 1999*, DOE/EIA-0484(99) (Washington, DC, March 1999). *IEO2000*: EIA, World Energy Projection System (2000).

Figure 13. Energy Consumption in the Developing World, 1995-2020

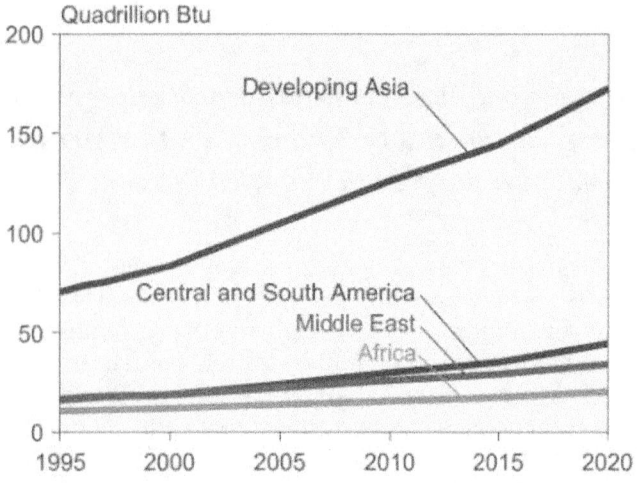

Sources: **1995-1997:** Energy Information Administration (EIA), *International Energy Annual 1997*, DOE/EIA-0219(97) (Washington, DC, April 1999). **Projections:** EIA, World Energy Projection System (2000).

recovery in 1999, and as a result showed strong energy demand growth this year. In particular, South Korea's energy consumption surpassed its pre-crisis levels. South Korea, along with Thailand, Malaysia, and Singapore, as well as Taiwan—an economy that did not fall prey to the recession in much of the rest of southeast Asia—all showed continued strength in electronics exports, which helped drive the recovery. Imports of electronics by the industrialized nations surged in 1999, largely to procure Y2K-related information technology [2].

Political instability is keeping Indonesia from enjoying the economic recovery. The violence that followed the vote for independence in East Timor has helped to unnerve international support in the country. In the wake of the East Timor violence, Australian companies have halted multi-million dollar wheat exports and cotton to Indonesia, stopped processing Indonesian crude oil, and curtailed air freight shipments between the two countries. Thailand's rice exports to Indonesia fell sharply from ranges of 100,000 to 150,000 tons at the beginning of 1999 to almost zero in the fall [3]. GDP growth in Indonesia is expected to remain negative until at least 2000.

On November 15, 1999, the Clinton Administration reached an accord with the Chinese government that is expected to pave the way for China's entrance into the World Trade Organization (WTO) [4]. The United States and China have negotiated this issue on and off for 13 years. By entering the WTO, China will be integrated into the global trading system, committing to economic reforms and opening Chinese markets to foreign investment and reducing import tariffs [5]. This may provide

the incentive to boost economic growth in the country. Beginning in 1997, GDP growth in China slowed from the double-digit growth rates experienced through much of the first part of the decade, although annual growth still remained at nearly 8 percent between 1997 and 1999. A drop in China's exports and foreign investment in 1999 and falling retail prices have reignited fears of the devaluation of the yuan that first arose during the height of the Asian economic recession. Nevertheless, in the long term, China is expected to maintain among the highest rates of growth in energy demand worldwide, increasing by 4.3 percent per year through 2020.

The forecast for *IEO2000* is more optimistic than last year's forecast with regard to prospects for growth among the transitional economies of the former Soviet Union (FSU). Russia and Ukraine, the region's largest economies, have experienced stronger than expected recovery from the August 1998 monetary crisis in Russia, which also affected most of the other economies in the region. In Russia, 1999 turned out to be the strongest year of economic growth since the late 1980s, with growth expected to be between 2.0 and 2.2 percent [6]. Stronger oil prices in 1999 are the primary reason for the country's GDP increase [7].

Current events do, however, add to the considerable uncertainty of FSU energy markets. For instance, the banking scandal revolving around a Russian money laundering scheme uncovered at the Bank of New York caused the International Monetary Fund (IMF) to delay delivery of the $640 million tranche standby credit in September 1999—part of the $4.5 billion IMF credit approved for Russia in July 1999. Important events in the country, including the upcoming presidential elections following the early resignation of President Boris Yeltsin and the ongoing Russian war in Chechnya, may also significantly affect the nation's economic situation and energy markets. Nevertheless, in the *IEO2000* reference case, economic recovery is expected to take hold in the FSU over the next several years. By 2005, the region is projected to begin experiencing strong GDP growth rates as it recovers from the social and economic upheavals of the early 1990s.

In Central and South America, the negative influence of the economic crises in Asia and Russia was muted somewhat by the economic strength of the United States—the region's biggest source of investment [8]. Recessions gripped many countries in the region in 1999, including Argentina, Chile, Colombia, Venezuela, and Ecuador. Brazil, the region's largest economy, experienced a relatively mild recession. Slowing reforms, trade conflicts among the Mercosur member countries, and rising global interest rates add to the uncertainty about short-term growth in the region.

In North America, the continued strength of the U.S. economy has, in large part, helped the economies of Mexico and Canada, its trading partners under the North American Free Trade Agreement (NAFTA). Higher oil prices have also bolstered Mexico's economy. The United States has had economic growth on the order of 4 percent annually for the past several years, and both Mexico and Canada have enjoyed GDP growth in excess of 3 percent. Most forecasters expect the U.S. and Canadian economies to slow somewhat in the short term, and over the 1997 to 2020 forecast horizon North American energy demand is projected to grow by 1.2 percent per year. Mexico's energy demand alone is expected to grow at by 3.0 percent per year (Figure 14).

In January 1999, 11 members of the European Union (Austria, Belgium, Finland, France, Germany, Italy, Luxemburg, the Netherlands, and Portugal) began the process of phasing in the single currency of the European Monetary Union, the euro, with an eye toward introducing euro notes and coins by January 1, 2002 [9]. This marks another milestone toward the increasingly unified European market. In terms of energy markets in Western Europe, the EU's Gas and Electricity Directives are resulting in rapid liberalization of the gas and electricity industries in most of the member countries. Since February 1999, the Electricity Directive has taken effect in 14 of the 15 member states, with only France not yet complying with the directive requirements, but expected to do so by the end of 2000 [10]. In the *IEO2000* reference case, energy consumption in Western Europe grows at just under 1 percent per year between 1997 and 2020, about the same as in last year's forecast.

Japan remains problematic in the short-term. While positive GDP growth in 1999—jumping to almost 8 percent in the first quarter of the year—indicated that the country's long-term recession is finally abating, at least half of the growth was directly attributed to the government's fiscal stimulus package, rather than consumer demand [11]. The Japanese government's November 1998 economic stimulus plan included a number of tax cuts and spending and lending measures, placing around $195 billion into the economy to boost consumer spending [12]. However, consumer spending has not improved substantially despite the package; there has been little in the way of corporate restructuring; and the large number of bad bank loans has not declined even

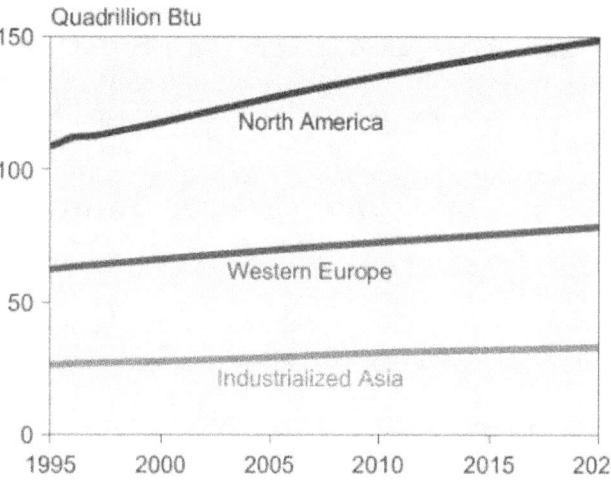

Figure 14. Energy Consumption in the Industrialized World, 1995-2020

Sources: **1995-1997**: Energy Information Administration (EIA), *International Energy Annual 1997*, DOE/EIA-0219(97) (Washington, DC, April 1999). **Projections**: EIA, World Energy Projection System (2000).

with the large, $517 billion government-provided bank rescue package [13]. As a result, sustained economic growth is not expected to return to the country before 2001.

In the industrialized countries, one of the main sources of uncertainty for the long-term energy consumption forecast is the potential impact of the Kyoto Climate Change Protocol. The *IEO2000* reference case does not take into account the impact of the Protocol, because the forecast is based on government policies in effect as of October 1999. As of January 2000, 83 countries and the European Community had signed the treaty, and only 22 of those signatories had ratified the treaty. The ratifications to date do not include any of the Annex I countries, that would be required to limit or reduce their greenhouse gas emissions under the terms of the Protocol.[3] The Protocol enters into force "on the ninetieth day after the date on which not less than 55 Parties to the Convention, incorporating Annex I Parties which accounted in total for at least 55 percent of the total carbon dioxide emissions for 1990 from that group, have deposited their instruments of ratification, acceptance, approval, or accession" [14].

Were the industrialized Annex I countries[4] to reduce their carbon emissions levels by reducing fossil fuel

---

[3]The Annex I countries under the Kyoto Protocol are Australia, Austria, Belgium, Bulgaria, Canada, Croatia, Czech Republic, Denmark, Estonia, European Community, Finland, France, Germany, Greece, Hungary, Iceland, Ireland, Italy, Japan, Latvia, Lithuania, Luxembourg, Monaco, Netherlands, New Zealand, Norway, Poland, Portugal, Romania, Russia, Slovakia, Slovenia, Spain, Sweden, Switzerland, Ukraine, United Kingdom of Great Britain and Northern Ireland, and the United States. Turkey and Belarus are Annex I countries that have not ratified the Framework Convention on Climate Change and did not commit to quantifiable emissions targets under the Protocol.

[4]Excluding the transitional Annex I countries of Eastern Europe and the former Soviet Union where emissions levels are likely to be lower than their Protocol targets in 2010.

consumption alone, energy demand might have to be reduced by between 30 and 60 quadrillion Btu in 2010.[5] It is likely, however, that other strategies, such as fuel switching, conservation measures, reforestation, emissions trading, joint implementation programs, and/or others, would also be employed to meet the Kyoto obligations.

Outlook by Primary Energy Source

As noted above, the *IEO2000* reference case represents a "business as usual" set of projections. Under these circumstances, potential international energy policies that have not yet been enacted, such as the Kyoto Protocol, are not taken into account. This year's projections suggest that use of energy from every source except nuclear power will increase over the forecast period (Figure 15).

Oil remains the dominant fuel throughout the forecast period, as it has been for more than two decades. In the industrialized world, the greatest increment in oil use is projected for the transportation sector. Oil is increasingly displaced by natural gas in the forecast for power generation in the industrialized countries, as gas becomes the fuel of choice for new power generation because it burns more cleanly and efficiently than oil. In the developing world, the projected increase in oil use occurs for all end uses. Natural gas consumption also increases in the developing world, but the infrastructure required for widespread gas use remains to be established.

Natural gas is projected to be the fastest growing primary energy source worldwide, maintaining growth of 3.1 percent annually over the 23-year projection period, twice as high as the rate for coal. Natural gas consumption rises from 82 to 167 trillion cubic feet between 1997 and 2020, with much of the increment projected to accommodate new electricity generation over the forecast. Gas is increasingly seen as the desired alternative for electric power, given the efficiency of combined-cycle gas turbines relative to coal- or oil-fired generators, and because it burns more cleanly than either coal or oil, making it a more attractive choice for countries interested in reducing greenhouse gas emissions.

Worldwide coal use increases by 2.3 billion short tons (44 percent) over the projection period. Substantial declines in coal use are projected for Western Europe and the EE/FSU, where natural gas is increasingly used to displace coal for electric power generation and for other industrial and building sector uses. Increases projected for the developing world, however, more than balance the decrement in coal use projected for Europe. The largest increases are projected for China and India,

where coal supplies are plentiful. Those two countries alone account for more than 90 percent of the projected rise in coal use worldwide over the forecast period (Figure 16).

Worldwide nuclear power consumption is expected to increase from 2,268 billion kilowatthours in 1997 to 2,464 billion kilowatthours in 2010 before declining to 2,136

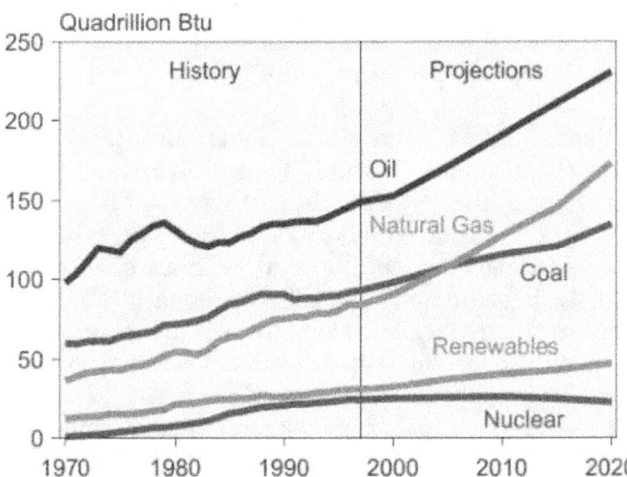

Figure 15. **World Energy Consumption by Fuel Type, 1970-2020**

Sources: **History:** Energy Information Administration (EIA), Office of Energy Markets and End Use, International Statistics Database and *International Energy Annual 1997*, DOE/EIA-0219(97) (Washington, DC, April 1999). **Projections:** EIA, World Energy Projection System (2000).

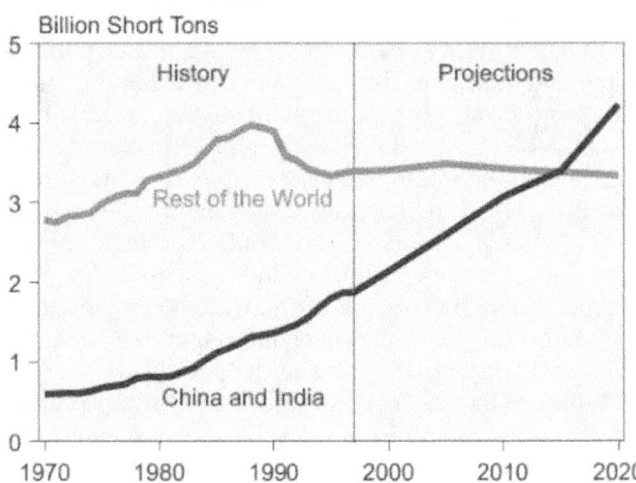

Figure 16. **World Coal Consumption by Region, 1970-2020**

Sources: **History:** Energy Information Administration (EIA), Office of Energy Markets and End Use, International Statistics Database and *International Energy Annual 1997*, DOE/EIA-0219(97) (Washington, DC, April 1999). **Projections:** EIA, World Energy Projection System (2000).

[5]This range was calculated by removing consumption of the most carbon-intensive fossil fuel possible, coal, and the least carbon-intensive fuel possible, natural gas with the understanding that it would probably be impractical to only reduce consumption by reducing coal use, so that a combination of fossil fuels would have to be reduced.

billion kilowatthours at the end of the forecast period. The projected decline in nuclear generation is attributed to the industrialized world and the EE/FSU, where older reactors are increasingly expected to be retired and few new reactors (with the notable exceptions of France and Japan) are planned to replace them. Should the Kyoto Protocol be enacted, however, it is possible that licenses for non-carbon emitting nuclear facilities would be extended and the decline of nuclear generation forestalled as industrialized countries attempt to reach their greenhouse gas targets. Most of the growth in nuclear capacity in the *IEO2000* reference case is expected to occur in the developing world (particularly in developing Asia), where nuclear power consumption increases by 3.7 percent per year between 1997 and 2020, while world nuclear generation declines overall by 0.3 percent per year (Figure 17).

The world's use of energy from hydroelectricity and other renewable energy sources is projected to grow by 1.9 percent annually in the reference case. The growth of renewables continues to be constrained by low fossil fuel prices, which discourage the development of renewable energy sources. Nevertheless, renewable energy sources retain an 8-percent share of total world energy consumption throughout the *IEO2000* projection period. Renewable energy could get a boost if the Kyoto Protocol were ratified, in that signatory nations might use non-carbon-emitting energy sources to reduce their use of fossil fuels.

Several initiatives—including the renewable portfolio standard proposed in the Clinton Administration's Comprehensive Electricity Competition Act (CECA)

and commitments by the European Union to advance wind-generated electricity—could help to advance the use of alternative renewable energy technologies. In addition, in the developing world, renewable energy sources are increasingly viewed as a way to bring electrification to remote, rural areas. Recent projects include a $25 billion commitment from Brazil for renewable generation in rural areas and World Bank loans for installing wind and solar energy systems in rural northwestern provinces of China.

Outlook for Carbon Emissions

The *IEO2000* projections indicate that, if fossil fuel consumption grows to the levels projected in the reference case, global carbon emissions will rise to 8.1 billion metric tons per year by 2010 and 10.0 billion metric tons by 2020 (Figure 18). Much of the increase in carbon emissions is expected in the developing countries, where emerging economies are expected to produce the highest growth rates for energy consumption. Developing countries alone account for 84 percent of the projected increment in the world's carbon emissions between 1990 and 2010 and 79 percent of the increment between 1990 and 2020 (Figure 19). Continued heavy reliance on coal and other fossil fuels projected for the developing countries ensures that, even if the Annex I countries were to adopt the terms of the Kyoto Protocol, worldwide emissions would still grow substantially over the forecast horizon.

Oil consumption is projected to account for the largest increment in worldwide carbon emission levels, emitting an additional 1.6 billion metric tons to the

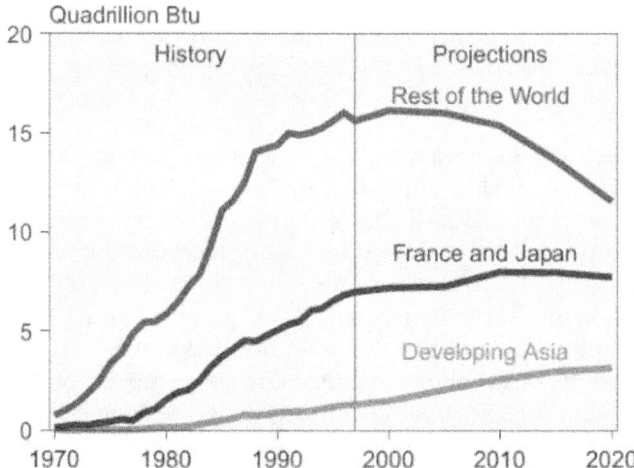

Figure 17. **World Nuclear Energy Consumption by Region, 1970-2020**

Sources: **History:** Energy Information Administration (EIA), Office of Energy Markets and End Use, International Statistics Database and *International Energy Annual 1997*, DOE/EIA-0219(97) (Washington, DC, April 1999). **Projections:** EIA, World Energy Projection System (2000).

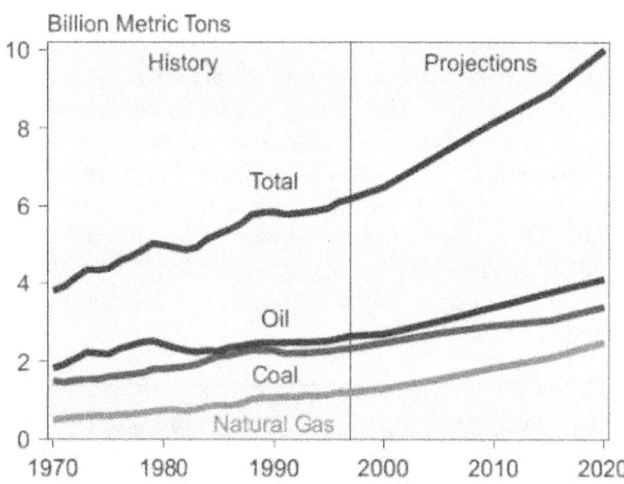

Figure 18. **World Carbon Emissions by Fuel Type, 1970-2020**

Sources: **History:** Energy Information Administration (EIA), Office of Energy Markets and End Use, International Statistics Database and *International Energy Annual 1997*, DOE/EIA-0219(97) (Washington, DC, April 1999). **Projections:** EIA, World Energy Projection System (2000).

Figure 19. World Carbon Emissions by Region, 1990-2020

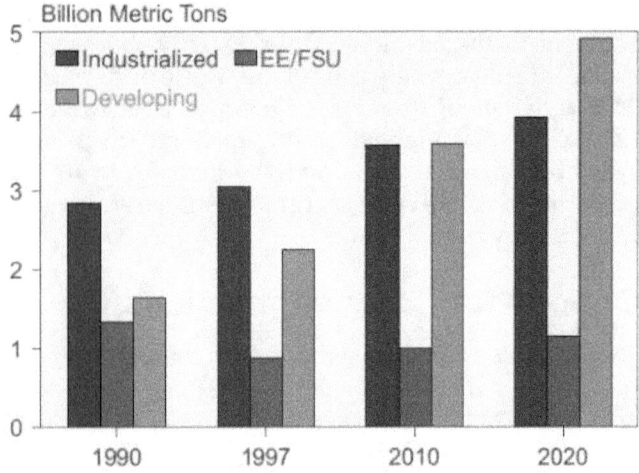

Sources: **1990 and 1997**: Energy Information Administration (EIA), Office of Energy Markets and End Use, International Statistics Database and *International Energy Annual 1997*, DOE/EIA-0219(97) (Washington, DC, April 1999). **Projections**: EIA, World Energy Projection System (2000).

Figure 20. Carbon Emissions in the Annex I Countries by Fuel Type, 1990-2020

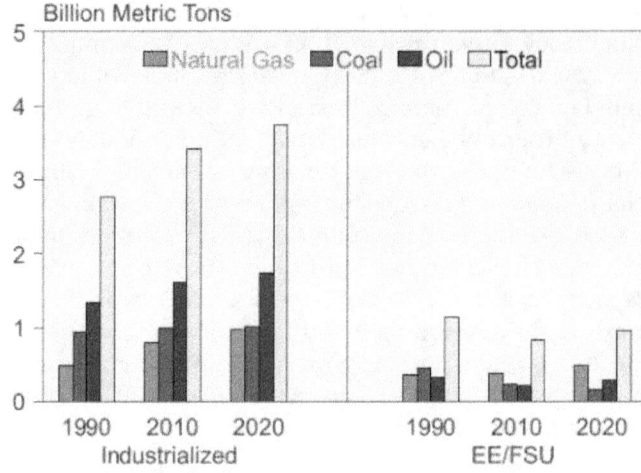

Sources: **1990**: Energy Information Administration (EIA), Office of Energy Markets and End Use, International Statistics Database and *International Energy Annual 1997*, DOE/EIA-0219(97) (Washington, DC, April 1999). **Projections**: EIA, World Energy Projection System (2000).

atmosphere by 2020 relative to the 1990 level. Natural gas follows with a 1.4 billion metric ton increment and coal with a 1.1 billion metric ton increment. Although natural gas use increases at a faster rate than oil use, because gas is a less carbon-intensive fuel than either oil or coal, its contribution to the increase in carbon emissions over the forecast is smaller.

The Kyoto Protocol, if ratified and implemented, could influence future patterns of energy consumption. However, because the Protocol has not been enacted, the *IEO2000* reference case projections do not take into account its potential impact. As a result, carbon emissions in the industrialized Annex I countries are projected to grow by 971 million metric tons between 1990 and 2020 (Figure 20). About half of the increment is attributed to natural gas consumption and 40 percent to oil consumption. The industrialized countries rely heavily on oil for transportation sector uses, where there are few economical alternatives. Only 9 percent of the increase in carbon emissions for the industrialized Annex I countries is expected to come from coal, as a result of decreasing coal consumption in Western Europe and only moderate increases in the other industrialized countries.

In this year's forecast separate projections for the Annex I and non-Annex I portions of Eastern Europe (EE) and the former Soviet Union (FSU) are presented, to provide a better estimate of the "hot air" credits potentially available from the EE/FSU region for carbon emissions trading under the Kyoto Protocol. Carbon emissions fell by 383 million metric tons in the Annex I transitional economies of the EE/FSU between 1990 and 1997, from 1,135

million metric tons to 751 million metric tons. Emissions in the Annex I portion of the EE/FSU are expected to rise to 835 million metric tons by 2010 and to 962 million metric tons by 2020, remaining below 1990 levels even at the end of the forecast.

In its *International Energy Outlook 1999 (IEO99)*, EIA projected that the EE/FSU region as a whole would have the potential to provide 374 million metric tons of credits to the Annex I emissions reduction effort in 2010. In *IEO2000*, with separate forecasts for the Annex I and non-Annex I EE/FSU nations, only 318 million metric tons of credits are projected for the Annex I EE/FSU countries. For the entire EE/FSU region in this year's projections, 364 million metric tons of credits would be available in 2010—slightly lower than the *IEO99* projection because of the stronger economic growth expected for the FSU in *IEO2000*.

Having a more realistic estimate of the credits potentially available from the Annex I EE/FSU countries is important because it allows analysts to see what level of effort would be required for the industrialized Annex I countries to meet their Kyoto Protocol targets. According to the *IEO2000* reference case projections, without trading, the industrialized Annex I countries would have to reduce their combined emissions in 2010 by 836 million metric tons (or 24 percent) relative to the reference case projection (Table 2); however, because EE/FSU Annex I emissions in total are projected to be about 318 million metric tons below their Protocol targets, the Annex I nations altogether would need to reduce emissions by only 519 million metric tons (or 12 percent) relative to the 2010 projection.

Table 2. Carbon Emissions in the Annex I countries, 1990 and 2010, and Effects of the Kyoto Protocol in 2010
(Million Metric Tons)

Region and Country	1990 Emissions	2010 Baseline Projection	2010 Kyoto Protocol Target	Reduction From 2010 Baseline	Percent Change	
					From 1990	From 2010 Baseline
Annex I Industrialized Countries						
North America	1,472	1,947	1,370	577	-7	-30
United States	1,345	1,787	1,251	536	-7	-30
Canada	127	160	119	41	-6	-26
Western Europe	934	1,016	860	156	-8	-15
Industrialized Asia	364	457	354	103	-3	-23
Japan	274	331	257	74	-6	-22
Australasia.	90	126	97	29	7	-23
Total	2,769	3,420	2,584	836	-7	-24
Annex I Transitional Economies						
Former Soviet Union	854	591	853	-261	-0	44
Eastern Europe	281	244	300	-56	7	23
Total	1,135	835	1,153	-318	2	38
Total Annex I Countries	3,904	4,255	3,737	519	-4	-12

Sources: **1990:** Energy Information Administration (EIA), *Emissions of Greenhouse Gases in the United States 1998*, DOE/EIA-0573(98) (Washington, DC, October 1999); and EIA, *International Energy Annual 1997*, DOE/EIA-0219(97) (Washington, DC, April 1999). **2010:** EIA, World Energy Projection System (2000).

Alternative Growth Cases

One of the major measures used to gauge uncertainty in the *IEO* forecast is the expected rate of future economic growth. *IEO2000* includes a high economic growth case and a low economic growth case, in addition to the reference case. The reference case projections are determined by establishing a set of regional assumptions about the economic growth paths—measured by gross domestic product (GDP)—and energy elasticity (the relationship between changes in energy consumption and changes in GDP). The two alternative growth cases, based on alternative ideas about possible economic growth paths, are derived to provide users with a way to quantify the range of uncertainty relative to the reference case (see Appendix A, Table A3, for reference case GDP assumptions).

For the high and low economic growth cases, different assumptions are made about the range of possible economic growth rates among the industrial, transitional EE/FSU, and developing economies as defined in the *IEO*. For the industrialized countries, one percentage point is added to the reference case GDP growth rates for the high economic growth case, and one percentage point is subtracted from the reference case GDP growth rates for the low economic growth case. Outside the industrialized world and excluding China and the EE/FSU, reference case GDP growth rates are increased and decreased by 1.5 percentage points for the high and low economic growth case estimates, respectively.

Because China experienced particularly high, often double-digit growth in GDP throughout much of the 1990s, it has the potential for a larger downturn in economic growth. In contrast, the EE/FSU region suffered a severe economic collapse in the early part of the decade and has been trying to recover from it, with mixed success. The EE/FSU nations have the potential for substantial increases in economic growth, should their current political and institutional problems moderate enough for the recovery of a considerable industrial base. As a result of these uncertainties, 3.0 percentage points are subtracted from the reference case GDP assumptions for China in the low economic growth case and 1.5 percentage points are added in the high economic growth case. For the EE/FSU region, 1.5 percentage points are subtracted from the reference case assumptions in the low economic growth case, and 3.0 percentage points are added in the high economic growth case.

The *IEO2000* reference case shows total world energy consumption reaching 608 quadrillion Btu in 2020, with the industrialized world consuming 260 quadrillion Btu, the transitional EE/FSU countries 76 quadrillion Btu, and the developing world 272 quadrillion Btu. In the high economic growth case, total world energy use is projected at 723 quadrillion Btu in 2020, 115 quadrillion Btu higher than in the reference case (Figure 21). Under the assumptions of the low economic growth case, worldwide energy consumption in 2020 would be 111 quadrillion Btu lower than in the reference case. Thus, a substantial range of 226 quadrillion Btu results between

Figure 21. Total World Energy Consumption in Three Cases, 1970-2020

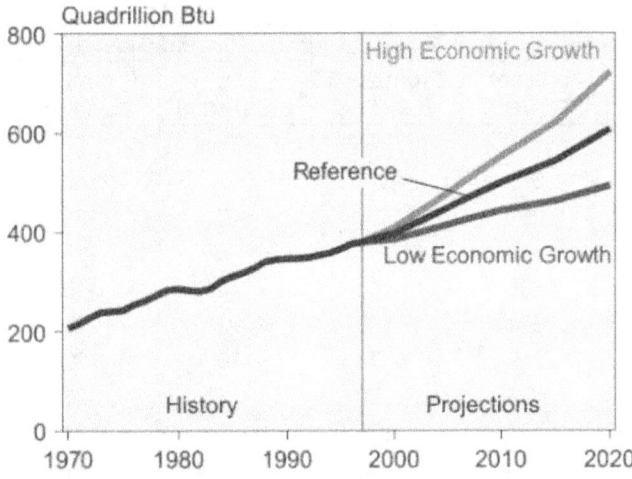

Sources: **History:** Energy Information Administration (EIA), Office of Energy Markets and End Use, International Statistics Database and *International Energy Annual 1997*, DOE/EIA-0219(97) (Washington, DC, April 1999). **Projections:** EIA, World Energy Projection System (2000).

high and low economic growth cases, almost 40 percent of the total reference case consumption projected for 2020. The variations in projected carbon emissions among the cases in 2020 are similarly wide: as low as 8,053 million metric tons (1,956 million metric tons less than the reference case projection) in the low economic growth case and as high as 11,950 million metric tons (1,940 million metric tons higher than in the reference case projection) in the high economic growth case.

Trends in Energy Intensity

Another way of quantifying the uncertainty surrounding a long-term forecast is to consider the relationship between energy consumption and GDP growth over time. Economic growth and energy demand are linked, but the strength of that link varies among regions and stages of economic development. In industrialized countries, history shows the link to be a relatively weak one. That is, energy demand lags behind economic growth. In developing countries, demand and economic growth have, in the past, been more closely correlated, with energy demand growth tending to track the rate of economic expansion. This trend may, however, be moderating in many parts of the developing world. For instance, between 1996 and 1997, GDP in developing Asia increased by 7.3 percent, but energy consumption increased by only 2.5 percent per year; and in Central and South America, GDP increased by 5.1 percent but energy use by only 3.1 percent annually [15].

The historical behavior of energy intensity—the ratio of energy use to GDP—in the FSU is problematic. The EE/FSU economies have always maintained higher levels of energy intensity than either the industrialized

or the developing countries. In the FSU, however, energy consumption grew more quickly than GDP until 1990, when the collapse of the Soviet Union created a situation in which both income and energy use were declining. GDP fell more rapidly than energy use, however, and as a result energy intensity increased. Over the forecast horizon, energy intensity is expected to improve in the region as the economies begin to recover from the economic and social declines of the early 1990s. Nevertheless, energy intensity in the EE/FSU in 2020 is expected to be almost double that in the developing world and five times that in the industrialized world (Figure 22).

The stage of economic development and the standard of living of individuals in a given region strongly influence the link between economic growth and energy demand. Advanced economies with high living standards have relatively high energy use per capita, but they also tend to be economies where per capita energy use is relatively stable or changes very slowly. In the industrialized nations, increases in energy use tend to correlate with employment and population growth. With the penetration of modern appliances and motorized personal transportation equipment already high in the industrialized countries, increments in personal income tend to result in spending on goods and services that are not energy intensive. To the extent that spending is directed to energy-consuming goods, it involves more often than not purchases of new equipment to replace old capital stock. The new stock is often more efficient than the equipment it replaces, resulting in a weaker link between income and energy demand. In developing countries, standards of living, while rising, tend to be low relative to those in more advanced economies.

Figure 22. World Energy Intensity by Region, 1970-2020

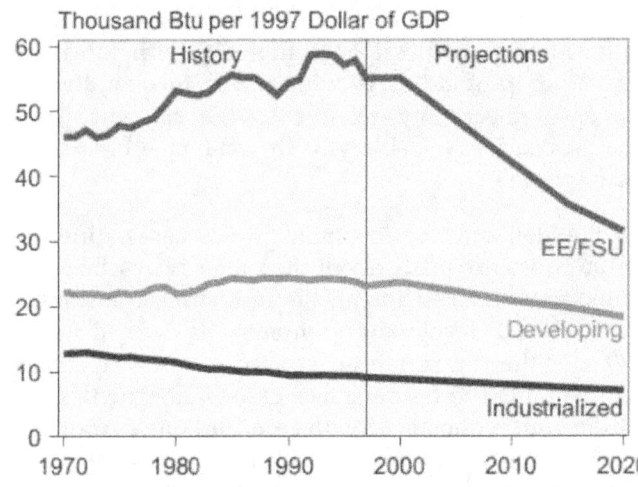

Sources: **History:** Energy Information Administration (EIA), Office of Energy Markets and End Use, International Statistics Database and *International Energy Annual 1997*, DOE/EIA-0219(97) (Washington, DC, April 1999). **Projections:** EIA, World Energy Projection System (2000).

Changing growth patterns of energy intensity could have dramatic impacts on energy consumption in the projection period, particularly among the developing countries. For instance, when energy intensities are assumed to decline in the developing countries by 60 percent—equal to the most rapid annual rate of decline observed between 1990 and 1997—energy consumption in the developing world is projected to be 140 quadrillion Btu in 2020, about 132 quadrillion Btu less than the reference case estimate of 272 quadrillion Btu. When energy intensities in the developing world are assumed to increase by 136 percent—the most rapid annual rate of increase observed between 1990 and 1997—energy consumption in the developing world is projected to be 812 quadrillion Btu in 2020, nearly three times the reference case projection.

Forecast Comparisons

Another way to examine the uncertainty associated with the *IEO2000* projections is to compare them with those derived by other forecasters. Four organizations provide forecasts comparable to those in *IEO2000*. The International Energy Agency (IEA) provides "business as usual" projections in its *World Energy Outlook 1998* out to the year 2020. Standard & Poor's Platt's, formerly DRI/McGraw-Hill (DRI), also provides energy forecasts by fuel to 2020 in its *World Energy Service: World Outlook 1999.* Petroleum Economics, Ltd. (PEL) and Petroleum Industry Research Associates (PIRA) also publish world

energy forecasts, but only to the years 2015 and 2010, respectively.

Regional breakouts among the forecasting groups vary, complicating the comparisons. For example, *IEO2000* includes Mexico in North America, but all the other services include Mexico in Latin America. As a result, for purposes of this comparison, the national-level *IEO2000* projections for Mexico have been moved from North America to Central and South America to provide forecasts for "Latin America" that match the other series. DRI and PIRA include only Japan in industrial Asia. DRI and *IEO2000* include Turkey in Middle East, but IEA includes Turkey, as well as the Czech Republic and Hungary, in "OECD Europe." Although most of the differences in regional groupings involve fairly small countries, they contribute to the variations among forecasts.

All the forecasts provide projections out to the year 2010 (Table 3). The growth rates for energy consumption among the reference case forecasts for the 1995-2010 time period are similar, between 2.1 and 2.3 percent per year, and all fall within the range of variation defined by the *IEO2000* low and high economic growth cases.

Regionally, the area of greatest uncertainty among the forecasts is the EE/FSU, with forecasts ranging from 0.4-percent annual growth in energy consumption between 1995 and 2010 (PEL) to 1.5-percent annual growth (PIRA). As a result, the PEL forecast for the

Table 3. Comparison of Energy Consumption Growth Rates by Region, 1995-2010
(Average Annual Percent Growth)

Region	IEO2000 Low Growth	IEO2000 Reference	IEO2000 High Growth	IEO99	DRI	IEA	PIRA	PEL
Industrialized Countries. . . .	0.9	1.2	1.6	1.3	1.5	1.3	1.1	1.2
United States and Canada. . .	1.1	1.4	1.7	1.4	1.5	1.1	1.3	1.3
Western Europe	0.6	1.0	1.4	1.2	1.6	1.5	1.0	1.3
Industrialized Asia	0.6	1.1	1.6	1.1	1.0	1.5	0.9	0.8
EE/FSU	0.6	1.0	1.7	0.9	0.6	1.4	1.5	0.4
Developing Countries	2.3	3.7	4.8	3.8	3.6	3.7	4.1	3.9
Asia	2.4	4.0	4.9	3.9	3.7	4.2	4.0	4.1
China	2.3	4.5	5.4	4.2	3.5	4.0	4.5	4.2
Other Asia[a]	2.4	3.4	4.3	3.6	3.9	4.5	4.6	4.1
Middle East	1.9	3.2	4.5	3.4	3.6	2.0	3.5	3.7
Africa.	1.6	2.5	3.4	2.5	2.6	2.7	3.1	2.7
Latin America	2.6	3.8	5.2	4.2	4.0	3.3	3.4	3.6
Total World	1.3	2.1	2.8	2.2	2.3	2.2	2.3	2.1

[a]Other Asia includes India and South Korea.

Sources: *IEO2000*: Energy Information Administration (EIA), World Energy Projection System (2000). *IEO99*: EIA, *International Energy Outlook 1999, DOE/EIA-0484(99) (Washington, DC, March 1999), Table A1, p. 141.* **DRI**: Standard & Poor's DRI, *World Energy Service: World Outlook* (Lexington, MA, January 1999), p. 5. **IEA**: International Energy Agency, *World Energy Outlook 1998* (Paris, France, November 1998), Business As Usual Case, pp. 412-463. **PIRA**: PIRA Energy Group, *Retainer Client Seminar* (New York, NY, October 1999), Tables II-4, II-6, and II-7. **PEL**: Petroleum Economics, Ltd., *Oil and Energy Outlook to 2015* (London, United Kingdom, February 2000).

EE/FSU region falls outside the range defined by the *IEO2000* low and high growth cases: PEL's projected growth rate of 0.4 percent per year is well below the *IEO2000* low growth case projection of 0.6 percent per year. The *IEO2000* reference case projects that energy use in the EE/FSU will increase by about 1.0 percent per year between 1995 and 2010.

The regions that comprise the developing world also are subject to a fair amount of uncertainty. Other Asia (including India and South Korea) shows the greatest variation among the forecasts for the developing world. The projections of energy consumption growth for Other Asia from 1995 to 2010 vary from 3.4 percent per year in the *IEO2000* reference case to the PIRA forecast of 4.6 percent per year, which exceeds the *IEO2000* high economic growth case projection of 4.3 percent per year. For China, the forecasts for 1995-2010 average annual growth in energy use range from 3.5 percent per year (DRI) to 4.5 percent per year (*IEO2000*, PIRA, and PEL), but none of the forecasts falls outside the range of the *IEO2000* low and high economic growth cases.

Only *IEO2000* and PEL provide forecasts for energy use in 2015, the end of the PEL forecast time horizon (Table 4). As was the case for the 1995-2010 projections, the two forecasts for 1995 to 2015 project the same growth rates for worldwide energy consumption—2.0 percent per year. The *IEO2000* forecast shows higher expectations for growth in energy use in the industrialized world and the EE/FSU but lower expectations for the developing world (except for China and Latin America) than in the PEL forecast. The largest variation is seen for China and the EE/FSU region. *IEO2000* projects an average annual

increase in EE/FSU consumption of 1.1 percent between 1995 and 2015, compared with PEL's projection of 0.6 percent per year; and *IEO2000* projects an average annual increase in China's consumption of 4.2 percent between 1995 and 2015, compared with PEL's projection of 3.8 percent per year.

IEO2000, IEA, and DRI all provide energy consumption projections for 2020 (Table 5). Again, expectations among the forecast services for the growth of world total energy consumption are similar, ranging from 2.0-percent annual growth (*IEO2000* and IEA) to 2.2 percent (DRI). For the EE/FSU, the projected growth rates range from 0.9 percent per year (DRI) to 1.5 percent per year (IEA), with *IEO2000* at 1.3 percent per year.

For some of the world's developing regions, the three forecasts are similar. For example, all three expect energy consumption in Africa to grow by 2.6 percent per year between 1995 and 2020, and all three expect similar growth rates for a combined developing Asia (3.7 percent per year in *IEO2000*, 3.5 percent in IEA, and 3.6 percent per year in DRI). Within developing Asia, however, there are differences among the forecasts in the expectations for China relative to the rest of developing Asia. Both IEA and DRI project slower growth in energy use for China than for "Other Asia" over the 1995-2020 time period, whereas *IEO2000* projects the reverse.

One key reason for the differences among the various forecasts is that they are based on different expectations about future economic growth rates. *IEO2000*, PIRA, and PEL all provide GDP growth rates for the 1995-2010 time period (Table 6), and all have similar expectations

Table 4. Comparison of Energy Consumption Growth Rates by Region, 1995-2015
(Average Annual Percent Growth)

Region	IEO2000 Low Growth	IEO2000 Reference	IEO2000 High Growth	IEO99	PEL
Industrialized Countries	0.8	1.1	1.5	1.2	1.0
United States and Canada	1.0	1.3	1.5	1.3	1.1
Western Europe	0.5	1.0	1.3	1.1	1.1
Industrialized Asia	0.5	1.0	1.4	1.0	1.0
EE/FSU	0.7	1.1	1.7	1.0	0.6
Developing Countries	2.0	3.5	4.5	3.6	3.7
Asia	2.1	3.7	4.6	3.8	3.9
China	2.0	4.2	5.1	4.1	3.8
Other Asia[a]	2.1	2.9	4.0	3.4	3.9
Middle East	1.8	3.0	4.2	3.2	3.4
Africa	1.5	2.5	3.5	2.4	2.5
Latin America	2.4	3.7	5.0	4.0	3.6
Total World	1.2	2.0	2.7	2.1	2.0

[a]Other Asia includes India and South Korea.

Sources: *IEO2000*: Energy Information Administration (EIA), World Energy Projection System (2000). *IEO99*: EIA, *International Energy Outlook 1999*, DOE/EIA-0484(99) (Washington, DC, March 1999), Table A1, p. 141. **PEL**: Petroleum Economics, Ltd., *Oil and Energy Outlook to 2015* (London, United Kingdom, February 2000).

Table 5. Comparison of Energy Consumption Growth Rates by Region, 1995-2020
(Average Annual Percent Growth)

Region	IEO2000			IEO99	DRI	IEA
	Low Growth	Reference	High Growth			
Industrialized Countries	0.7	1.0	1.4	1.1	1.2	1.0
United States and Canada	0.9	1.1	1.4	1.2	1.2	0.8
Western Europe	0.5	0.9	1.3	1.1	1.3	1.1
Industrialized Asia	0.4	0.9	1.3	1.0	1.5	1.2
EE/FSU	0.9	1.3	2.0	1.1	0.9	1.5
Developing Countries	2.1	3.5	4.6	3.5	3.5	3.5
Asia	2.1	3.7	4.6	3.7	3.6	3.9
China	2.1	4.2	5.1	4.1	3.3	3.6
Other Asia[a]	2.1	3.1	4.0	3.3	3.8	4.2
Middle East	1.8	3.0	4.2	3.0	3.4	2.6
Africa	1.6	2.6	3.6	2.3	2.6	2.6
Latin America	2.5	3.8	5.2	3.9	3.9	3.2
Total World	1.2	2.0	2.8	2.1	2.2	2.0

[a]Other Asia includes India and South Korea.

Sources: **IEO2000**: Energy Information Administration (EIA), World Energy Projection System (2000). **IEO99**: EIA, *International Energy Outlook 1999, DOE/EIA-0484(99) (Washington, DC, March 1999)*, Table A1, p. 141. **DRI**: Standard & Poor's DRI, *World Energy Service: World Outlook* (Lexington, MA, January 1999), p. 5. **IEA**: International Energy Agency, *World Energy Outlook 1998* (Paris, France, November 1998), Business As Usual Case, pp. 412-463.

Table 6. Comparison of Economic Growth Rates by Region, 1995-2010
(Average Annual Percent Growth in Gross Domestic Product)

Region	IEO2000			IEO99	PIRA	PEL[a]
	Low Growth	Reference	High Growth			
Industrialized Countries	1.5	2.3	3.2	2.3	2.5	—
United States and Canada	1.8	2.7	3.6	2.6	2.8	2.9
Western Europe	1.5	2.4	3.3	2.5	2.4	2.5
Industrialized Asia	0.6	1.5	2.3	1.5	1.6	1.4
EE/FSU	1.7	3.0	5.0	2.6	3.1	—
Former Soviet Union	0.9	2.2	3.8	1.3	—	1.4
Eastern Europe	3.0	4.3	6.9	4.5	—	3.3
Developing Countries	3.2	4.8	6.0	4.8	4.8	—
Asia	3.8	5.5	6.8	5.3	5.3	—
China	4.6	7.2	8.5	6.9	6.3	7.0
Other Asia[b]	3.4	4.7	6.0	4.5	4.2	3.3
Middle East	2.1	3.4	4.7	4.0	3.6	3.1
Africa	2.4	3.7	5.0	3.6	3.6	3.5
Latin America	2.8	4.0	5.3	4.3	3.6	3.3
Total World	1.8	2.9	3.9	2.9	3.6	2.9

[a]North America includes only the United States. Industrialized Asia includes only Japan.
[b]Other Asia includes India and South Korea.

Sources: **IEO2000**: Energy Information Administration (EIA), World Energy Projection System (2000). **IEO99**: EIA, *International Energy Outlook 1999, DOE/EIA-0484(99) (Washington, DC, March 1999)*, Table A1, p. 141. **PIRA**: PIRA Energy Group, *Retainer Client Seminar* (New York, NY, October 1999), Tables II-4, II-6, and II-7. **PEL**: Petroleum Economics, Ltd., *Oil and Energy Outlook to 2015* (London, United Kingdom, February 2000).

for economic growth in the industrialized world: relatively higher growth is projected for the United States and Canada and for Western Europe than is projected for industrialized Asia. The GDP assumptions for the industrialized regions in *IEO2000* are similar to those in *IEO99*. (DRI is not included in the comparisons of economic growth projections, because *IEO2000* adopts the DRI forecast for modeling purposes. As a result, the DRI and *IEO2000* economic growth expectations are exactly the same, and differences in energy consumption forecasts must be attributed to other characteristics of the two modeling systems.)

Expectations for economic growth in the EE/FSU region over the 1995-2010 time period are also nearly the same across forecasts for the total region, ranging from 3.0 percent per year (*IEO2000*) to 3.1 percent (PIRA). PEL does not provide GDP growth rate assumptions for the total EE/FSU region, but the PEL forecast is less optimistic than the others about the potential growth of both Eastern Europe and the FSU, and thus its projected growth rates for the entire region are likely to be lower. In any case, where separate estimates are provided, there is general consensus among the forecasts that Eastern Europe will enjoy substantially better economic growth than the FSU between 1995 and 2010.

All the forecasters have similar views about developing Asia's economic growth between 1995 and 2010. China is expected to have the highest GDP growth in all the forecasts, ranging from 6.3 percent per year (PIRA) to 7.2 percent per year (*IEO2000*). All the GDP growth estimates fall within the range defined by the *IEO2000* high

and low economic growth cases. The PEL estimates may be somewhat exaggerated relative to the other sources, however, because GDP growth rates are only provided between 1997 and 2010 in most cases, and between 1998 and 2005 for Africa. Given the importance of 1997 as the first year of the recession in southeast Asia, this may skew the PEL rates.

Two projection series, *IEO2000* and IEA, provide economic growth estimates for the 1995-2020 period (Table 7). Expectations for GDP growth in the industrialized regions are similar in the two forecasts, but the IEA forecast shows lower economic growth rates than *IEO2000* for North America and Western Europe and higher growth rates for industrialized Asia. IEA is slightly less optimistic than *IEO2000* about the rate of recovery in the EE/FSU. This may be explained by the regional accounting of the Czech Republic and Hungary—two of Eastern Europe's stronger economies—which the IEA does not include in EE/FSU. They are included in Western Europe (the OECD Europe designation) in the *World Energy Outlook 1998*.

Projections vary not only with respect to levels of total energy demand and economic growth but also with respect to the composition of primary energy inputs. Four of the forecasts—*IEO2000*, IEA, PIRA, and DRI—provide energy consumption projections by fuel in 2010. Table 8 shows a summary of the projections for growth in energy use by fuel from each of the forecasts between 1995 and 2010. Unfortunately, DRI does not provide separate projections for nuclear and "other" energy sources, but instead provides a forecast for "primary electricity,"

Table 7. Comparison of Economic Growth Rates by Region, 1995-2020
(Average Annual Percent Growth in Gross Domestic Product)

Region	IEO2000			IEO99	IEA
	Low Growth	Reference	High Growth		
Industrialized Countries	1.3	2.2	3.1	2.2	—
United States and Canada	1.5	2.4	3.3	2.2	2.1
Western Europe	1.4	2.3	3.2	2.4	2.0
Industrialized Asia	0.7	1.6	2.5	1.9	1.8
EE/FSU	2.4	3.8	5.7	2.9	3.3
Developing Countries	3.0	4.7	6.0	4.8	—
Asia	3.4	5.3	6.6	5.3	—
China	3.7	6.5	7.9	6.7	5.5
Other Asia[a]	3.2	4.6	5.9	4.6	4.2-4.5
Middle East	2.3	3.6	5.0	4.1	2.7
Africa	2.4	3.8	5.2	3.6	2.5
Latin America	2.8	4.1	5.4	4.4	3.3
Total World	1.7	2.8	3.9	2.9	3.1

[a]Other Asia includes India and South Korea.

Sources: **IEO2000:** Energy Information Administration (EIA), World Energy Projection System (2000). **IEO99:** EIA, *International Energy Outlook 1999, DOE/EIA-0484(99) (Washington, DC, March 1999), Table A1, p. 141.* **IEA:** International Energy Agency, *World Energy Outlook 1998* (Paris, France, November 1998), Business As Usual Case, pp. 412-463.

the combination of consumption from the two energy sources.

In terms of oil consumption, all the forecasts expect similar growth worldwide. Oil demand is projected to increase by between 2.0 percent per year (*IEO2000*, PIRA, and IEA) and 2.2 percent per year (DRI). Most of the forecasting sources expect strongest growth in natural gas over the 1995-2010 time period, the only exception being the IEA forecast, where growth in hydroelectricity and other forms of renewable energy equals the expected growth in gas use. Moreover, the optimism of the IEA regarding renewable energy is demonstrated by the fact it is the only energy source projection among all of the forecasts that exceeds the high economic growth case projection from *IEO2000*. *IEO2000* remains slightly more pessimistic than the other forecasting sources regarding the growth of coal consumption over the 15-year projection period.

PEL and *IEO2000* provide by-fuel forecast values for the year 2015 (Table 9). The two forecasts have similar views about oil, nuclear, and renewable energy consumption between 1995 and 2015, with some differences in the forecasts for natural gas and coal. *IEO2000* expects the stronger growth in natural gas use to displace coal consumption, particularly for electric power generation. PEL, on the other hand, expects coal use to substantially keep pace with natural gas over the 1995-2015 time horizon.

IEO2000, IEA, and DRI are the only forecasts that provide estimates for 2020 (Table 10). The projected growth rates among the fuels and forecast services largely parallel those for the 1995-2010 time period. The most noticeable difference is the expectation over all three services that nuclear energy consumption will either begin to decline or at least slow considerably between 2010 and 2020.

Table 8. Comparison of World Energy Consumption Growth Rates by Fuel, 1995-2010
(Average Annual Percent Growth)

Fuel	IEO2000 Low Growth	IEO2000 Reference	IEO2000 High Growth	IEO99	DRI	IEA	PIRA	PEL
Oil	1.2	2.0	2.7	2.0	2.2	2.0	2.0	2.1
Natural Gas	2.5	3.1	3.9	3.5	3.0	2.8	3.3	2.9
Coal	0.6	1.7	2.4	1.6	2.0	2.2	2.4	1.8
Nuclear	0.3	0.7	1.1	0.5	—[a]	0.6	0.7	0.6
Renewable/Other	1.2	2.0	2.7	2.2	—[a]	2.8	1.7	2.0
Total.	**1.3**	**2.1**	**2.8**	**2.2**	**2.3**	**2.2**	**2.3**	**2.1**
Primary Electricity.	0.8	1.4	2.0	1.5	1.8	1.3	1.2	1.4

[a]DRI reports nuclear and hydroelectric power together as "primary electricity."

Sources: *IEO2000*: Energy Information Administration (EIA), World Energy Projection System (2000). *IEO99*: EIA, *International Energy Outlook 1999*, DOE/EIA-0484(99) (Washington, DC, March 1999), Table A1, p. 141. **DRI**: Standard & Poor's DRI, *World Energy Service: World Outlook* (Lexington, MA, January 1999), p. 5. **IEA**: International Energy Agency, *World Energy Outlook 1998* (Paris, France, November 1998), Business As Usual Case, pp. 412-463. **PIRA**: PIRA Energy Group, *Retainer Client Seminar* (New York, NY, October 1999), Tables II-4, II-6, and II-7. **PEL**: Petroleum Economics, Ltd., *Oil and Energy Outlook to 2015* (London, United Kingdom, February 2000).

Table 9. Comparison of World Energy Consumption Growth Rates by Fuel, 1995-2015
(Average Annual Percent Growth)

Fuel	IEO2000 Low Growth	IEO2000 Reference	IEO2000 High Growth	IEO99	PEL
Oil	1.2	2.0	2.7	1.9	2.1
Natural Gas	2.4	3.1	3.7	3.4	2.7
Coal	0.4	1.4	2.2	1.6	1.6
Nuclear	-0.2	0.3	0.7	0.1	0.2
Renewable/Other	1.0	1.8	2.5	2.1	2.0
Total.	**1.2**	**2.0**	**2.7**	**2.1**	**2.0**
Primary Electricity.	0.5	1.2	1.8	1.3	1.3

Sources: *IEO2000*: Energy Information Administration (EIA), World Energy Projection System (2000). *IEO99*: EIA, *International Energy Outlook 1999*, DOE/EIA-0484(99) (Washington, DC, March 1999), Table A1, p. 141. **PEL**: Petroleum Economics, Ltd., *Oil and Energy Outlook to 2015* (London, United Kingdom, February 2000).

Table 10. Comparison of World Energy Consumption Growth Rates by Fuel, 1995-2020
(Average Annual Percent Growth)

Fuel	IEO2000 Low Growth	IEO2000 Reference	IEO2000 High Growth	IEO99	DRI	IEA
Oil	1.2	1.9	2.7	1.8	2.2	1.9
Natural Gas	2.4	3.1	3.8	3.3	2.9	2.6
Coal	0.4	1.6	2.4	1.7	2.1	2.1
Nuclear	-0.6	-0.2	0.3	-0.3	—[a]	-0.0
Renewable/Other	1.0	1.8	2.5	2.0	—[a]	2.5
Total	1.2	2.0	2.8	2.1	2.2	2.0
Primary Electricity	0.4	1.1	1.7	1.2	1.1	0.9

[a]DRI reports nuclear and hydroelectric power together as "primary electricity."

Sources: *IEO2000*: Energy Information Administration (EIA), World Energy Projection System (2000). *IEO99*: EIA, *International Energy Outlook 1999*, DOE/EIA-0484(99) *(Washington, DC, March 1999), Table A1, p. 141*. **DRI**: Standard & Poor's DRI, *World Energy Service: World Outlook* (Lexington, MA, January 1999), p. 5. **IEA**: International Energy Agency, *World Energy Outlook 1998* (Paris, France, November 1998), Business As Usual Case, pp. 412-463.

Performance of Past *IEO* Forecasts for 1990 and 1995

In an effort to measure how well the *IEO* projections have estimated future energy consumption trends over the series' 14-year history, we present a comparison of *IEO* forecasts produced for the years 1990 and 1995. The forecasts are compared with actual data published in EIA's *International Energy Annual 1997* [16], as part of EIA's commitment to provide users of the *IEO* with a set of performance measures to assess the forecasts produced by this agency.

The *IEO* has been published since 1985. In *IEO85*, mid-term projections were derived only for the world's market economies. That is, no projections were prepared for the centrally planned economies (CPE) of the Soviet Union, Eastern Europe, Cambodia, China, Cuba, Laos, Mongolia, North Korea, and Vietnam. The *IEO85* projections extended to 1995 and included forecasts of energy consumption for 1990 and 1995 and primary consumption of oil, natural gas, coal, and "other fuels." *IEO85* projections were also presented for several individual countries and subregions: the United States, Canada, Japan, the United Kingdom, France, West Germany, Italy, the Netherlands, other OECD Europe, other OECD (Australia, New Zealand, and the U.S. Territories), OPEC, and other developing countries. Beginning with *IEO86*, nuclear power projections were published separately from the "other fuel" category.

The regional aggregation has changed from report to report. In 1990, the report coverage was expanded for the first time from coverage of only the market economies to coverage of the entire world. Projections for China, the former Soviet Union, and other CPE countries were provided separately.

Historical data for total regional energy consumption in 1990 show that the *IEO* projections from those early years were consistently lower than the actual data for the market economies. For the four editions of the *IEO* printed between 1985 and 1989 (no *IEO* was published in 1988) in which 1990 projections were presented, total projected energy consumption in the market economies ran between 2 and 5 percent below the actual amounts published in the *International Energy Annual 1997* (Figure 23).

In addition, market economy projections for 1995 in the 1985 through 1993 *IEO* reports (EIA did not release forecasts for 1995 after the 1993 report) were consistently

Figure 23. Comparison of *IEO* Forecasts with 1990 Energy Consumption in Market Economies

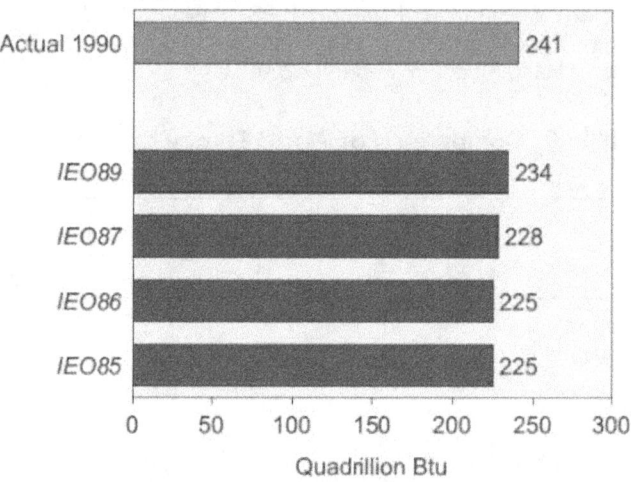

Sources: **History:** Energy Information Administration (EIA), *International Energy Annual 1997*, DOE/EIA-0219(97) (Washington, DC, April 1999). **Projections:** EIA, International Energy Outlook, DOE/EIA-0484 (Washington, DC, various years).

lower than the historical 1995 data (Figure 24). Most of the difference is attributed to those market economy countries outside the Organization for Economic Cooperation and Development (OECD). Through the years, EIA's economic growth assumptions for OPEC and other market economy countries outside the OECD have been low. The 1993 forecast was, as one might expect, the most accurate of the forecasts for 1995, but its projection

for OPEC and the other market economy countries was still more than 10 percent below the actual number.

IEO90 marked the first release of a worldwide energy consumption forecast. Since *IEO90*, the forecasts for worldwide energy demand have been between 2 and 5 percent higher than the actual amounts consumed (Figure 25). Much of the difference can be explained by the unanticipated collapse of the Soviet Union economies in the early 1990s. The *IEO* forecasters could not foresee the extent to which energy consumption would fall in this region. In *IEO90*, total energy consumption in the FSU was projected to reach 67 quadrillion Btu in 1995. The projection was reduced steadily in the next three *IEO* reports, but even in 1993 energy demand for 1995 in the FSU region was still projected to be 53 quadrillion Btu, as compared with actual 1995 energy consumption of 43 quadrillion Btu, some 10 quadrillion Btu (or about 5 million barrels of oil per day) less than projected in *IEO93*.

Considering the forecasts for the year 1995 strictly in terms of depicting future trends associated with the fuel mix, the *IEO* reports have performed well. Each *IEO* since 1990 has projected the fuel mix within 2 percentage points of the actual 1995 mix. The earliest *IEO*s tended to be too optimistic about the growth of coal use in the market economies[6] (Figure 26), and not optimistic enough about the recovery of oil consumption after the declines in the early 1980s that followed the price shocks caused by oil embargoes in 1973 and 1974 and the 1979-1980 revolution in Iran (Figure 27). The *IEO85* and *IEO86* reports projected that oil would account for only

Figure 24. Comparison of *IEO* Forecasts with 1995 Energy Consumption in Market Economies

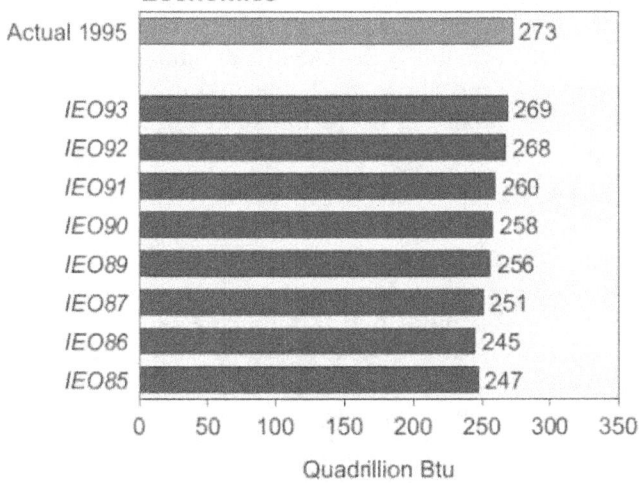

Sources: **History:** Energy Information Administration (EIA), *International Energy Annual 1997*, DOE/EIA-0219(97) (Washington, DC, April 1999). **Projections:** EIA, International Energy Outlook, DOE/EIA-0484 (Washington, DC, various years).

Figure 25. Comparison of *IEO* Forecasts with 1995 World Energy Consumption

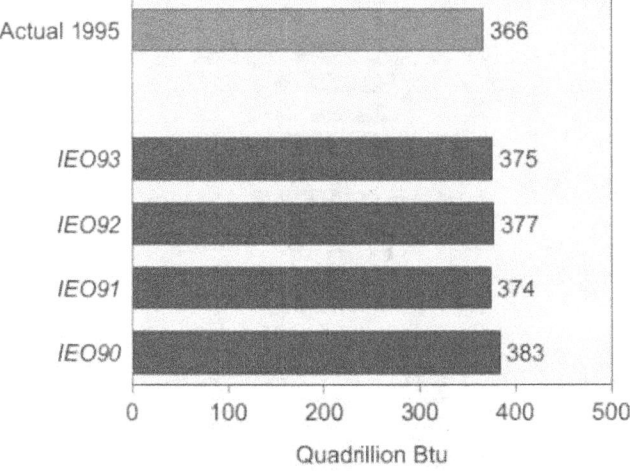

Sources: **History:** Energy Information Administration (EIA), *International Energy Annual 1997*, DOE/EIA-0219(97) (Washington, DC, April 1999). **Projections:** EIA, International Energy Outlook, DOE/EIA-0484 (Washington, DC, various years).

Figure 26. Comparison of *IEO* Forecasts with 1995 Coal Consumption in Market Economies

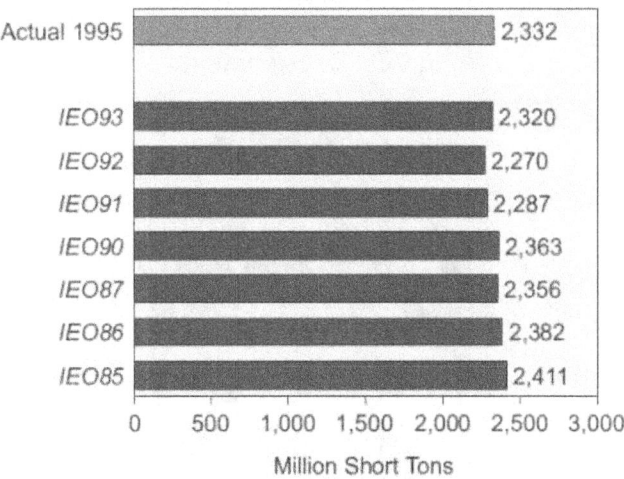

Sources: **History:** Energy Information Administration (EIA), *International Energy Annual 1997*, DOE/EIA-0219(97) (Washington, DC, April 1999). **Projections:** EIA, International Energy Outlook, DOE/EIA-0484 (Washington, DC, various years).

[6]Projections for West Germany and later unified Germany have been removed from the values considered here because of the lack of continuity in the coal data series after reunification.

about 40 percent of total energy consumption for the market economies in 1995, whereas oil actually accounted for 45 percent of the total in 1995.

The forecasts for world coal consumption that appeared in the *IEO*s printed between 1990 and 1993 were consistently high, between 4 and 17 percent higher than actual coal use (Figure 28), largely because of overestimates for the former Soviet Union and Eastern Europe—regions that experienced substantial declines in coal consumption during the years following the collapse of the Soviet Union. Most of the by-fuel projections for the FSU were greater than the actual consumption numbers, with the

exception of hydroelectricity and other renewable resources (Figure 29). Natural gas use did not decline as much as oil and coal use because gas is a plentiful resource in the region and was used extensively to fuel the domestic infrastructure, but even the *IEO* estimates for 1995 natural gas use were 16 to 22 percent higher than the actual use.

The EIA projections for total energy consumption in China were below the actual 1995 consumption level in *IEO90* (by 12 percent) and *IEO91* (by 7 percent) but higher in *IEO92* (by 7 percent) and about the same in *IEO93*. The underestimates in the earlier *IEO*s balanced, in part, the overestimates for the EE/FSU countries; however, even the 3- to 16-percent underestimate of projected 1995 coal use in China could not make up for the 30- to 55-percent overestimate of FSU coal use. In terms of other fuels, EIA consistently overestimated China's gas consumption and underestimated its oil

Figure 27. Comparison of *IEO* Forecasts with 1995 Oil Consumption in Market Economies

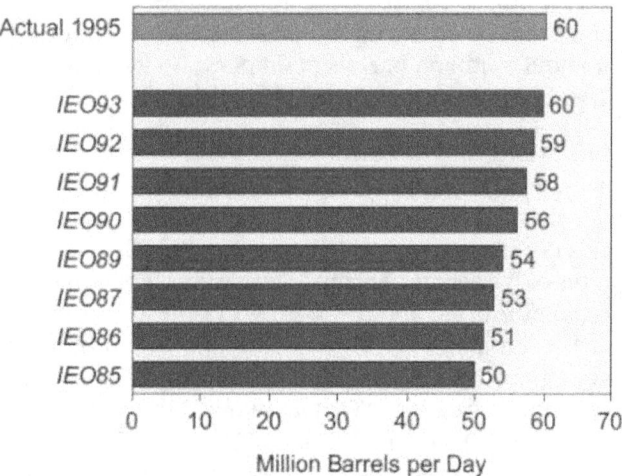

Million Barrels per Day

Sources: **History:** Energy Information Administration (EIA), *International Energy Annual 1997*, DOE/EIA-0219(97) (Washington, DC, April 1999). **Projections:** EIA, International Energy Outlook, DOE/EIA-0484 (Washington, DC, various years).

Figure 28. Comparison of *IEO* Forecasts with 1995 World Coal Consumption

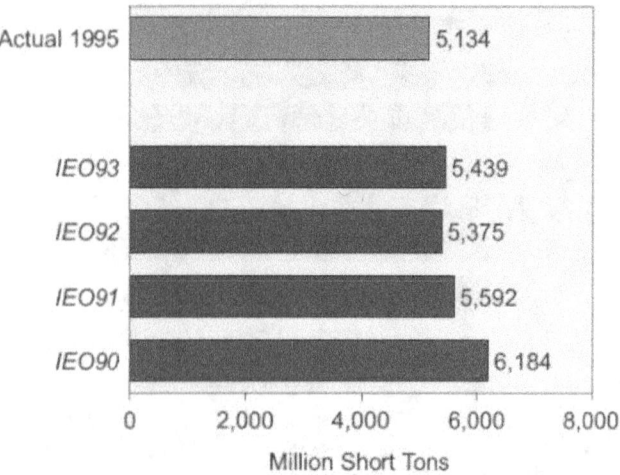

Million Short Tons

Sources: **History:** Energy Information Administration (EIA), *International Energy Annual 1997*, DOE/EIA-0219(97) (Washington, DC, April 1999). **Projections:** EIA, International Energy Outlook, DOE/EIA-0484 (Washington, DC, various years).

Figure 29. Comparison of *IEO* Forecasts with 1995 Energy Consumption in the Former Soviet Union by Fuel Type

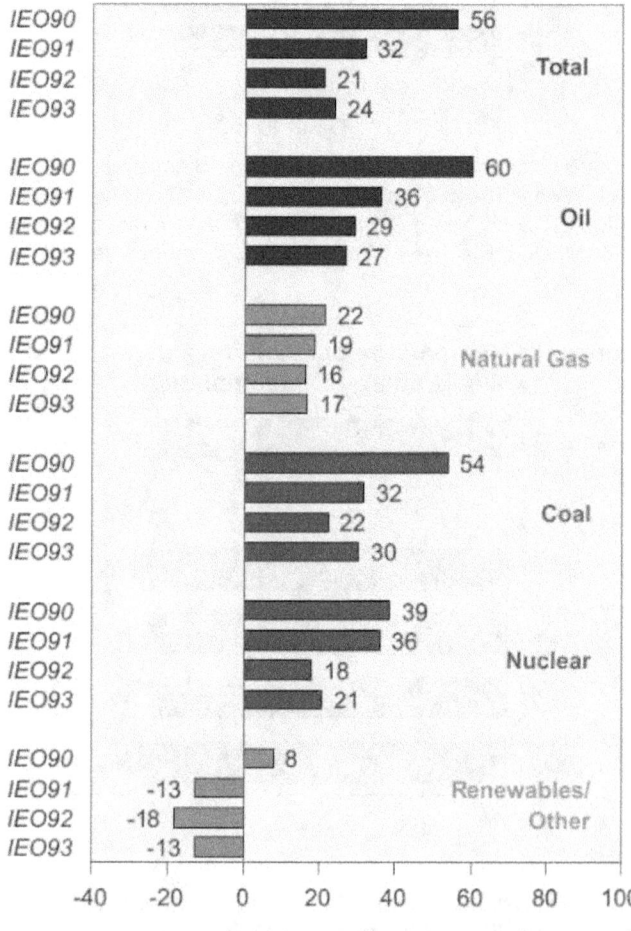

Percent Difference From Actual 1995

Sources: **History:** Energy Information Administration (EIA), *International Energy Annual 1997*, DOE/EIA-0219(97) (Washington, DC, April 1999). **Projections:** EIA, International Energy Outlook, DOE/EIA-0484 (Washington, DC, various years).

consumption. Nuclear power forecasts were fairly close for China, within 5 percent of the actual consumption (Figure 30). It is noteworthy, however, that consumption of natural gas and nuclear power was quite small in 1995, so that any variation between actual historical consumption and the projections results in a large percentage difference. EIA consistently underestimated economic growth in China. As late as 1993, EIA expected GDP in China to grow by about 7.3 percent per year during the decade of the 1990s, whereas it actually grew by 10.7 percent per year between 1990 and 1995.

The comparison of *IEO* projections and historical data in the context of political and social events underscores the importance of these events in shaping the world's energy markets. Such comparisons also point out how important a model's assumptions are to the derivation of accurate forecasts. The political and social upheaval in Eastern Europe and the former Soviet Union was not predictable, and it dramatically affected the accuracy of the projections for the region. If higher economic growth rates had been assumed for China, more accurate forecasts for that region might have been achieved. It is important for users of the *IEO* or any other projection series to realize the limitations of the forecasts. Failing an ability to predict future volatility in social, political, or economic events, the projections should be used as a plausible path or trend for the future and not as a precise prediction of future events.

References

1. Energy Information Administration, *Short-Term Energy Outlook* (February 2000), Table A3, web site www.eia.doe.gov/emeu/steo/pub/contents.html.

2. Cambridge Energy Research Associates, *Global Energy Watch: Afloat But Adrift? Energy Markets and the Undercurrent of Change* (Cambridge, MA, November 1999), p. 6.

3. WEFA Group, *Asia Monthly Monitor* (Eddystone, PA, October 1999), p. IN.1

4. E. Eckholm and D.E. Sanger, "China and U.S. Sign Landmark Trade Deal," *The New York Times on the Web, Late News* (November 15, 1999), web site www.nytimes.com.

5. S. Faison, "Reformer's Comeback: New Power Against Opponents of Open Markets," *The New York Times* (November 16, 1999), p. A11.

6. "Monthly Report: Russia, Poland, Czech Republic, Hungary," *PlanEcon Report*, Vol. 15, No. 19 (October 22, 1999), p. 7.

7. PlanEcon, *Review and Outlook for the Former Soviet Republics* (Washington, DC, October 1999), p. iii.

8. S. Romero, "Rally in Latin Markets Hints at Regional Revival," *The New York Times* (November 6, 1999).

9. "Euro: One Currency for Europe: Quest—Questions and Answers" (August 31, 1999), web site http://europe.eu.int/euro/html/entry.html.

10. Cambridge Energy Research Associates, *Global Energy Watch: Afloat But Adrift? Energy Markets and the Undercurrent of Change* (Cambridge, MA, November 1999), pp. 22-23.

11. Standard & Poor's DRI, *The U.S. Economy: 1999/8* (Lexington, MA, August 1999), p. 44.

12. S. WuDunn, "Japan Raises Plan To Revive Economy Up to $195 Billion," *The New York Times on the Web* (November 16, 1998), web site www.nytimes.com.

13. S. WuDunn, "Japan's Parliament Passes $517 Billion Bank Rescue Package" *The New York Times on the Web* (October 17, 1998), web site www.nytimes.com.

Figure 30. Comparison of *IEO* Forecasts with 1995 Energy Consumption in China by Fuel Type

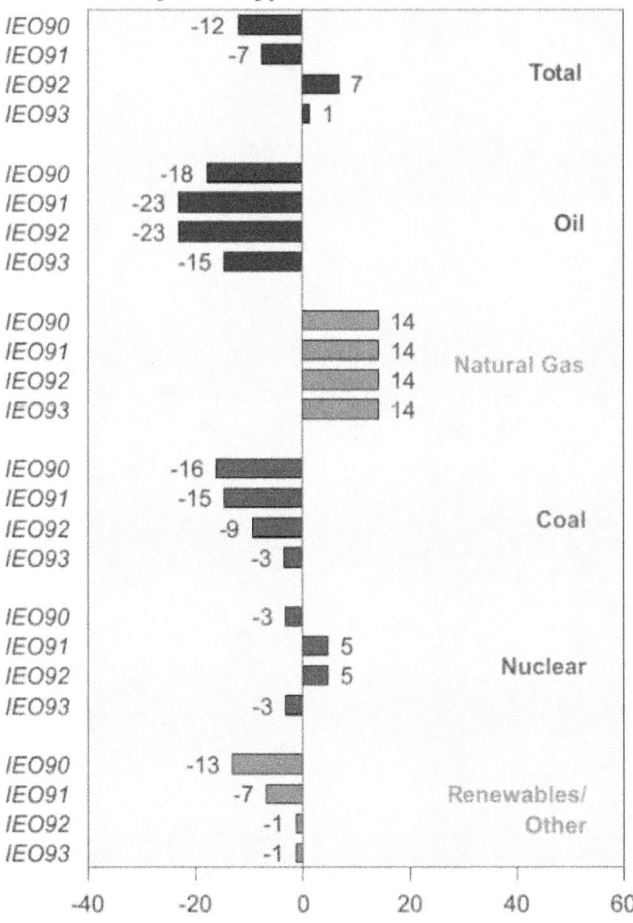

Percent Difference From Actual 1995

Sources: **History:** Energy Information Administration (EIA). *International Energy Annual 1997*, DOE/EIA-0219(97) (Washington, DC, April 1999). **Projections:** EIA, International Energy Outlook, DOE/EIA-0484 (Washington, DC, various years).

14. United Nations Framework Convention on Climate Change, "The Convention and Kyoto Protocol" (October 12, 1999), web site www.unfcc/de/resource/convkp.html.

15. Energy Information Administration, *International Energy Annual 1997*, DOE/EIA-0219(97) (Washington, DC, April 1999), and Standard and Poor's DRI, *World Economic Outlook*, Vol. 1 (Lexington, MA, Third Quarter 1999).

16. Energy Information Administration, *International Energy Annual 1997*, DOE/EIA-0219(97) (Washington, DC, April 1999).

World Oil Markets

The IEO2000 projections reflect a change in short-term expectations for world oil prices. In the long term, OPEC production cutbacks are expected to be relaxed, and prices are projected to rise gradually through 2020 as the oil resource base is expanded.

The crude oil market rebounded dramatically in 1999. Prices rose from the low monthly average of $9.39 per barrel (nominal U.S. dollars) in December 1998 to $24.44 in December 1999, an increase of almost $15 a barrel. Prices were influenced by the successful adherence to announced cutbacks in production by members of the Organization of Petroleum Exporting Countries (OPEC) as well as several non-OPEC countries, notably, Mexico and Norway. In addition, the price decline in 1998 significantly dampened the annual production growth that non-OPEC suppliers had provided since the mid-1990s, and petroleum demand in Southeast Asia began to recover from the severe recession of 1997-98.

Oil consumption rose in 1999 by slightly more than 1 million barrels per day with industrialized nations accounting for about one-half of the increase. Before the 1998 recession, oil demand in developing Asia (including China) had grown at a robust annual rate of about 8.0 percent between 1991 and 1997. As the Asian economies began recovering in 1999, oil demand grew in China by 5.1 percent and in the rest of Asia grew by 2.0 percent. With economic problems in Brazil and political uncertainty in Colombia, Ecuador, and Venezuela, oil demand in Latin America did not increase in 1999. Persistent economic problems in Russia caused declines in oil demand in both 1998 and 1999 for the former Soviet Union (FSU); however, oil demand in the FSU is expected to show slight growth in 2000 [1].

On March 23, 1999, OPEC (not including Iraq) agreed to production cutbacks totaling 1.7 million barrels per day. Four non-OPEC suppliers (Mexico, Norway, Russia, and Oman) pledged an additional 0.4 million barrels per day. Two earlier OPEC ministerial meetings in 1998 had yielded plans for oil production cutbacks that were never successfully realized, but the active encouragement by non-OPEC producers may have lent an air of seriousness to the more recent OPEC pledges. Since the March 23rd meeting, OPEC's production management efforts have been successful, and their target of raising prices above $20 per barrel has been met. A September 22-23, 1999, OPEC ministerial meeting yielded no additional production cutbacks, but there was agreement to hold the current course. The question now is when OPEC will raise the production targets for its members.

At the beginning of 1999, constraints on worldwide oil supplies were becoming evident as the low oil price environment prevailed. Stripper production in the United States was in decline. Exploration and development spending was being slashed. Rig utilization rates, especially for onshore equipment, had drastically fallen. Announced spending plans worldwide were reduced. Oil-producing countries faced severe fiscal deficits, causing national oil companies to cut capital spending. Private-sector restructuring led to mergers involving leading multinational oil companies. The oil market pessimism prevalent at the beginning of 1999 was not evident, however, by the end of the year.

Incorporating the recent price turbulence into the construction of an intermediate- and long-term oil market outlook is difficult and raises the following questions: Will prices remain above $20 per barrel even when the production targets of OPEC producers are raised and significant increases in non-OPEC production are once again expected? Will sustained and robust economic growth in developing countries continue, given the severe setback to the Asian economies in 1998? Will technology guarantee that oil supply development moves forward even in a low world oil price environment?

Although oil prices more than doubled in real terms from 1998 to 1999, that development is not indicative of the trend in the *International Energy Outlook 2000* (*IEO2000*) reference case. In the short term, oil prices in are expected to continue at the levels seen during the later months of 1999 into 2000. As OPEC production cutbacks are relaxed and non-OPEC production increases are realized, however, prices are expected to fall back slightly from the 2000 level, then increase gradually out to 2020. When the economic recovery in Asia is complete, demand growth in developing countries throughout the world is expected to be sustained at robust levels. Worldwide oil demand reaches almost 113 million barrels per day by 2020 in the reference case, requiring an increment to world production capability of almost 40 million barrels per day relative to current capacity. OPEC producers are expected to be the major beneficiaries of increased production requirements, but non-OPEC supply is expected to remain competitive, with major increments to supply coming from offshore resources, especially in the Caspian Basin and deepwater West Africa.

Over the past 25 years, oil prices in real 1998 dollars have ranged from $12.10 to $63.30 per barrel. In the future, one can expect volatile behavior to recur principally because of unforeseen political and economic circumstances. Tensions in the Middle East, for example, could give rise to serious disruptions of normal oil production and trading patterns. On the other hand, significant excursions from the reference price trajectory are not likely to be long sustained. High real prices deter consumption and encourage the emergence of significant competition from marginal but potentially important sources of oil and non-oil energy supplies. Persistently low prices have the opposite effects (see box on page 28).

Limits to long-term oil price escalation include substitution of other fuels (such as natural gas) for oil, marginal sources of conventional oil that become reserves when prices rise, and nonconventional sources of oil that become reserves at still higher prices (see box in the natural gas chapter of this report, pages 45-46). Advances in exploration and production technologies are likely to bring down prices when such additional oil resources become part of the reserve base. The *IEO2000* low and high world oil price cases suggest that the projected trends in growth for oil production are sustainable without severe oil price escalation. There are oil market analysts, however, who find this viewpoint to be overly optimistic, based on what they consider to be a significant overestimation of both proven reserves and ultimately recoverable resources.

Highlights of the *IEO2000* projection for the world oil market are as follows:

- The reference case price projection shows an oil price increase of more than $4 per barrel from 1999 to 2000, a decline of slightly less than $3 per barrel in 2001, and then an 0.4-percent average annual increase through 2020.

- Deepwater exploration and development initiatives are generally expected to be sustained worldwide, with offshore West Africa emerging as a major future source of oil production. Technology and resource availability can sustain large increments in oil production capability at the reference case prices. The low price environment of 1998 and early 1999 did slow the pace of development in some prospective production areas, especially, the Caspian Basin region.

- Economic development in Asia is crucial to long-term growth in oil markets. The evolution of Asian oil demand projected in the reference case would strengthen economic ties between the Middle East and Asian markets.

- Although OPEC's share of world oil supply is projected to increase significantly over the next two decades, competitive forces among energy producers are expected to remain strong enough to forestall efforts to escalate real oil prices significantly. The competitive forces operate within OPEC, between OPEC and non-OPEC sources of supply, and between oil and other sources of energy (particularly, natural gas).

- The uncertainties associated with the *IEO2000* reference case projections are significant. Changes in the prospects for sustained economic recovery in developing Asia, Japan's economic turnaround, China's economic reforms, and economic recovery in Brazil, other Latin American economies, and the FSU could lead to oil market behavior quite different from that portrayed in the *IEO2000* projections.

Growth in Oil Demand

World petroleum consumption projections are slightly lower in *IEO2000* than in last year's forecast in the early years (about 1 percent in 2005), due to the much higher oil prices expected in the near term, as well as the lingering effects of the economic slowdown in Asia, Central and South America, and Russia. World oil consumption is expected to increase by 1.1 million barrels per day in 1999 [2], exceeding the increase of 0.5 million barrels per day in 1998 but lower than the average annual increase of nearly 1.6 million barrels per day from 1994 to 1997.

Oil provides a larger share of world energy consumption than any other energy source, at 39 percent of the total in 1997. Petroleum is used heavily in the transportation sector and is also used to provide heat and power as well as industrial feedstocks. World oil consumption is projected to increase by a total of 39.8 million barrels per day (an average rate of 1.9 percent per year), from 73 million barrels per day in 1997 to 112.8 million barrels per day in 2020 (Figure 31). Between 1970 and 1997 oil use rose by a total of 26.2 million barrels per day, an average annual increase of 1.7 percent; and the 1970-1997 growth might have been still larger without the price shocks of 1973-1974 and 1979-1980. Oil's share of the energy market is expected to decline only slightly over the forecast period.

The industrialized countries, currently the largest consumers of petroleum, are expected to remain the largest users through 2020. Oil consumption in the industrialized countries is projected to rise from 43.1 million barrels per day in 1997 to 54.5 million barrels per day in 2020. The developing countries, however, are expected to make the largest contribution to the increment in oil demand, an increase of 24.7 million barrels per day from 1997 to 2020 (Figure 32), representing 62 percent of the growth in worldwide petroleum consumption.

Figure 31. World Oil Consumption by Region, 1970-2020

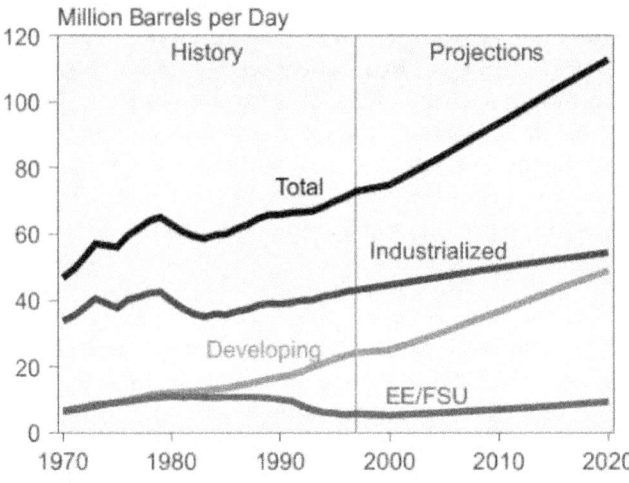

Million Barrels per Day

Sources: **History:** Energy Information Administration (EIA), Office of Energy Markets and End Use, International Statistics Database and *International Energy Annual 1997*, DOE/EIA-0219(97) (Washington, DC, April 1999). **Projections:** EIA, World Energy Projection System (2000).

Petroleum consumption in developing countries was just over one-half (56 percent) of the total consumption in industrialized countries in 1997 but is projected to reach 90 percent of that in the industrialized countries by 2020.

Regionally, developing Asia and North America increased oil use by the largest amount from 1970 to 1997, and they are expected to lead the increases in consumption in the forecast period (Figure 33). Substantial growth is also expected in Central and South America,

Figure 32. Increments in Oil Consumption by Region, 1997-2020

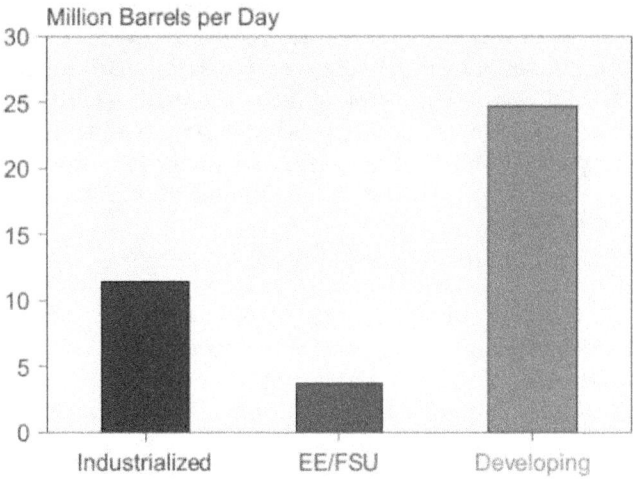

Million Barrels per Day

Source: Energy Information Administration, World Energy Projection System (2000).

the Middle East, and Africa. Consumption in Eastern Europe and the former Soviet Union (EE/FSU) was lower in 1997 than in 1970, as a result of political and economic difficulties during the 1990s, primarily in the FSU. Petroleum consumption in the FSU is expected to remain flat for the next few years and then start to rise after 2000. The EE/FSU total is projected to increase by 3.7 million barrels per day between 1997 and 2020.

Oil demand is driven by economic growth, along with rising population. The industrialized countries consume oil at much higher levels per capita than the developing countries, such as China and Brazil (Figure 34), but there are also large differences among the industrialized

Figure 33. Increments in Oil Consumption by Region, 1970-1997 and 1997-2020

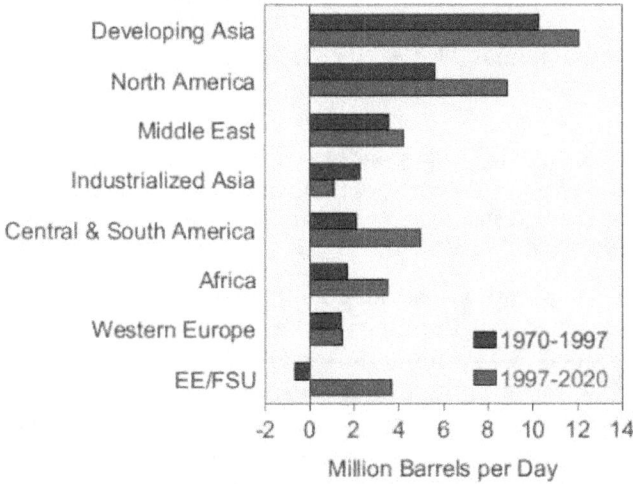

Million Barrels per Day

Sources: **1970 and 1997:** Energy Information Administration (EIA), Office of Energy Markets and End Use, International Statistics Database and *International Energy Annual 1997*, DOE/EIA-0219(97) (Washington, DC, April 1999). **2020:** EIA, World Energy Projection System (2000).

Figure 34. Per Capita Oil Use by Selected Country and Region, 1970-2020

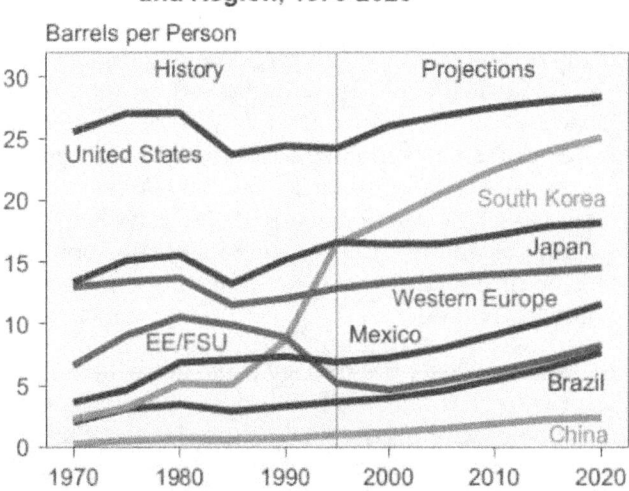

Barrels per Person

Sources: **History:** Energy Information Administration (EIA), Office of Energy Markets and End Use, International Statistics Database and *International Energy Annual 1997*, DOE/EIA-0219(97) (Washington, DC, April 1999). **Projections:** EIA, World Energy Projection System (2000).

Are Low World Oil Prices Sustainable?

Crude oil prices in 1998 were at their lowest sustained levels in more than a decade. While members of the Organization of Petroleum Exporting Countries (OPEC) agreed upon incremental production cutbacks twice in 1998, the strategy did not seem to have much impact on plummeting prices. December 1998 saw the average U.S. imported refiner acquisition cost fall to $9.39 per barrel.

The OPEC strategy was not working for three reasons: (1) lack of discipline among OPEC members in adhering to the agreed-upon production cutbacks, (2) the return of Iraq as a significant oil exporter under a United Nations Security Council resolution that allowed Iraq a certain level of revenues for humanitarian purposes, and (3) the surprising resilience of non-OPEC oil supply even in a low world oil price environment.

The OPEC ministerial meeting held on March 23, 1999, yielded pledges from members for 1.7 million barrels per day in production cutbacks, and four non-OPEC suppliers (Mexico, Norway, Oman, and Russia) pledged an additional 0.4 million barrels per day. Whether it was the active encouragement provided by the non-OPEC producers, the recovery of some Asian oil demand after the 1998 recession, the discernible slowing in the growth of non-OPEC supply, or some combination thereof, oil prices had risen by about $15 per barrel by the end of 1999.

The significant upturn in prices surprised many oil market analysts who were convinced that an extended period of sustained low oil prices was both feasible and probable. Is there a compelling argument that OPEC should keep the market awash in Persian Gulf crude oil in order to discourage the exploration and development of higher-cost oil reserves? If one argues that OPEC members are generally focused on total revenues and market share, then higher production at lower prices can certainly accommodate such goals. The answer to the question lies in what happened after oil prices fell by half in the early 1980s: technology and the significant lowering of exploration, development, and extraction costs sustained, and even increased, non-OPEC production.

In order for a low price strategy to be successful, significant volumes of non-OPEC output would have to be uneconomical to produce under sustained low prices. To examine the possible effects of such a price environment on OPEC and non-OPEC producers, an alternative case was developed for IEO2000, assuming that the 1998 world oil price remained constant in nominal terms through 2005. In the *IEO2000* reference case, the world oil price is projected to be $20.49 per barrel in

2005. In the constant nominal price case, the 2005 price projection is $10.71 per barrel. The difference of almost $10 per barrel leads to a level of oil demand in the constant nominal price case that is more than 5.1 million barrels per day above the reference case level in 2005. The demand increase is not particularly surprising; however, it is of interest to see which oil suppliers are expected to provide the increment in production.

The figure below shows the marginal operating costs of OPEC and non-OPEC oil production (including crude oil, natural gas liquids, and liquids from other hydrocarbons). Non-OPEC costs are typically higher than those of OPEC producers: only slightly more than 5 percent of OPEC production has operating costs higher than $5 per barrel, compared with more than one-third of non-OPEC production. Amazingly, about three-fourths of OPEC production costs less than $3 per barrel. Even in the constant nominal price case, however, much of the world's oil could still be produced at a profit. Assuming constant 1998 prices through 2005, total production in 2005 would be only about 2 to 3 million barrels per day lower than it was in 1997.

Marginal Operating Costs of Oil Production, 1997

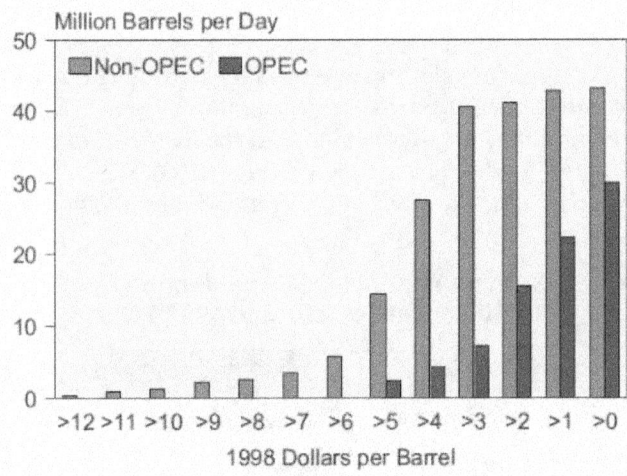

Source: Petroleum Economics, Ltd., *The Outlook for Fuel Oil, Petroleum Products, and Crude Markets in an Environment of Sustained Low Prices and Asian Uncertainty* (Sugarland, TX, September 1998).

The exploration, development, and operating costs associated with bringing proven crude oil reserves into production are shown in the figure opposite. Again, OPEC enjoys a significant advantage. Even in the constant nominal price case, Persian Gulf OPEC producers would be able to bring new fields into production at a 50 percent or higher rate of return on investment. In contrast, more than two-thirds of proven non-OPEC reserves can only be brought into production at an estimated cost of more than $10 per barrel. It can only be

(continued on page 29)

Are Low World Oil Prices Sustainable? (Continued)

concluded that new production from non-OPEC fields would be extremely vulnerable to a sustained period of low world oil prices as in the constant nominal price case.

In the *IEO2000* reference case, non-OPEC oil production grows by 0.8 percent annually between 1997 and 2005, and OPEC production grows by 2.6 percent a

Costs of Developing Proven, Recoverable Oil Reserves

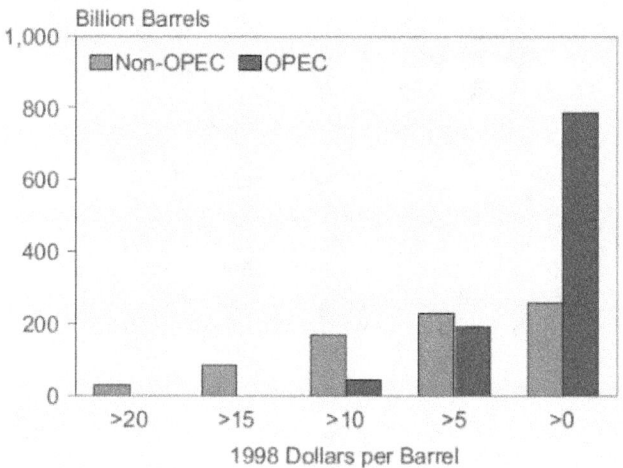

Note: Reserve estimates as of January 1, 1999.
Source: Petroleum Economics, Ltd., *The Outlook for Fuel Oil, Petroleum Products, and Crude Markets in an Environment of Sustained Low Prices and Asian Uncertainty* (Sugarland, TX, September 1998).

year. In the constant nominal price case, however, non-OPEC production grows by only 0.1 percent per year, while the annual increase in OPEC production doubles to 5.2 percent. The projected OPEC market share in 2005—less than 44 percent in the reference case—exceeds 49 percent in the constant nominal price case. Non-OPEC areas particularly vulnerable to sustained low prices include North America, Latin America, the North Sea, West Africa, China, and the former Soviet Union. Overall, projected non-OPEC production in 2005 is more than 2.5 million barrels per day lower in the constant nominal price case than in the *IEO2000* reference case.

This analysis suggests that the continuation of high levels of production by OPEC members, accompanied by a sustained low world oil price, could indeed force some non-OPEC producers out of competition in oil supply markets. History has shown, however, that quota agreements among OPEC producers have often deteriorated into various forms of cheating. In addition, the inability to gauge future technological advances and cost-cutting measures accurately, as well as the difficulty of holding together such a diverse group of oil producers, could make a low price strategy a risky one for OPEC. In fact, reasonable arguments can be made that any artificial (non-market) means of production management by OPEC might achieve short-term objectives but will not optimize revenues or stabilize market share in the long run.

---

countries. Per capita oil use in the United States, for example, is much higher than in Japan or Western Europe. Consumption per capita is projected to increase at a rapid pace in developing countries, but in most cases the levels remain much lower than those of the industrialized countries. One notable exception is South Korea, where rising per capita incomes allow per capita oil consumption to reach the levels of the industrialized countries, or even surpass them, in the forecast period.

In most countries oil intensity (oil consumed per dollar of GDP), decreases over time (Figure 35). The industrialized countries, especially Western Europe and Japan, tend to have lower levels of oil intensity, reflecting their more energy-efficient, fuel-diverse, and service-oriented economies. Intensity levels in developing countries are projected to decline at a faster rate, however, as energy efficiency improvements penetrate the economies.

The transportation sector is the primary user of petroleum, consuming 49 percent of the oil used in the world in 1997. The patterns of consumption between the industrialized and developing countries are quite different, however. In the heat and power segments of the markets

Figure 35. Oil Intensity by Selected Country and Region, 1970-2020

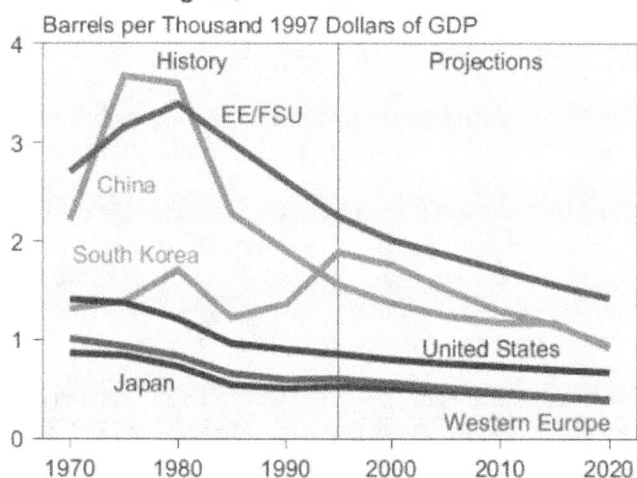

Sources: **History:** Energy Information Administration (EIA), Office of Energy Markets and End Use, International Statistics Database and *International Energy Annual 1997*, DOE/EIA-0219(97) (Washington, DC, April 1999). **Projections:** EIA, World Energy Projection System (2000).

in industrialized countries, nonpetroleum energy sources were able to compete with and substitute for oil throughout the 1980s; and by 1990, oil consumption in other sectors was less than in the transportation sector. Most of the expected gains in worldwide oil use occur in the transportation sector. Of the total increase (11.4 million barrels per day) projected for the industrialized countries from 1997 to 2020, 10.7 million barrels per day is attributed to the transportation sector (Figure 36), where few alternatives are economical until late in the forecast.

In the developing countries, the transportation sector also shows the fastest projected growth in oil use, rising nearly to the level of nontransportation oil consumption by 2020. In the developing world, however, in contrast to the industrialized countries, oil use for purposes other than transportation is projected to contribute 42 percent of the total increase in petroleum consumption. The growth in nontransportation petroleum consumption in developing countries is caused in part by the substitution of petroleum products for noncommercial fuels (such as wood burning for home heating and cooking) as incomes rise and the energy infrastructure matures.

Industrialized Countries

The largest increases in oil consumption among the industrialized countries from 1997 to 2020 are projected for North America (Figure 37). The largest absolute increase is projected for the United States (6.5 million barrels per day), and the most rapid growth is expected in Mexico (3.3 percent per year). Mexico's projected economic and population growth rates are the highest among the industrialized countries and regions in the forecast, accompanied by strong growth in both transportation and nontransportation oil consumption. North America as a whole is projected to contribute 22 percent of the increase in worldwide oil use.

Growth in petroleum consumption in Western Europe from 1997 to 2020 is expected to be considerably below the average annual growth rate of 1.4 percent per year from 1985 to 1997. The projected increase of 1.5 million barrels per day amounts to an annual average growth rate of 0.4 percent. Outside the transportation sector, oil use is projected to decline as natural gas makes inroads into the heat and power sectors of the market.

Industrialized Asia is projected to add 1.1 million barrels per day to its petroleum consumption between 1997 and 2020, and more than half the increase is projected for Japan, the world's second-largest petroleum-consuming country. After growing by 2.7 percent per year from 1985 to 1996, petroleum consumption in Japan is projected to slow to 0.4 percent per year from 1997 to 2020, as the country reaches saturation in terms of per capita motor vehicle use. Australasia's consumption of

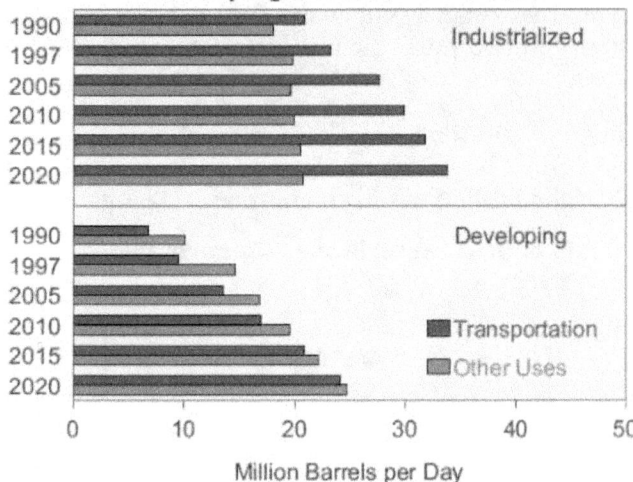

Figure 36. Oil Consumption for Transportation and Other Uses in Industrialized and Developing Nations, 1990-2020

Million Barrels per Day

Sources: **1990 and 1997:** Energy Information Administration (EIA), Office of Energy Markets and End Use, International Statistics Database and *International Energy Annual 1997*, DOE/EIA-0219(97) (Washington, DC, April 1999). **Projections:** EIA, World Energy Projection System (2000).

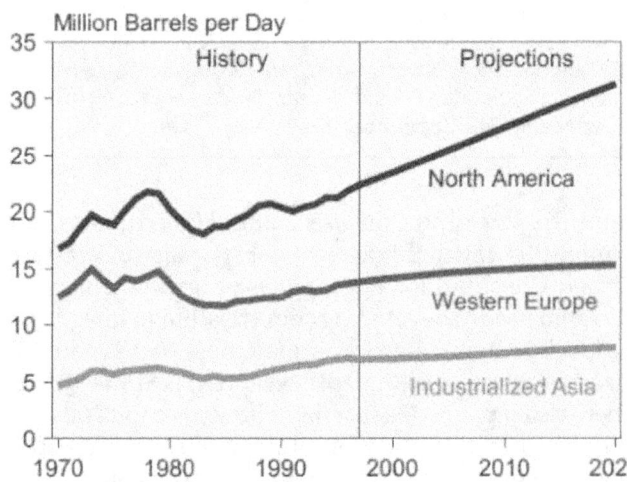

Figure 37. Oil Consumption in the Industrialized World by Region, 1970-2020

Sources: **History:** Energy Information Administration (EIA), Office of Energy Markets and End Use, International Statistics Database and *International Energy Annual 1997*, DOE/EIA-0219(97) (Washington, DC, April 1999). **Projections:** EIA, World Energy Projection System (2000).

petroleum is projected to increase by 0.5 million barrels per day over the forecast period (1.4 percent per year).

Developing Countries

Petroleum consumption in developing countries is projected to more than double, increasing from 24.2 million barrels per day in 1997 to 49.0 million barrels per day in 2020 (3.1 percent average annual growth). Although the region overall experienced a slowdown in oil demand with the recession that began in mid-1997, developing Asian economies affected by the recession are already beginning to show strong recovery, and *IEO2000* expects the region to contribute 30 percent of the world-wide increase in petroleum consumption over the next two decades (Figure 38). China alone is projected to provide 14 percent of the world increase in oil demand, and China and India combined are expected to add 8.0 million barrels per day to oil demand from 1997 to 2020, as compared with 6.5 million barrels per day for the United States.

China has the highest projected growth rate for oil consumption among the nations of the world at 4.1 percent per year, followed closely by Brazil and India at 4.0 and 3.7 percent per year, respectively. Road infrastructure projects currently planned in China [3] are expected to contribute to more rapid growth in transportation petroleum consumption. South Korea's petroleum consumption, after a fourfold increase (1.7 million barrels per day) from 1985 to 1997, dropped by more than 0.4 million barrels per day in 1998 [4] as a result of the economic and financial turmoil that spread throughout Asia. The country's demand for petroleum is expected to recover as economic conditions improve and is projected to grow at a more modest rate of 2.0 percent per year rate from 1997 to 2020.

Strong growth is also expected in Central and South America, with a projected increase of nearly 5 million barrels per day for the region as a whole. Oil consumption in Brazil is projected to increase by 4.0 percent per year from 1997 to 2020. Recent financial and economic

difficulties are expected to slow petroleum consumption growth in the near term, but rapid growth is expected to return in both the transportation and end-use sectors of the market. Petroleum consumption in the rest of Central and South America is expected to nearly double over the forecast period. Again, much of the increase projected to occur in the transportation sector.

Substantial increases in oil consumption are also expected in the Middle East (4.2 million barrels per day) and Africa (3.5 million barrels per day). Much of the increase will be used to fuel electricity generation in African nations, where the infrastructure needed to support the use of other fuels for power generation still is lacking.

Eastern Europe and the Former Soviet Union

As a result of political and economic turmoil in the FSU, oil consumption in 1996 was 55 percent below its 1987 level (Figure 39). In 1997, however, there was an increase of 0.2 million barrels per day over 1996. Oil use in the FSU is expected to remain at about the 1997 level through 2000 and then rise through the rest of the forecast period. The reference case projection for 2020 is 7.6 million barrels per day, still below the peak level of 9.1 million barrels per day for the FSU in 1982.

Petroleum consumption also declined in the early 1990s in Eastern Europe but has been rising slowly since 1995. A trend of slow growth is expected to continue, with petroleum consumption rising to 1.8 million barrels per day by 2020, the same level as 1989. In contrast to the FSU, all the increase in Eastern Europe is projected for the transportation sector. Petroleum consumption in

Figure 38. Oil Consumption in the Developing World by Region, 1970-2020

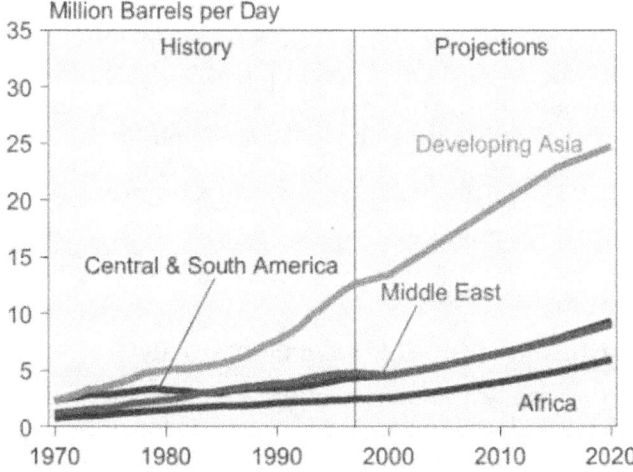

Sources: **History:** Energy Information Administration (EIA), Office of Energy Markets and End Use, International Statistics Database and *International Energy Annual 1997*, DOE/EIA-0219(97) (Washington, DC, April 1999). **Projections:** EIA, World Energy Projection System (2000).

Figure 39. EE/FSU Oil Consumption by Region, 1970-2020

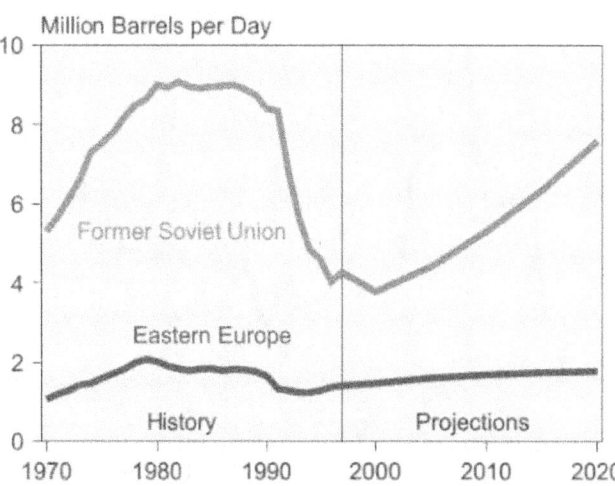

Sources: **History:** Energy Information Administration (EIA), Office of Energy Markets and End Use, International Statistics Database and *International Energy Annual 1997*, DOE/EIA-0219(97) (Washington, DC, April 1999). **Projections:** EIA, World Energy Projection System (2000).

other sectors declines slightly in the projections as natural gas is substituted for oil.

World Oil Price

The near-term price trajectory in the *IEO2000* reference case is considerably different from that in last year's *International Energy Outlook (IEO99)*. In *IEO99*, the rebound from the plummeting oil prices of 1998 and early 1999 was expected to occur gradually out to 2005, based on a recent series of unsuccessful attempts by OPEC member nations to adhere to announced production cutbacks. The *IEO2000* reference case incorporates the dramatic 1999 price increases that have followed the latest, so far successful, pledges by OPEC and some non-OPEC producers. In both outlooks, the reference case price trajectory beyond 2005 shows a gradual increase of about 0.4 percent per year out to 2010, reaching $21.00 per barrel (in constant 1998 U.S. dollars) in 2010 and $22.04 in 2020. Three possible long-term price paths are shown in Figure 40.

Oil demand rises significantly over the projection period in all three *IEO2000* price scenarios. In the high and low world oil price cases, the increases are 34 million barrels per day and 44 million barrels per day, respectively. The assumed size of proven worldwide reserves (more than 1 trillion barrels) and U.S. Geological Survey estimates of ultimately recoverable oil imply that resources are not a key constraint on world oil demand to 2020. More important are the political, economic, and environmental circumstances that could shape developments in oil supply and demand.

Figure 40. World Oil Prices in Three Cases, 1970-2020

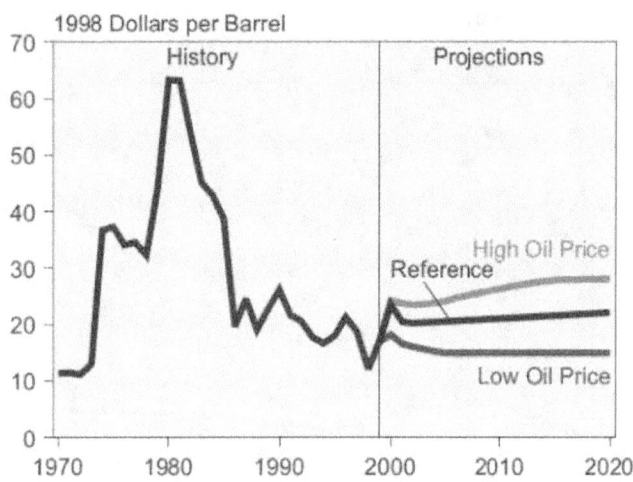

Sources: **History:** Energy Information Administration (EIA). *Annual Energy Review 1998*, DOE/EIA-0384(98) (Washington, DC, July 1999). **Projections:** 1999-2001—EIA, *Short-Term Energy Outlook*, on-line version (February 7, 2000), web site www.eia.doe.gov/emeu/steo/pub/contents.html. 2002-2020—EIA, *Annual Energy Outlook 2000*, DOE/EIA-0383(2000) (Washington, DC, December 1999).

The Composition of World Oil Supply

The *IEO2000* reference case projects an increase in world oil supply of almost 40 million barrels per day over the projection period. Gains in production are expected for both OPEC and non-OPEC producers; however, less than one-third of the production rise is expected to come from non-OPEC areas. Over the past two decades, the growth in non-OPEC oil supply has resulted in an OPEC market share of 41 percent, substantially under its historic high of 52 percent in 1973. New exploration and production technologies, aggressive cost-reduction programs by industry, and attractive fiscal terms to producers by governments all contribute to the outlook for continued growth in non-OPEC oil production.

While the long-term outlook for non-OPEC supply remains optimistic, the low oil price environment of 1998 and early 1999 had a definite impact on exploration and development activity. By the end of 1998, North American drilling activity had fallen by more than 25 percent from its level a year earlier. Worldwide, only the Middle East region registered no decline in drilling activity during 1998. In general, onshore drilling had fallen more sharply than offshore. Worldwide, offshore rig utilization rates were generally sustained at levels better than 80 percent of capacity [5].

The reference case projection indicates that more than two-thirds of the increase in demand over the next two decades will be met by increases in production by members of OPEC rather than by non-OPEC suppliers. OPEC production in 2020 is projected to be more than 25 million barrels per day higher than it was in 1998 (Figure 41). The *IEO2000* estimates of OPEC production capacity out to 2005 are slightly less than those projected in *IEO99*, reflecting a shift toward non-OPEC supply projects in the current high price environment. Some analysts suggest that OPEC might pursue significant price escalation through conservative capacity expansion decisions rather than undertake ambitious production expansion programs. This outlook discounts such suggestions, in light of the generous return on investment that OPEC producers (especially those in the Persian Gulf region) receive even in a relatively low world oil price environment.

Expansion of OPEC Production Capacity

It is generally acknowledged that OPEC members with large reserves and relatively low costs for expanding production capacity can accommodate sizable increases in petroleum demand. In the *IEO2000* reference case, the production call on OPEC suppliers grows at a robust annual rate of 3.1 percent (Table 11 and Figure 42). OPEC capacity utilization is expected to increase sharply after 2000, reaching 95 percent by 2015 and

Figure 41. World Oil Production in the Reference Case by Region, 1970-2020

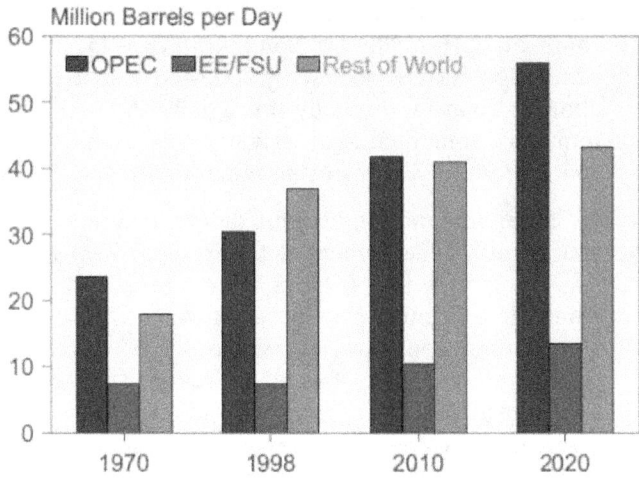

Sources: **History:** Energy Information Administration (EIA), *International Petroleum Monthly*, DOE/EIA-0520(99/12) (Washington, DC, December 1999). **Projections:** EIA, World Energy Projection System (2000).

Table 11. OPEC Oil Production, 1990-2020
(Million Barrels per Day)

Year	Reference Case	High Oil Price	Low Oil Price
History			
1990	24.5	—	—
1998	30.4	—	—
Projections			
2000	30.8	30.2	31.5
2005	36.5	34.5	40.1
2010	41.7	38.3	47.3
2015	48.3	43.4	55.6
2020	55.9	50.1	64.9

Note: Includes the production of crude oil, natural gas plant liquids, refinery gain, and other liquid fuels.
Sources: **History:** Energy Information Administration (EIA), *International Petroleum Monthly*, DOE/EIA-0520(99/12) (Washington, DC, December 1999), Table 1.4. **Projections:** EIA, World Energy Projection System (2000).

Figure 42. OPEC Oil Production in Three Oil Price Cases, 1970-2020

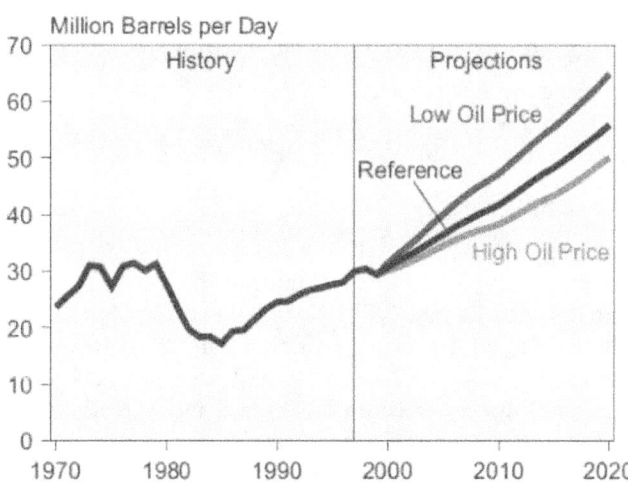

Sources: **History:** Energy Information Administration (EIA), *International Petroleum Monthly*, DOE/EIA-0520(99/12) (Washington, DC, December 1999). **Projections:** EIA, World Energy Projection System (2000).

remaining there for the duration of the projection period.

Iraq's role in OPEC will be particularly interesting to observe over the next half-dozen years. During 1999, Iraq expanded its production capacity to 2.8 million barrels per day in order to reach the slightly more than $5.2 billion in oil exports allowed by United Nations Security Council resolutions. Such expansion was required in the low price environment of early 1999. For the purposes of the *IEO2000* reference case, Iraq is assumed to maintain its current oil production capacity of 2.8 million barrels per day into the year 2000 and to export an average of 1.5 to 1.7 million barrels per day. The Security Council resolutions are assumed to remain in place through 2001.

Iraq has indicated a desire to expand its production capacity aggressively, to about 6 million barrels per day, once U.N. sanctions are lifted. Preliminary discussions with potential outside investors (including France, Russia, and China) about exploration projects have already taken place. Such a significant increase in Iraqi oil exports would offset a significant portion of the price stimulus associated with current OPEC production cutbacks.

Given the requirements for OPEC production capacity expansion implied by the *IEO2000* estimates, much attention has been focused on the oil development, production, and operating costs of individual OPEC producers. With Persian Gulf producers enjoying a reserve-to-production ratio in excess of 85 years, substantial capacity expansion is obviously feasible.

Production costs in Persian Gulf OPEC nations are less than $1.50 per barrel, and the capital investment

required to increase their production capacity by 1 barrel per day is less than $5,000 [6]. Assuming the *IEO2000* low price trajectory, total development and operating costs over the entire projection period, expressed as a percentage of gross oil revenues, would be less than 18 percent. Thus, Persian Gulf OPEC producers can expand capacity at a cost that is a relatively small percentage of projected gross revenues.

For OPEC producers outside the Persian Gulf, the cost to expand production capacity by 1 barrel per day is considerably greater, exceeding $10,000 in some member nations. Yet even those producers can still expect margins in excess of 32 percent on investments to expand production capacity in the low price case over the long term [7]. Venezuela has the greatest potential for capacity expansion and has aggressive plans to increase its production capacity to 4.6 million barrels per day by 2005 from the current level of 3.4 million barrels per day. It is unclear, however, whether the current political climate will support the outside investment required for any substantial expansion of production capacity. Tables D1-D10 in Appendix D show the ranges of production potential for both OPEC and non-OPEC producers.

The reference case projection implies aggressive efforts by OPEC member nations to apply or attract investment capital to implement a wide range of production capacity expansion projects. If those projects are not undertaken, world oil prices could escalate; however, the combination of potential profitability and the threat of competition from non-OPEC suppliers argues for the pursuit of an aggressive expansion strategy for OPEC.

In *IEO2000*, OPEC members outside the Persian Gulf are expected to continue increasing their production. The outlook for Nigeria's offshore production potential is optimistic, although development of production capacity there is unlikely before 2005. In addition, increased optimism about production potential in Algeria, Indonesia, and Venezuela supports the possibility of reducing the Persian Gulf share of OPEC oil exports

Non-OPEC Supply

Growing non-OPEC oil supplies played a significant role in the erosion of OPEC's market share over the past two decades, as non-OPEC supply became increasingly diverse. North America dominated non-OPEC supply in the early 1970s, the North Sea and Mexico evolved as major producers into the 1980s, and much of the new production in the 1990s has come from the developing countries of Latin America, the non-OPEC Middle East, and China. In the *IEO2000* reference case, non-OPEC supply from proven reserves is expected to increase steadily, from 44.5 million barrels per day in 1998 to 56.6 million barrels per day in 2020 (Table 12).

There are several important differences between the *IEO2000* production profiles and those published in *IEO99*:

- The U.S. production decline is slightly less severe in the *IEO2000* projections as a result of higher near-term oil prices, technological advances, and lower costs for deep exploration and production in the Gulf of Mexico.

- The rebound in near-term oil prices coupled with enhanced subsea and recovery technologies delays the *IEO99* estimated peak for North Sea production by a year to the 2004-2005 time period and slightly tempers the production decline out to 2020.

- Resource development in the Caspian Basin region was significantly delayed in the *IEO99* projection in view of the prospects for a prolonged low price environment. In *IEO2000*, Caspian output rises to almost 2.5 million barrels per day by 2005 and increases by about 7.1 percent annually through 2020. There still remains a great deal of uncertainty regarding export routes from the Caspian Basin region.

- *IEO99* anticipated significant delays in the exploration and development activities for deepwater projects worldwide. Although there remained considerable optimism about deepwater prospects, significant output from such projects was not anticipated until oil prices returned to a range of $18 to $20 per barrel. With the current rebound in prices, output from deepwater projects in the U.S. Texas Gulf, the North Sea, West Africa, the South China Sea, Colombia, and the Caspian Basin is accelerated in *IEO2000* by 1 to 3 years.

In the *IEO2000* reference case, North Sea production peaks in 2004 at more than 7.2 million barrels per day. Production from Norway, Western Europe's largest producer, is expected to peak at about 3.7 million barrels per day in 2003 and then gradually decline to about 2.9 million barrels per day by the end of the forecast period with the maturing of some of its larger and older fields. The United Kingdom sector is expected to produce about 3.1 million barrels per day by 2005, followed by a decline to 2.6 million barrels per day by 2020.

Two non-OPEC Middle East producers are expected to increase output gradually through 2005. Enhanced recovery techniques are expected to increase current output in Oman by more than 150,000 barrels per day, with only a gradual production decline anticipated after

Table 12. Non-OPEC Oil Production, 1990-2020
(Million Barrels per Day)

Year	Reference Case	High Oil Price	Low Oil Price
History			
1990.	42.2	—	—
1998.	44.5	—	—
Projections			
2000.	45.2	45.4	45.0
2005.	47.1	47.7	45.9
2010.	51.5	52.6	49.9
2015.	54.8	56.5	52.7
2020.	56.6	58.6	54.3

Note: Includes the production of crude oil, natural gas plant liquids, refinery gain, and other liquid fuels.
Sources: **History:** Energy Information Administration (EIA), *International Petroleum Monthly*, DOE/EIA-0520(99/12) (Washington, DC, December 1999), Table 1.4. **Projections:** EIA, World Energy Projection System (2000).

2005. Current oil production in Yemen could increase by at least 100,000 barrels per day within the next couple of years, and those levels would show little decline throughout the forecast period. Syria is expected to hold its production flat through the first half of the decade, but with little in the way of new resource potential, its production declines by about one-third from 2005 to 2020.

Oil producers in the Pacific Rim are expected to increase production significantly with the use of enhanced exploration and extraction technologies. Deepwater fields offshore from the Philippines have improved the reserve picture there, and production is expected to reach almost 250,000 barrels per day by 2005. Vietnam's long-term production potential also is still viewed with considerable optimism, although exploration activity has been slower than originally anticipated. Output levels from Vietnamese fields are expected to exceed 500,000 barrels per day by 2020.

Australia has significantly added to its proven reserves recently, and it is likely that Australia will become a million barrel per day producer by 2005. Papua New Guinea also continues to add to its reserve posture and is expected to achieve production volumes approaching 200,000 barrels per day by 2005, followed by only a modest decline over the rest of the forecast. India, too, is expected to show some modest production increase early in the decade and only a modest decline in output thereafter. Malaysia shows little potential for any significant new finds, and its output is expected to peak at around 825,000 barrels per day in the early 2000s, followed by a gradual decline to about 625,000 barrels per day by 2020. Exploration and test-well activity have pointed to some production potential for Bangladesh and Mongolia, but significant output is not expected before 2005.

Oil producers in Central and South America have significant potential for increasing output over the next decade. Brazil has just recently become a million barrel per day producer and has considerable production potential waiting to be tapped. Its production is expected to rise throughout the forecast period, topping 1.7 million barrels per day by 2020. Colombia's current economic downturn has delayed its bid to join the relatively short list of million barrel per day producers, but it is expected to top 1.2 million barrels per day within 5 years and show little decline through 2020. The oil sectors in both countries would benefit significantly from a more favorable climate for attracting foreign investment.

Argentina is expected to increase its production volumes by at least 100,000 barrels per day over the next 2 years, and by 2005 it is also likely to become a million barrel per day producer. Although the current political situation in Ecuador is in transition, there is still optimism that Ecuador will increase production by more than 100,000 barrels per day within the next few years.

Several West African producers (Angola, Cameroon, Chad, Congo, Gabon, Ivory Coast) are expected to reap the benefits of substantial exploration activity, especially considering the recent rebound in oil prices. Angola is expected to become a million barrel per day producer within 5 years. Given the excellent exploration results, Angola could produce volumes of up to 1.8 million barrels per day well into the later years of the forecast period. The other West African producers with offshore tracts are expected to increase output by up to 300,000 barrels per day for the duration of the forecast period. North African producers Egypt and Tunisia produce mainly from mature fields and show little promise of adding to their reserve posture, and their production volumes are expected to fall gradually throughout the forecast. Sudan and Equatorial Guinea, which have dramatically increased their production recently, are expected to be producing moderate volumes by 2005 and Eritrea, Somalia, and South Africa after 2005.

In North America, falling U.S. output is expected to be more than offset by production increases in Canada and Mexico. Canada's output is projected to increase by about 200,000 barrels per day over the next 2 years, mainly from Newfoundland's Hibernia oil project, which could produce more than 150,000 barrels per day at its peak sometime in the next several years. After 2001, Canada is expected to gradually add an additional 600,000 barrels per day in output from a combination of frontier area offshore projects and oil from tar sands. Higher near-term prices, technological advances, and lower costs for deepwater exploration and production in the Gulf of Mexico temper the projected decline in U.S. production. Mexico is expected to adopt energy policies that encourage the efficient development of its vast resource base, and production volumes approaching 4 million barrels per day are projected from 2010 through 2020.

With the rebound in near-term oil prices, oil production in the FSU is expected to reach 7.6 million barrels per day by 2005—a level that could be significantly higher if the outlook for investment in Russia were not so pessimistic. The long-term production potential for the FSU is still regarded with considerable optimism, especially for the resource-rich Caspian Basin region. Oil production in the region is projected to exceed 13.1 million barrels per day by 2020 in the *IEO2000* reference case, implying export volumes in excess of 7.5 million barrels per day. In China, oil production is expected to increase gradually to 3.6 million barrels per day by 2020, but China's import requirements will be as large as its domestic production by 2010 and will continue to grow as its petroleum consumption increases.

The estimates for non-OPEC production potential presented in this outlook are based on such parameters as numbers of exploration wells, finding rates, reserve-to-production ratios, advances in both exploration and extraction technologies, and the sensitivity to changes in the world oil price. A critical component of the forecasting methodology is the constraint placed on the exploration and development of undiscovered resources. For the purpose of the three *IEO2000* world oil price cases, no more than 15 percent of the mean United States Geological Survey estimate of undiscovered oil was allowed to be developed over the forecast period. Tables D1-D10 in Appendix D show the ranges of production potential for both OPEC and non-OPEC producers.

The expectation in the late 1980s and early 1990s was that non-OPEC production in the longer term would be stagnant or decline gradually in response to resource constraints. The relatively insignificant cost of developing oil resources in OPEC countries (especially those in the Persian Gulf region) was considered such an overwhelming advantage that non-OPEC production potential was viewed with pessimism. In actuality, however, despite a relatively low price environment, non-OPEC production has risen every year since 1993, adding almost 4 million barrels per day between 1993 and 1997. It is expected that non-OPEC producers will continue to increase output, producing an additional 7 million barrels per day by 2010. Three factors are generally given credit for the impressive resiliency of non-OPEC production: development of new exploration and production technologies, efforts by the oil industry to reduce costs, and efforts by producer governments to promote exploration and development by encouraging outside investors with attractive fiscal terms.

Alternative Non-OPEC Supply Cases

The only variable affecting the estimates of non-OPEC production potential in the three *IEO2000* world oil price cases is the world oil price assumption. As a result, the range in non-OPEC supply is modest, varying by slightly less than 4.3 million barrels per day at the end of the forecast period. In fact, however, improved technology and a better understanding of the underlying resource potential have been major factors sustaining non-OPEC supply in the recent past. To examine the effects of those factors, two additional cases—the high and low non-OPEC supply cases—were developed for *IEO2000*.

Both non-OPEC supply cases are based on the reference case world oil price assumption, and are considered feasible alternatives to the projections of non-OPEC supply in the reference case. The high non-OPEC supply projections would, of course, be more likely if oil prices were higher, and the projections in the low case would be more likely if prices were lower. Figure 43 compares

OPEC and non-OPEC production estimates in the reference case with those in the two alternative non-OPEC supply cases. The alternative cases used reference case assumptions except for the following departures.

High Non-OPEC Supply Case:

- Due to increased optimism regarding the offshore production potential in the FSU, Latin America, West Africa, and the South China Sea, undiscovered oil in those regions is assumed to be 15 percent greater than the estimates in the reference case by 2020.

- One-third of the world's (non-OPEC, non-U.S.) undiscovered oil is considered economical to develop over the forecast period—almost 65 billion barrels more than in the reference case.

- Technology improvements over the forecast period are assumed to be transferrable worldwide. In the reference case, there is an assumed 5-year lag for technology transfer to nonindustrialized countries.

Figure 43. OPEC and Non-OPEC Oil Production in Three Cases, 1990-2020

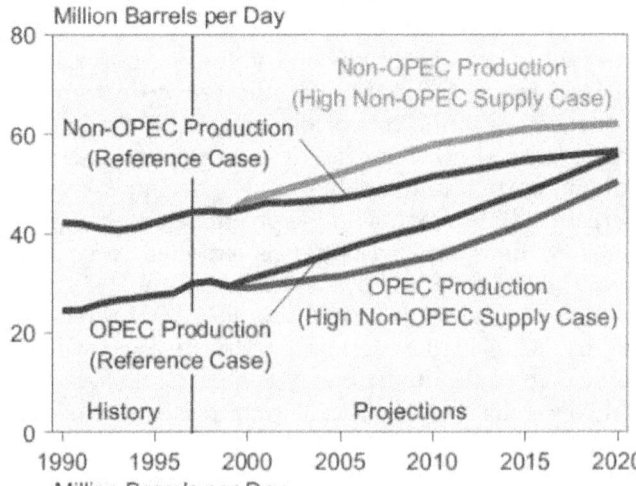

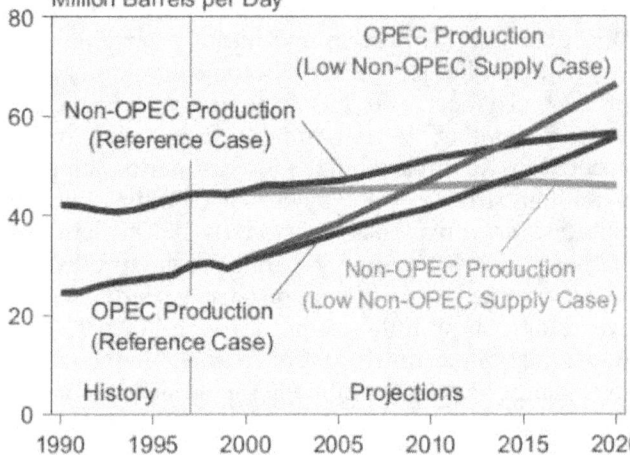

Sources: **History:** Energy Information Administration (EIA), *International Energy Annual 1997*, DOE/EIA-0219(97) (Washington, DC, April 1999). **Projections:** EIA, World Energy Projection System (2000).

- A reserve-to-production ratio of 10 years (slightly less than the current non-OPEC ratio) is used as a lower bound for production estimates, as compared with 15 years in the reference case.

Low Non-OPEC Supply Case:

- The amount of oil production from undiscovered reserves in deepwater areas is assumed to be 25 percent less than the reference case estimate as a result of persistent low oil prices and the finding of more natural gas deposits than oil deposits.

- Only one-fifth of the undiscovered oil in non-OPEC areas is considered economical to develop over the forecast period.

- There are assumed to be no significant technology improvements over the forecast period, and worldwide oil recovery rates are assumed to average only 35 percent. The reference case assumes a gradual increase in worldwide recovery rates to 45 percent by 2020.

- Russia's oil production is assumed to be one-third of that estimated in the reference case.

The high non-OPEC supply case assumptions result in 1.6-percent annual growth in non-OPEC production over the forecast period, as compared with a 1.2-percent growth rate in the reference case. Non-OPEC oil production reaches a peak of 62.1 million barrels per day in the high case in 2020, compared with a peak of 56.6 million barrels per day in the reference case. Figure 44 compares peak production levels for six non-OPEC regions in the reference, high non-OPEC supply, and low non-OPEC supply cases.

In the reference case, OPEC production peaks at 55.9 million barrels per day, and the OPEC share of

Figure 44. Non-OPEC Oil Production by Region in Three Cases, 2020

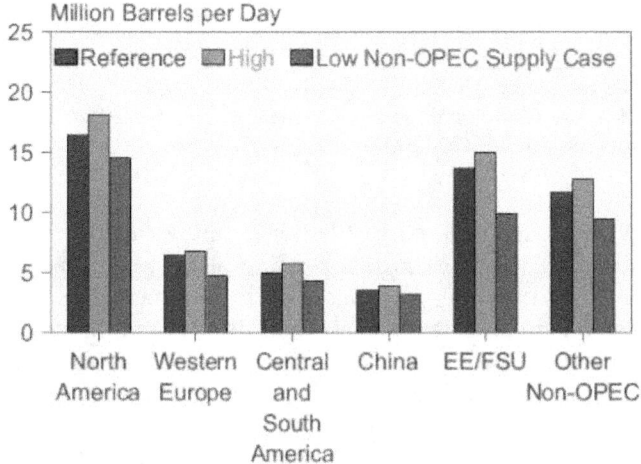

Source: Energy Information Administration, World Energy Projection System (2000).

worldwide production reaches almost 50 percent by 2020. In the high non-OPEC supply case, OPEC production peaks at 50.4 million barrels per day and never assumes a market share above 45 percent. The low non-OPEC supply case projects only modest 0.2-percent annual growth in non-OPEC production over the forecast period. Non-OPEC production peaks in 2015 at 46.7 million barrels per day. OPEC production reaches 66.4 million barrels per day in 2020, with about a 59-percent share of the world market.

Worldwide Petroleum Trade in the Reference Case

In 1997, industrialized countries imported 16.5 million barrels of oil per day from OPEC producers. Of that total, 10.3 million barrels per day came from the Persian Gulf region. Oil movements to industrialized countries represented almost two-thirds of the total petroleum exported by OPEC member nations and more than 63 percent of all Persian Gulf exports (Table 13). By the end of the forecast period, OPEC exports to industrialized countries are estimated to be about 5.3 million barrels per day higher than their 1997 level, and more than half the increase is expected to come from the Persian Gulf region.

Despite such a substantial increase, the projected share of total petroleum exports in 2020 that goes to the industrialized nations is considerably lower than their 1997 share, at almost 56 percent. Their share of all Persian Gulf exports falls even more dramatically, to almost 37 percent. The significant shift in the balance of OPEC export shares between the industrialized and non-industrialized nations is a direct result of the robust economic growth anticipated for the developing nations of the world, especially those of Asia. OPEC petroleum exports to developing countries are expected to increase by almost 16 million barrels per day over the forecast period, with more than half the increase going to the developing countries of Asia. China, alone, will most likely import about 5.3 million barrels per day from OPEC by 2020, virtually all of which is expected to come from Persian Gulf producers.

North America's petroleum imports from the Persian Gulf are expected to more than double over the forecast period (Figure 45). At the same time, more than half of North America's imports in 2020 are expected to be from Atlantic Basin producers and refiners, with significant increases in crude oil imports anticipated from Latin American producers, including Venezuela, Brazil, Colombia, and Mexico. West African producers, including Nigeria and Angola, are also expected to increase their export volumes to North America. Caribbean Basin refiners are expected to account for most of the increase in North American imports of refined products.

Table 13. Worldwide Petroleum Trade in the Reference Case, 1997 and 2020
(Million Barrels per Day)

	Importing Region							
	Industrialized				Nonindustrialized			
Exporting Region	North America	Western Europe	Asia	Total	Pacific Rim	China	Rest of World	Total
	1997							
OPEC								
Persian Gulf	2.0	3.5	4.8	10.3	4.2	0.5	1.3	**6.0**
North Africa.	0.3	1.9	0.0	2.2	0.0	0.0	0.1	**0.1**
West Africa	0.8	0.6	0.0	1.5	0.2	0.0	0.1	0.3
South America	1.8	0.2	0.0	2.0	0.1	0.0	1.1	**1.2**
Asia.	0.1	0.0	0.5	0.6	0.2	0.0	0.0	**0.2**
Total OPEC	**5.0**	**6.3**	**5.3**	**16.5**	**4.8**	**0.5**	**2.6**	**7.9**
Non-OPEC								
North Sea.	0.8	5.4	0.0	6.3	0.0	0.0	0.0	**0.1**
Caribbean Basin	2.7	0.4	0.0	3.2	0.2	0.0	2.2	**2.4**
Former Soviet Union	0.0	2.6	0.0	2.6	0.1	0.0	0.1	**0.2**
Other Non-OPEC.	2.7	1.9	0.5	5.0	7.3	0.5	1.2	**9.0**
Total Non-OPEC	**6.3**	**10.3**	**0.5**	**17.1**	**7.6**	**0.5**	**3.6**	**11.6**
Total Petroleum Imports	**11.2**	**16.6**	**5.9**	**33.7**	**12.4**	**0.9**	**6.2**	**19.5**
	2020							
OPEC								
Persian Gulf	4.1	3.7	5.5	13.3	9.0	5.3	8.8	**23.1**
North Africa.	0.3	2.1	0.0	2.4	0.1	0.0	0.1	**0.3**
West Africa	0.8	0.9	0.2	1.9	0.1	0.0	0.0	**0.1**
South America	3.5	0.4	0.1	4.1	0.2	0.0	0.1	**0.2**
Asia.	0.1	0.0	0.1	0.2	0.1	0.0	0.0	**0.1**
Total OPEC	**8.7**	**7.1**	**6.0**	**21.8**	**9.6**	**5.3**	**9.0**	**23.8**
Non-OPEC								
North Sea.	0.6	5.1	0.0	5.7	0.1	0.0	0.0	**0.1**
Caribbean Basin	4.2	0.4	0.1	4.7	0.2	0.0	3.1	**3.2**
Former Soviet Union	0.5	4.4	0.2	5.1	2.5	0.5	0.3	**3.2**
Other Non-OPEC.	3.2	2.0	0.2	5.4	0.1	0.7	2.7	**3.6**
Total Non-OPEC	**8.4**	**12.0**	**0.5**	**20.9**	**8.1**	**1.2**	**6.1**	**10.1**
Total Petroleum Imports	**17.1**	**19.1**	**6.5**	**42.7**	**17.7**	**6.5**	**15.1**	**33.9**

Notes: Totals may not equal sum of components due to independent rounding.
Sources: **1997:** Energy Information Administration (EIA), Energy Markets and Contingency Information Division. **2020:** EIA, Office of Integrated Analysis and Forecasting, IEO2000 WORLD Model run IEO00.B20 (2000).

With a moderate decline in North Sea production, Western Europe is expected to import increasing amounts from Persian Gulf producers and from OPEC member nations in both northern and western Africa. Substantial imports from the Caspian Basin are also expected. Industrialized Asian nations are expected to increase their already heavy dependency on Persian Gulf oil. The developing countries of the Pacific Rim are expected to increase their total petroleum imports between 1995 and 2020 by almost 43 percent.

Worldwide crude oil distillation refining capacity was about 80.3 million barrels per day at the beginning of 1999. To meet the projected growth in international oil demand in the reference case, worldwide refining capacity would have to increase by more than 40 million barrels per day by 2020. Substantial growth in distillation capacity is expected in the Middle East, Central and South America, and especially in the Asia Pacific region. Refiners in North America and Europe, while making only modest additions to their distillation capacity, are

Figure 45. Imports of Persian Gulf Oil by Importing Region, 1997 and 2020

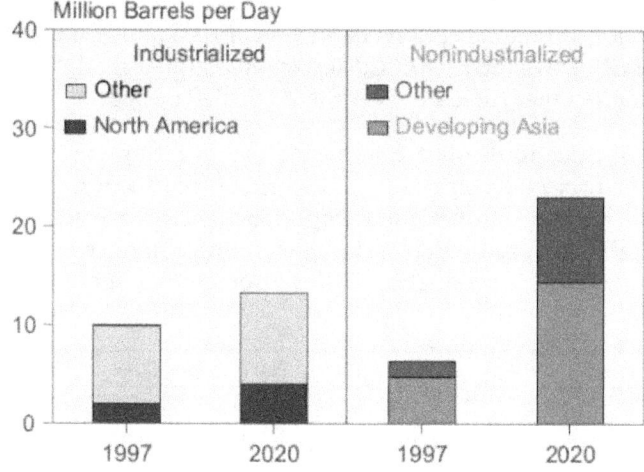

Sources: **History:** Energy Information Administration (EIA), *International Petroleum Monthly*, DOE/EIA-0520(99/12) (Washington, DC, December 1999). **Projections:** EIA, World Energy Projection System (2000).

expected to continue improving product quality and enhancing the usefulness of the heavier portion of the barrel through investment in downstream capacity. Future investments by developing countries are also expected to include more advanced configurations designed to meet the anticipated increase in demand for lighter products, especially transportation fuels.

Other Views of Prices and Production

Several oil market analysis groups produce world oil price and production forecasts. Table 14 compares the *IEO2000* world oil price projections with similar forecasts from Standard & Poor's Platt's (DRI), the International Energy Agency (IEA), Petroleum Economics, Ltd. (PEL), Petroleum Industry Research Associates (PIRA), the Gas Research Institute (GRI), National Resources Canada (NRCan), WEFA Energy (WEFA), and Deutsche Banc Alex.Brown (DBAB).

The collection of forecasts includes a wide range of price projections. The volatility of world oil prices in the late 1990s has helped to define this wide range with differing views about whether oil prices will sustain the higher levels achieved in 1999 with the recovery of many southeast Asian economies and the production quotas achieved by the OPEC member countries. Prices for 2005 range from DRI's $15.70 per barrel (constant 1998 U.S. dollars) to NRCan's $20.97 per barrel. NRCan released its last world oil price projections in 1997 but revisited oil prices in a December 1999 publication, *Canada's Emissions Outlook: An Update*, and decided that the forecasts were valid without revision. The IEA forecast, which is closest to *IEO2000* ($20.47 vs. $20.49), was also released some time ago, in November 1998 in the IEA's

Table 14. Comparison of World Oil Price Projections, 2000-2020
(1998 Dollars per Barrel)

Forecast	2005	2010	2015	2020
IEO2000				
Reference Case	20.49	21.00	21.53	22.04
High Price Case	24.16	26.31	27.86	28.04
Low Price Case	14.90	14.90	14.90	14.90
DRI (October 1999)	15.70	16.66	18.58	19.94
IEA (November 1998)	20.47	20.47	30.10	30.10
PEL (February 2000)	16.78	14.78	—	—
PIRA (October 1999)	19.75	20.64	—	—
WEFA (October 1999)	16.54	18.62	19.28	19.77
GRI (January 2000)	17.90	17.90	17.90	—
NRCan (April 1997)	20.97	20.97	20.97	20.97
DBAB (December 1999)	17.57	17.86	17.84	18.20

Notes: *IEO2000* projections are for average landed imports to the United States. DRI, GRI, WEFA, and DBAB projections are for composite refiner acquisition prices. PEL projections are for Brent crude oil. PIRA projections are for West Texas Intermediate crude oil at Cushing.

Sources: **IEO2000**: Energy Information Administration, *Annual Energy Outlook 2000*, DOE/EIA-0383(2000) (Washington, DC, December 1999). **DRI**: Standard & Poor's Platt's, *Oil Market Outlook: Long Term Focus* (Lexington, MA, October 1999), p. 2. **IEA**: International Energy Agency, *World Energy Outlook 1998* (Paris, France, 1998), p. 84. **PEL**: Petroleum Economics, Ltd., *Oil and Energy Outlook to 2015* (London, United Kingdom, February 2000). **PIRA**: PIRA Energy Group, *Retainer Client Seminar* (New York, NY, October 1999), Table II-3. **WEFA**: WEFA Group, *U.S. Energy Outlook 1999* (Eddystone, PA, February 1998), p. 1.12. **GRI**: Gas Research Institute, *2000 Data Book of the GRI Baseline Projections of U.S. Energy Supply and Demand to 2015*, Vol. 2 (Washington, DC, January 1999), p. PRC-1. **NRCan**: Natural Resources Canada, *Canada's Energy Outlook, 1996-2020*, Annex C2 (Ottawa, Ontario, Canada, April 1997). **DBAB**: Deutsche Banc Alex.Brown, Inc., "World Oil Supply and Demand Estimates," e-mail from Adam Sieminski (December 20, 1999).

World Energy Outlook 1998—the last *Outlook* released with oil prices.

IEO2000 expects oil prices higher than those in most of the other forecasts, in the range of $20 to $21 per barrel through 2005, as does PIRA at $19.75 per barrel in 2005. Recent forecasts from DRI, DBAB, and GRI all show lower prices, in the range of $16 to $18 per barrel in 2005. PEL (in 2010) and IEA (after 2010) may be considered outliers among the sets of projections. PEL's price projection is slightly lower than the *IEO2000* low price path in 2010, when the PEL time series ends ($14.78 from PEL compared with $14.90 in the *IEO2000* low world oil price case). Similarly, after 2010, IEA price expectations exceed the *IEO2000* high price path by about $2 per barrel. With the exceptions of NRCan in 2005 and IEA after 2010, however, the *IEO2000* prices are higher than those in the other forecasts over the 2005-2020 period.

Oil price forecasts are influenced by differing views of the projected composition of world oil production. Two factors are of particular importance: (1) expansion of OPEC oil production and (2) the timing of a recovery in EE/FSU oil production. All the forecasters agree that

Table 15. Comparison of World Oil Production Forecasts

Forecast	Percent of World Total			Million Barrels per Day		
	OPEC	EE/FSU	Rest of World	OPEC	EE/FSU	Rest of World
History						
1997	40	10	50	29.9	7.4	36.8
Projections						
2000						
IEO2000.	40	10	50	30.6	7.6	37.5
DRI[a]	42	8	50	31.1	6.3	36.9
PEL	41	10	49	31.2	7.5	36.5
PIRA.	37	10	53	28.8	7.5	40.9
DBAB	42	10	48	31.1	7.6	36.1
2005						
IEO2000.	45	9	46	38.2	8.0	38.9
DRI	45	8	47	38.0	6.6	39.2
PEL	43	10	47	36.0	8.2	39.8
PIRA.	39	10	51	33.4	8.5	44.1
DBAB	47	10	43	39.5	8.8	36.4
2010						
IEO2000.	45	11	44	42.0	10.5	40.9
DRI	47	7	45	44.7	7.0	43.0
IEA[b]	47	11	42	43.8	10.2	38.7
PEL	47	10	43	44.1	8.9	40.3
PIRA.	42	11	47	39.9	10.1	44.8
DBAB	49	11	40	46.0	10.3	37.5
2015						
IEO2000.	46	12	41	47.6	12.5	42.3
DRI	48	7	45	50.0	7.4	47.6
DBAB	50	12	38	52.3	12.0	39.3
2020						
IEO2000.	50	12	38	55.5	13.5	43.1
DRI	47	6	46	55.1	7.6	53.8
IEA[b]	55	10	35	49.0	9.4	31.5
DBAB	52	12	36	59.1	13.9	41.6

[a]In the DRI projections, EE/FSU includes only the former Soviet Union.
[b]In the IEA projections, OPEC includes only Middle East OPEC.
Note: Percentages may not add to 100 due to independent rounding.
Sources: *IEO2000*: Energy Information Administration, World Energy Projection System (2000) and "DESTINY" International Energy Forecast Software (Dallas, TX: Petroconsultants, 2000). **DRI:** Standard & Poor's Platt's, *Oil Market Outlook: Long Term Focus* (Lexington, MA, October 1999). **IEA:** International Energy Agency, *World Energy Outlook 1998* (Paris, France, November 1998), p. 101 and p. 117. **PEL:** Petroleum Economics, Ltd., *Oil and Energy Outlook to 2015* (London, United Kingdom, February 2000). **PIRA:** PIRA Energy Group, *Retainer Client Seminar* (New York, NY, October 1999). **DBAB:** Deutsche Banc Alex.Brown, fax from Adam Sieminski (December 20, 1990).

recovery in the EE/FSU will be fairly slow. The share of EE/FSU production does not grow above 12 percent in any of the forecasts included in this comparison. DRI is the least optimistic about recovery in the region, and its projection never exceeds 8 percent. Indeed, DRI's forecast of Russia's share of world oil production (oil production estimates for the entire region are not available from DRI) falls to 7 percent in 2010 and to 6 percent in 2020. All the other forecasts expect production in the EE/FSU to make up about 11 percent of the world total by 2010. Both *IEO2000* and DBAB project a 12-percent share for EE/FSU production by 2015, which is maintained through 2020, whereas IEA projects 11 percent in 2010 and 10 percent in 2020.

The forecasts that provide projections through 2020 (*IEO2000*, DRI, DBAB, and IEA) all expect OPEC production in 2020 to be between 20 and 30 million barrels per day higher than it was in 1997. There is more variation in expectations among the four forecasts for non-OPEC suppliers. DRI expects a substantial increase of 17 million barrels per day of supply from other suppliers, whereas IEA expects a decrease of 5 million barrels per day. IEA projects that the "other" share of world oil production will fall to 35 percent by 2020, while the OPEC share increases to 55 percent. (The IEA estimate for the OPEC share is actually understated, because IEA does not publish oil production forecasts for the entire OPEC membership but only for "Middle East" OPEC. With non-Persian Gulf members supplying about 35 percent of OPEC's current production, it can be assumed

that the total OPEC share of world oil production is even higher in 2020 than the 55 percent shown in Table 15.) *IEO2000* and DBAB expect more moderate growth from "other," non-OPEC supply totaling about 5 to 6 million barrels per day between 1997 and 2020.

References

1. Energy Information Administration, *Short-Term Energy Outlook*, on-line version (February 7, 2000). web site www.eia.doe.gov/emeu/steo/pub/contents.html.

2. Energy Information Administration, *Short-Term Energy Outlook*, on-line version (February 7, 2000). web site www.eia.doe.gov/emeu/steo/pub/contents.html.

3. Standard and Poor's DRI, *China's Automotive Market: The Next Decade*" (Lexington, MA, December 1998), p. 37.

4. BP Amoco, *BP Amoco Statistical Review of World Energy* (London, UK, June 1999).

5. "Offshore Prospects Showing Delays in Present Environment," *Oil & Gas Journal*, Vol. 97, No. 2 (January 11, 1999), p. 18.

6. DRI/McGraw-Hill, *Oil Market Outlook* (Lexington, MA, July 1995), Table 1, p. 10.

7. Energy Information Administration, *Oil Production Capacity Expansion Costs for the Persian Gulf*, DOE/EIA-TR/0606 (Washington, DC, February 1996).

Natural Gas

Natural gas is the fastest growing primary energy source in the IEO2000 forecast. The use of natural gas is projected to more than double between 1997 and 2020, providing a relatively clean fuel for efficient new gas turbine power plants.

World natural gas consumption continues to grow, increasing its market share of total primary energy consumption. In the *International Energy Outlook 2000* (*IEO2000*), natural gas remains the fastest growing component of world energy consumption. Over the *IEO2000* forecast period from 1997 to 2020, gas use is projected to more than double in the reference case, reaching 167 trillion cubic feet in 2020 from the 1997 level of 82 trillion cubic feet (Figure 46). Over the 1997-2020 period, the role of natural gas in energy use is projected to increase in all regions except the Middle East and Africa, where its share remains relatively stable. The developing countries of Asia and of South and Central America will see the strongest growth rates in gas demand. Large incremental increases are also projected for industrialized countries, including the United States, and for the former Soviet Union (FSU).

In the *IEO2000* reference case, a slowly increasing share of world gas consumption is used in the electric power sector (rising from 29 percent in 1997 to 33 percent in 2020), and natural gas accounts for the largest increment in electricity generation (increasing by 33 quadrillion Btu). Not only do combined-cycle gas turbine power plants offer some of the highest commercially available

plant efficiencies, but natural gas is also attractive for environmental reasons. When it is burned, natural gas releases less sulfur dioxide, less particulate matter, and less carbon dioxide than does oil or coal.

For the industrialized countries, natural gas—compared with other fuels—is expected to provide the greatest incremental increase in energy consumption among the major fuels and has the fastest average annual growth in the forecast (2.1 percent per year, compared with 1.0 percent for oil). The percentage of gas used for power generation also grows from 20 percent in 1997 to over 30 percent in 2020. In 1997, natural gas consumption in the developing countries was a smaller portion of total energy use (14 percent) than the world average (22 percent). From that starting point, gas consumption in developing countries grows at a faster rate in the reference case than any other fuel (an average of 5.6 percent per year, compared with 3.1 percent for both oil and coal). Increments in gas use in the developing countries are expected to supply both power generation and other uses, such as town gas and fuel for industry.

In Central and South America, the power sector currently relies heavily on hydroelectric power, which accounts for about 12 percent of primary energy use (compared with a 2.5-percent share in the global energy mix in 1997). Because dependence on hydroelectric resources in the region has led to problems in maintaining electricity supply during times of drought, fuel diversification is now being pursued. In developing Asia, gas use is also desirable for environmental reasons and to diversify the energy mix away from heavy reliance on oil imports. The problem has been complicated, however, by greater distances between gas resources and market centers, leading to a combination of liquefied natural gas (LNG) and pipeline trade in the region.

There are also new efforts to develop natural gas use in the Middle East, although the projected growth rates and incremental increases there are smaller than in Asia or the Americas. Domestic resources may supply some of the increase in gas use in the Middle East (for example, in Saudi Arabia), but trade will also be important. In addition to Turkey and Israel, both Oman and the United Arab Emirates may become gas importers, using pipeline imports from Qatar to meet domestic demand while sustaining their own LNG export commitments.

Figure 46. World Natural Gas Consumption, 1970-2020

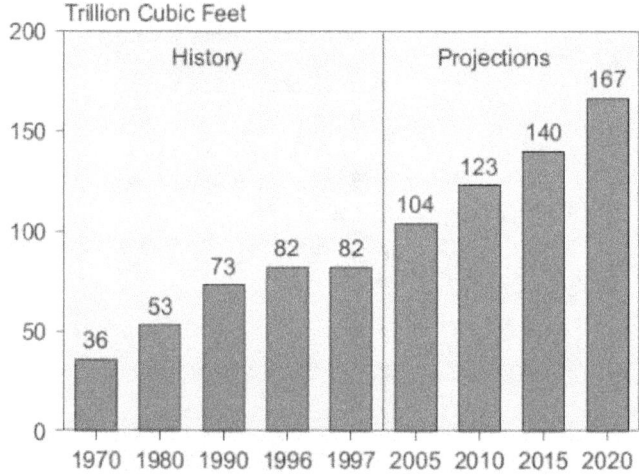

Sources: **History:** Energy Information Administration (EIA), Office of Energy Markets and End Use, International Statistics Database and *International Energy Annual 1997*, DOE/EIA-0219(97) (Washington, DC, April 1999). **Projections:** EIA, World Energy Projection System (2000).

Some important gas market developments in 1999 include:

- The completion of several major international pipelines and firming of plans for other new pipelines in Europe and South America. Steady growth in pipeline infrastructure is leading to increased trade, which can facilitate a more transparent (and mature) gas market. The 1999 completion of the Europipe II from Norway to Germany will lead to an expanded role for North Sea gas in Germany. On the southern side of Europe, Italy moved forward with plans to build a new pipeline for imports of Libyan gas. In South America, pipelines from Bolivia to Brazil and from Argentina to Chile (the GasAtacama and the Norandino) were completed in 1999.

- Completion in Asia of several major international pipelines and plans for additional lines. In Asia, the new pipeline from Myanmar to Thailand began building up deliveries to contracted volumes, more than a year behind schedule. Contracts were signed for two new pipelines that would carry Indonesian gas exports to Singapore, and plans moved forward for a pipeline from Papua New Guinea to Australia, with finalization of gas sales contracts.

- The completion of several LNG facilities. Three grassroots natural gas liquefaction facilities came on stream in 1999 in Trinidad and Tobago (Atlantic LNG), Nigeria (Bonny), and Qatar (Rasgas). An expansion, Indonesia's eighth train ("Train H") at its Bontang facility is also starting operations at the end of 1999 or early in 2000. Qatar has concluded agreements with India for the sale of 7.5 million metric tons of LNG with deliveries starting in 2003.

Reserves

Global gas reserves have more than doubled over the past 20 years, outpacing the 62-percent growth in oil reserves over the same period. *Oil & Gas Journal* estimated proven world gas reserves as of January 1, 2000, at 5,146 trillion cubic feet, an increase of 1.5 trillion cubic feet over the previous year's estimate (see box on pages 45-46).[7] Over the past 20 years, reserve estimates have grown rapidly in the FSU and in developing countries in the Middle East, South and Central America, and the Asia-Pacific region (Figure 47).

The largest incremental increases in reserves over the past year were nearly 4 trillion cubic feet for the Asia-Pacific region and more than 33 trillion cubic feet for Africa, mostly in Algeria and Egypt. Those reserve expansions were offset, however, by reported decreases in all other regions. Reserves in Mexico were reported to decline by more than 50 percent (from 63 to 30 trillion

cubic feet), and reserves in the United States and Western Europe also declined by 3 and 2 trillion cubic feet, respectively. Gas reserves reported by *Oil & Gas Journal* are compiled from voluntary survey responses and do not always reflect most recent changes. Some significant gas discoveries made in 1999, for example in Asia and the Middle East, are not reflected in the most recent estimates.

In regional terms, world gas reserves are more widely distributed than oil reserves. The Middle East, which holds nearly 65 percent of global oil reserves, accounts for only 34 percent of gas reserves (Figure 48). Thus,

Figure 47. World Natural Gas Reserves by Region, 1975-2000

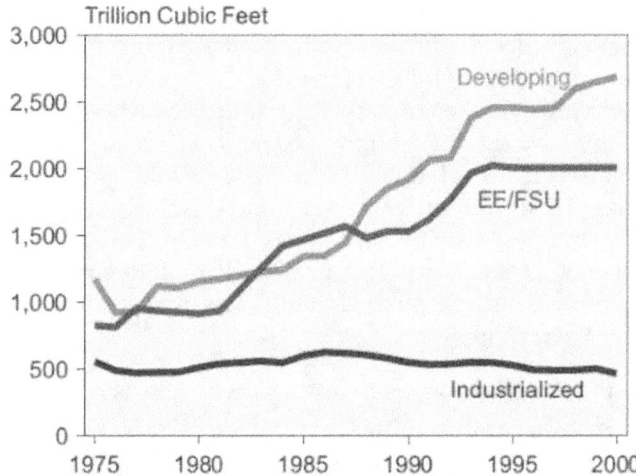

Sources: **1975-1993:** "Worldwide Oil and Gas at a Glance," *International Petroleum Encyclopedia* (Tulsa, OK: PennWell Publishing, various issues). **1994-2000:** *Oil & Gas Journal* (various issues).

Figure 48. World Natural Gas Reserves by Region as of January 1, 2000

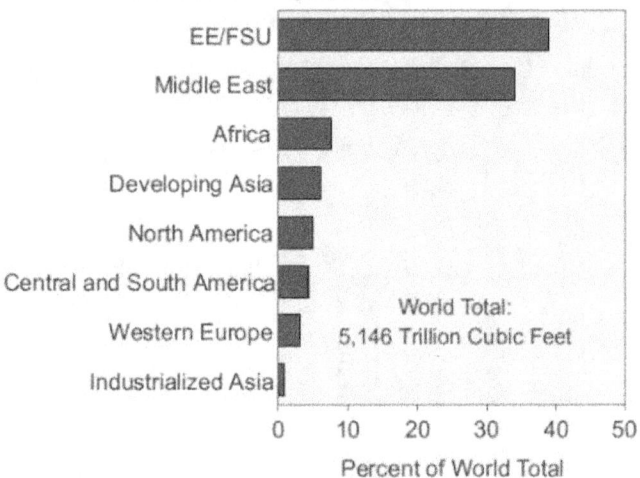

Source: "Worldwide Look at Reserves and Production," *Oil & Gas Journal*, Vol. 97, No. 51 (December 20, 1999), pp. 91-93.

---

[7]Proven reserves, as reported by the *Oil & Gas Journal*, are estimated quantities that can be recovered under present technology and prices. Figures reported for Canada and the FSU, however, include reserves in the probable category.

Taking Stock: Reporting Reserves and Selected Natural Gas Finds of 1999

Because the world relies on fossil fuels for more than 85 percent of its energy consumption, the amount of fuel available is of great interest. Estimates of reserves influence policy and business decisions, with impacts reaching individual consumers and even shaping society (as in the automobile culture that developed on perceptions of abundant oil). But the process of estimating and reporting reserves is complex and affects how reserve numbers should be used. The following paragraphs highlight some of the issues related to reserve estimates and their implications with respect to natural gas.

- *Natural gas reserves are a moving target.* Because the world consumes a portion of its gas reserves every day and new gas resources are discovered every year, estimating global reserves is difficult. In addition, the amount and type of information about reserves and gas consumption varies from country to country.

- *Reserve estimates are not updated as soon as new discoveries are made.* Although energy companies and governments often like to make public announcements of new discoveries, it usually takes time to assess the size of gas resources once they have been discovered. For example, 1999 was an important year

for natural gas exploration and discoveries, but none of the gas finds made last year is included in the recent reserves report of the *Oil and Gas Journal (O&GJ)* which is used in this chapter. The table below provides a sampling of significant gas finds in 1999.

- *How one defines "reserves" is important to the resulting estimate.* There are other reasons, as well, why discovered gas resources are not reported as natural gas reserves. The *O&GJ* defines "proven reserves" as those quantities that can be recovered under present technology and prices. Cedigaz, on the other hand, defines "proved reserves" as those corresponding to discoveries that are reasonably sure to be able to produce in present economic and technical conditions.[a] Both determinations are subjective and can vary depending on who is evaluating the resources.

- *Known but remote gas resources are often excluded from estimates of reserves.* It is a particularly complex issue to account for the large amounts of gas that are discovered far from demand centers ("remote gas"), because the cost of transporting the gas (unlike oil) is much higher; therefore, much of the remote gas

(continued on page 46)

A Sampling of Significant Natural Gas Resource Discoveries in 1999

Region/Country	1999 Discoveries
Industrialized	
Australia	Deepwater offshore discovery in the northwest near Gorgon gas field was made by WAPET consortium (including Mobil, Chevron, Texaco, and Shell).
Developing	
South America	
Peru	Another well (drilled by Shell and Mobil) in the Camisea region could yield large resources (2 trillion cubic feet), although it has not yet been flow-tested.
Asia	
Bangladesh	Unocal's third major discovery in Bangladesh, Moulavi Bazar field.
China	Major oil and gas finds by Phillips in Bohai Bay. Early hydrocarbon estimates of 2 billion barrels (gas not specified) and later reported at 4 billion barrels. Early in 2000 major gas find reported in remote Tarim Basin, with estimated gas reserves of 7 trillion cubic feet.
Indonesia	West Natuna discoveries. Others in east Kalimantan.
Thailand	Moderate-size finds made by Unocal and Chevron.
Africa	
Egypt	A very large offshore deepwater find along the Mediterranean coast could double oil and gas reserves (gas reserves currently at 35 trillion cubic feet).
Angola	Emerging as important new oil and gas province, with discoveries made in 1999 by Elf (new field has 3.5 billion barrel hydrocarbon reserve potential).
Nigeria	A major oil and gas field, named Erha, was found in deepwater offshore by an Exxon exploration affiliate.

Sources include newswires, newspapers, and assorted publications, such as *Business Wire, PR Newswire, Afx News, Oil & Gas Journal, Houston Chronicle,* and *Lloyd's List.*

[a]"Worldwide Look at Reserves and Production," *Oil & Gas Journal,* Vol. 97, No. 51 (December 20, 1999), pp. 91-93; and Cedigaz, *Natural Gas in the World—1999 Survey* (Paris, France, October 1999), p. 24.

some regions with limited oil reserves hold a greater portion of global gas stocks. The FSU, in particular, accounts for around 6 percent of world oil reserves but nearly 40 percent of proven gas reserves, most of which (33 percent of world reserves) is located in the Russian Federation. The Russian reserves are the largest in the world, more than double the second-largest reserve volume in Iran. Gas reserves are also more widely distributed than oil reserves within the Middle East, where Qatar, Iraq, Saudi Arabia, and the United Arab Emirates all have significant gas volumes (Table 16). Reserve-to-production (R/P) ratios exceed 100 years in the Middle East and Africa and are next highest in the FSU at 83.4 years. South and Central America also has a high ratio (71.5 years), but in North America and Europe R/P ratios are relatively low at 11.4 years and 18.3 years, respectively. The R/P average for natural gas in the world is 63.4 years, compared with 41 years for oil [1].

Table 16. World Natural Gas Reserves by Country as of January 1, 2000

Country	Reserves (Trillion Cubic Feet)	Percent of World Total
World	**5,146**	**100.0**
Top 20 Countries	**4,571**	**88.8**
Russian Federation	1,700	33.0
Iran	812	15.8
Qatar	300	5.8
United Arab Emirates	212	4.1
Saudi Arabia	204	4.0
United States	164	3.2
Algeria	160	3.1
Venezuela	143	2.8
Nigeria	124	2.4
Iraq	110	2.1
Turkmenistan	101	2.0
Malaysia	82	1.6
Indonesia	72	1.4
Uzbekistan	66	1.3
Kazakhstan	65	1.3
Canada	64	1.2
Netherlands	63	1.2
Kuwait	52	1.0
China	48	0.9
Mexico	30	0.6
Rest of World	**575**	**11.0**

Source: "Worldwide Look at Reserves and Production," *Oil & Gas Journal*, Vol. 97, No. 51 (December 20, 1999), pp. 91-93.

Regional Activity

North America

In the *IEO2000* reference case, natural gas consumption is projected to grow by 1.6 percent per year between 1997 and 2020 in Canada and the United States and by 2.4 percent in Mexico. Fuel use for electric power generation is largely responsible for the increases in all three countries. In the United States alone, natural gas consumption for electricity generation (excluding cogenerators) is projected to grow from 3.4 trillion cubic feet in 1997 to 9.3 trillion cubic feet in 2020. In projections from EIA's *Annual Energy Outlook 2000 (AEO2000)*, nearly 90 percent of new electricity generating capacity between 1997 and 2020 is combined-cycle or combustion turbine technology fueled by natural gas or both oil and gas; and while increases are also expected in the other U.S. demand sectors, the growth in gas use in the electric utility sector is by far the most significant [2]. For Canada, the Canadian Energy Research Institute (CERI) estimates that gas demand for electricity generation could nearly triple in the next decade, assuming the continued restructuring of the electricity sector that is currently either underway or anticipated in many provinces.

In 1998, approximately 55 percent of Canada's natural gas production was exported to the United States, and Canadian gas accounted for about 14 percent of U.S. consumption. Canada's exports have been growing steadily in response to increasing demand in the United States, more than tripling since 1985. By 2005, *AEO2000* projects that Canada's share of end-use consumption in the U.S. gas market will increase to 18.4 percent.

Currently, significant pipeline construction both within Canada and between the United States and Canada is underway to accommodate U.S. import demand. By the end of 2000, five major new natural gas pipeline projects and an upgrade on a sixth (Alliance, Millennium, NOVA, Northern Border, TransCanada, and Maritimes Northeast) are expected to be complete, allowing a considerable increase in trade between the two countries [3]. Most of the construction will provide access to supplies in western Canada, and the Maritimes and Northeast project will transport supplies from Canada's offshore Atlantic Sable Island fields to markets in New England. Gas fields with more than 6 trillion cubic feet of combined reserves near Sable Island and at Terra Nova are under development, and Cambridge Energy Research Associates has indicated that natural gas reserves off Nova Scotia may be five times what has already been discovered, with approximately 53 trillion cubic feet possible [4].

Considerable pipeline construction is also under way in the United States. Several major projects will provide access to new sources of both supply and demand and increase capacity along corridors where utilization rates are high during peak periods. Recently completed projects include Interstate's Pony Express project, the Trailblazer system expansion, the Transwestern Pipeline expansion, and the El Paso Natural Gas system expansion. The first two provide access to Wyoming and Montana production regions, and the last two provide access to New Mexico's San Juan Basin. Further expansions are underway that will increase flows from these areas to markets on the east and west coasts. U.S. pipeline capacity expansion is expected to slow after 2001, however, to less than 1 percent a year. Overall utilization of pipeline capacity is expected to increase significantly after 2001 as demand for natural gas to fuel electricity generation leads to increased flows during the summer months.

LNG imports are also becoming more economical for the United States. LNG imports are expected to increase more than fivefold between 1997 and 2020, from 0.08 trillion cubic feet per year to 0.39 trillion cubic feet per year. In the past, U.S. LNG imports have come predominantly from Algeria. New sources of supply include Australia, Trinidad and Tobago, and Qatar, and Abu Dhabi and Norway are potential sources.

Additions to U.S. LNG import capability include a 50-percent increase in offloading capacity at the Everett, Massachusetts, port facility; a projected reopening of the Southern Natural Gas Company LNG terminal at Elba Island, Georgia; and a potential reopening of Columbia LNG's Cove Point, Maryland, facility. Both Elba Island and Cove Point were closed over 15 years ago when LNG became too costly to compete with other sources of natural gas in the United States. Cove Point subsequently reopened in the early 1990s for peak-period service storage only. Preliminary approval to reopen Elba Island was granted by the Federal Energy Regulatory Commission (FERC) in December 1999, and Southern plans to begin importing up to 0.8 trillion cubic feet of LNG per year from Trinidad in 2002 [5]. Anticipating increased demand for LNG shipments, especially in the Northeast, Columbia LNG is hoping to recommission its Cove Point, Maryland, facility to provide LNG tanker unloading services. The terminal can deliver up to 1 billion cubic feet per day to Columbia's main system. If response is sufficient to a planned open season for customers to bid on capacity, as Columbia expects it to be, Cove Point will file with the FERC for authorization to recommission [6].

Although Mexico has considerable resources that could be developed, production is not expected to keep pace with rising internal demand, and Mexico is expected to remain a net importer of natural gas. As in the United States and Canada, most of the projected growth in demand is for electricity generation. A recent forecast by the Mexican government indicates that natural gas demand will grow by 9.2 percent a year between 1998 and 2007, with consumption for electricity generation increasing by 20 percent a year.

Another area of significant growth in Mexico's gas consumption is expected to be manufacturing and assembly plants located close to the U.S. border, where U.S. producers are in a much better position to satisfy the demand [7]. Although considerable investment is currently being made in the expansion of pipeline infrastructure, Mexico continues to have the problem, at least in the near term, of not being able to transport natural gas from southern producing regions to northern consuming regions in quantities sufficient to meet demand. *AEO2000* projects U.S. exports to Mexico to grow from 0.05 trillion cubic feet in 1997 to 0.24 trillion cubic feet in 2020, in the wake of Mexico's recent elimination of a 4-percent import tariff and an increase in pipeline capacity between the two countries.

Western Europe

Europe's gas reserves, which account for less than 5 percent of global resources, are located predominantly in the Netherlands, Norway, and the United Kingdom. Production in those three countries currently surpasses production in other regions with greater reserves, such as the Middle East. Nearly one-third of Europe's gas demand is met by supplies from outside the region, particularly pipeline imports from the FSU and Algeria, as well as LNG primarily from North Africa. Recent demand increases reflect rising gas use for power generation as well as in the industrial sector. Demand growth has been particularly strong in Greece, Portugal, Italy, Spain, Finland, Belgium, and Denmark. *IEO2000* projects growth in Western Europe's gas use averaging 2.9 percent per year, reaching 25.9 trillion cubic feet by 2020 (Figure 49).

European investments in infrastructure in 1998 included the completion of the Interconnector and at least four other significant pipeline projects, and 1999 saw the on-schedule completion and commissioning of the Europipe II. The 420-mile Europipe II, operated by Statoil, links Norway's west coast to Dornum in northwest Germany. With its commissioning, imports from Norway could supply up to 30 percent of Germany's natural gas use by 2010 [8]. In *IEO2000*, German gas consumption is expected to rise by an average of 2.9 percent per year, nearly doubling in 2020 from the 1997 level of 3.4 trillion cubic feet (Figure 50).

Elsewhere in Europe, pipeline projects such as a Swiss line scheduled for completion in October 1999 will improve and increase north-south gas flows. Future construction in Switzerland could eventually double capacity by 2003. A new leg of the Yamal-Europe pipeline was

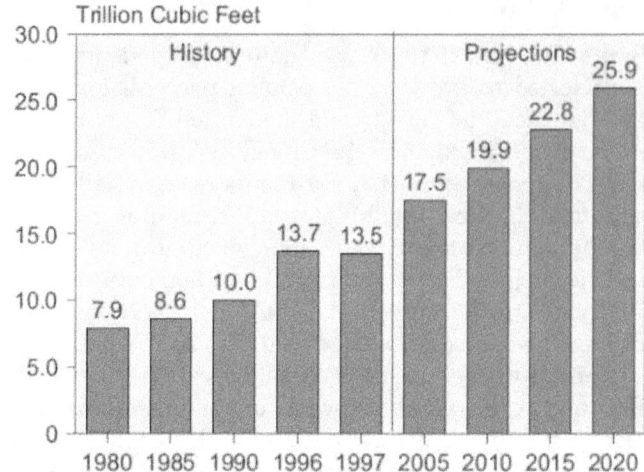

Figure 49. Natural Gas Consumption in Western Europe, 1970-2020

Sources: **History:** Energy Information Administration (EIA), Office of Energy Markets and End Use, International Statistics Database and *International Energy Annual 1997*, DOE/EIA-0219(97) (Washington, DC, April 1999). **Projections:** EIA, World Energy Projection System (2000).

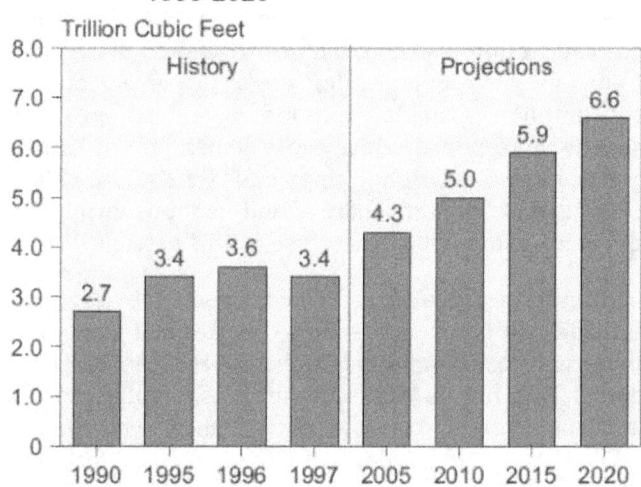

Figure 50. Natural Gas Consumption in Germany, 1990-2020

Sources: **History:** Energy Information Administration (EIA), Office of Energy Markets and End Use, International Statistics Database and *International Energy Annual 1997*, DOE/EIA-0219(97) (Washington, DC, April 1999). **Projections:** EIA, World Energy Projection System (2000).

also completed in Germany in September. Financed by Wingas and Gazprom investment, the 209-mile section stretches from Frankfurt-on-Oder to Rueckersdorf, Thueringen, and will transport up to 990 billion cubic feet of gas per year to German and West European consumers [9]. Germany's Ruhrgas started construction of a 71-mile pipeline from Mittelbrunn in Saarland to Esch in Luxembourg. With deliveries scheduled to start mid-2000, the $45 million project will transport 20 billion cubic feet of gas annually to a gas and steam power plant under a 15-year contract signed with Soteg [10].

In addition to the announced and planned mergers of such large international corporations as BP, Amoco, and ARCO and Mobil and Exxon, Europe also saw important mergers in 1999. TotalFina and Elf Aquitaine agreed in September to a friendly merger that would rival BP Amoco (which plans to merge with ARCO) in European gas production and marketing. In October, however, the European Union announced an investigation into the merger, citing concerns about potential dominance in the liquid petroleum gas (LPG) market [11]. In Germany, Veba and Viag will merge to create a group with electricity, natural gas, and water businesses and with the stated goal of pursuing growth abroad through targeted acquisitions [12].

Strong gas market growth continues in Italy and Spain. Italy's Eni announced that final agreements were reached with Libya's National Oil Corporation (NOC) to import gas via a new undersea pipeline. Plans call for imports of some 280 billion cubic feet per year starting as soon as 2003, with Italy now in the stage of awarding contracts [13]. Regional approval has been given to Edison-Mobil plans for a new Italian LNG receiving terminal in the northern Adriatic Po Delta (Figure 51) to be built on an artificial island; approval by the national government is still required. The terminal could be operational as early as 2003, with Egypt as the potential supplier of LNG [14].

Spain has agreed to buy more LNG from Shell-led facilities in Nigeria, enabling Shell to go forward with expansion plans there. Spanish gas demand is expected to grow rapidly after 2000, when a series of gas-fired, combined-cycle power plants are due for commissioning [15]. Competition has been increasing in the Spanish gas and power markets with the government's announcement of new operators allowed in each market. Five more companies now able to trade in gas include Enagas, Gas de Asturias, Gas de Euskadi, Iberdrola, and BP Amoco [16]. In Portugal, plans continue to develop the country's first LNG receiving terminal, to be sited on the Atlantic coast near the Sines oil terminal. Transgas Atlantico is seeking a turnkey contractor for the project and has issued an invitation to tender. Construction could begin by the end of 2000.

Eastern Europe and the Former Soviet Union

In most of Eastern Europe, natural gas consumption continued to decline in 1998, although there were increases in some countries. The Russian Federation continued to dominate world trade movements of natural gas, exporting 4.2 trillion cubic feet to Europe and to other FSU countries. The only other exports of natural gas from the FSU were 63.5 billion cubic feet delivered to Iran from Turkmenistan [17].

Nonpayment for gas supplies continues to be an issue throughout the FSU, both within and between countries,

and barter continues to be an accepted form of payment. Uzbekistan threatened to stop deliveries to Kyrgyzstan on November 15, 1999, if the mounting debt was not paid. Because of problems with Kyrgyzstan's hard currency, Uzbekistan had agreed to take partial payment for gas supplies in flour, but Kyrgyzstan had fallen considerably behind even in its "flour debt." As a result, the Uzbek gas transport company Uztransgaz indicated that it had no option but to cut off supplies [18]. Deliveries were temporarily halted but resumed in mid-December after the payment of $3 million, partly in cash and partly in goods. The gas currently being received is roughly half the amount received during 1998, and it is going mainly to homes in the northern part of the country [19].

Belarus, in debt to the Russian gas monopoly Gazprom, has major internal problems with consumers not paying their gas bills, which in turn make it difficult for the Belarusian government to pay Gazprom. The government gave internal consumers until January 1, 2000, to pay gas debts, and the state-owned gas transport company Beltranshaz proposed that partial payment of the debt to Gazprom be made in agricultural equipment [20].

The situation is similar in Ukraine, which owes Russia more than $1 billion for natural gas purchases. Ukraine's internal nonpayment problem is significant, with consumers owing Naftogaz Ukrainy about $3 billion. In an effort to secure payment from domestic customers, the government has been cracking down on nonpaying customers by curtailing supplies. A total of 13,700 were disconnected in 1998, including 3,650 industrial customers. After some initial problems involving cutoffs during the cold winter months, the government pledged to forbid cutoffs during the winter. Cutoffs resumed in the spring, however, and at the beginning of April 1999, 378 debtor firms were disconnected from natural gas supplies [21].

Ukraine has resorted to barter to satisfy its external debt, agreeing to deliver to Russia 11 bombers and 500 cruise missiles, valued at $285 million, by the end of 1999 as partial payment of the debt owed to Gazprom. The first of the bombers, all of which were inherited after the breakup of the Soviet Union, were delivered to Russia in November 1999 [22]. Ukraine is also in debt to Turkmenistan for gas supplies delivered as far back as 1993, and Turkmenistan cut off supplies in June 1999. The debt has since been restructured, with payments to be completed in December 2001. In October 1999, the debt exceeded $300 million [23].

Russia has also threatened to take action against Ukraine for reasons other than nonpayment. Ukraine is the main transit route for Russian gas to reach European markets, and Russia has long accused Ukraine of siphoning off gas during transit. Itera, the main supplier of Russian gas to Ukraine, threatened to stop supplying gas to

Figure 51. Italy's Natural Gas Infrastructure

Source: International Energy Agency, *Natural Gas Information 1998* (Paris, France, 1999), p. VI.27.

Ukraine by October 1, 1999, pending payment of debt. Gas exports are a major source of revenue for Russia, and over 90 percent of gas exported by Russia passes through Ukraine. According to a Gazprom spokesman, since ceasing gas flow to Ukraine would entail giving up the European market, it is unlikely that gas shipments to and through Ukraine will cease anytime soon [24].

Although Russian natural gas trade with Europe is currently dependent on Ukraine, Russia expects new export routes to be developed in the next few years. The first section of the Yamal-Europe pipeline, through Belarus and Poland, went into operation in 1999, and a second parallel section is in the planning stages. The new pipeline allows Russia to eliminate Ukraine from its route to Western Europe, which currently receives 25 percent of its natural gas from Gazprom. Gazprom is eager to increase exports to Western European customers, who pay on time and in U.S. dollars, in sharp contrast with domestic customers, who make only 20 percent of their payments in cash, if at all [25]. Although still awaiting final approval, the Blue Stream pipeline, which would traverse the Black Sea bed and transport Russian gas to Turkey and Southeast Europe, is expected to become operational in 2001. A third project under consideration is the construction of a pipeline through the Baltic Sea to Germany. If these projects are built and become fully operational, shipments of Russian gas through Ukraine will decline by about one-third [26].

The Blue Stream project to supply Russian gas to energy-hungry Turkey is in competition with another project, the Trans-Caspian project, which would supply Turkey and western markets with gas from Turkmenistan and Azerbaijan. Although it was initially intended to ship supplies from Turkmenistan alone, Azerbaijan entered the pictured after the discovery earlier this year of large volumes of natural gas at its offshore Shakh Deniz field. Reserves are said to be between 14 and 25 trillion cubic feet, and geologists have indicated that additional finds are likely. Given that Azerbaijan has indicated that this discovery alone would allow them to export 0.6 to 0.7 trillion cubic feet per year, it is doubtful that they will elect to play solely a transport role in the Trans-Caspian project. Azerbaijan is also exploring the possibility of exporting gas to Iran by way of an existing pipeline.

As of late November 1999, Turkey had still not made the decision whether to support the Blue Stream project or the U.S.-supported Trans-Caspian project. Turkey has indicated that it does not feel that the two projects are in competition with, or alternatives to, each other, because the amount of gas proposed from both sources will still fall short of satisfying Turkey's projected natural gas demand. The Trans-Caspian line would carry 0.6 trillion cubic feet of gas per year from Turkmenistan, and possibly Azerbaijan, to Turkey. According to its sponsors, it could ultimately gain another 0.5 trillion cubic feet of capacity to serve international markets. The Blue Stream pipeline would have an ultimate capacity of 1.1 trillion cubic feet. Turkey has projected that it will need 1.9 trillion cubic feet of gas by 2010, and needs to find and purchase as much natural gas as it can as soon as possible [27].

Even as Russia hopes to reduce its dependence on Ukraine, Ukraine wants to diversify its gas sources so that it is less dependent on Russia. Ukraine currently produces just about 25 percent of the gas it consumes, with most of the balance coming from Russia. Ukrainian officials met with a delegation from Afghanistan in September 1999 to discuss the possibility of receiving gas from Afghanistan [28], and Ukraine has announced plans to conduct talks with Kazakhstan about purchasing 0.2 trillion cubic feet of natural gas in 2000 [29], representing roughly 10 percent of the amount currently imported from Russia.

Another major consumer seeking to lessen its dependence on Russia is Poland. In the past Russia has met nearly all of Poland's natural gas needs. As a move towards diversification, however, Poland signed a 5-year gas supply contract with Norway in December 1999 for supplies beginning in 2001. Talks to determine how the gas might be transported to Poland are only at an initial stage, however [30]. Poland is anticipating large increases in natural gas demand between now and 2020 and is dependent on imports to meet most of the new demand. Natural gas industry restructuring, which will facilitate gas purchases from countries such as Norway, Holland, and Germany, is in the works, with an anticipated start date sometime in 2001. Accompanying the restructuring is a move toward privatization, which the government expects to be completed in 2005.

In spite of the uncertainties and problems currently facing the EE/FSU, *IEO2000* projects significant future growth in the region's natural gas markets. Consumption in the FSU is projected to grow at a rate of 2.1 percent a year between 1997 and 2020, with the strongest growth at the end of the forecast period from 2015 to 2020. The projected increase in Eastern Europe is steadier but considerably higher, at an overall rate of 5.6 percent per year. Total EE/FSU consumption nearly doubles, from 22.3 trillion cubic feet in 1997 to 41.2 trillion cubic feet at the end of the forecast period. The considerable effort, both internally and via foreign investment, that is going into the development of the region's natural gas infrastructure will be a significant factor in increasing future production, consumption, and export capabilities for natural gas.

Central and South America

Central and South American gas markets are small in terms of total volumes handled, but they continue to show strong growth with active upstream and downstream development. Between 1990 and 1997, the region's gas consumption grew by an average of more than 5 percent per year. Estimated reserves in the region account for less than 5 percent of global gas reserves, but much of the area has been underexplored, and discoveries are accompanying recent exploration activity. Production, consumption, and trade are also limited. Production and consumption of natural gas in the region were at about 2.9 trillion cubic feet in 1997. In 1999, natural gas trade extended outside the region with initiation of LNG exports from Trinidad and Tobago. The only international pipelines in the region before 1999 operated from Bolivia to Argentina and from Argentina to Chile. The *IEO2000* reference case projects that the region's gas use, facilitated by additional pipelines, will grow to 15.3 trillion cubic feet by 2020, at an average annual growth rate of 7.5 percent (Figure 52).

Figure 52. Natural Gas Consumption in Central and South America, 1990-2020

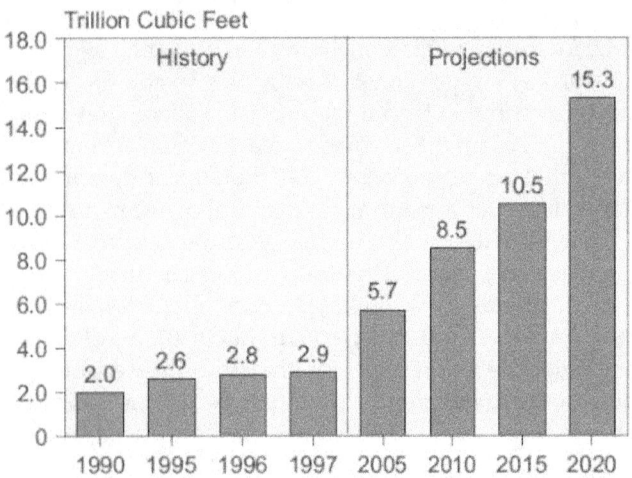

Sources: **History:** Energy Information Administration (EIA), Office of Energy Markets and End Use, International Statistics Database and *International Energy Annual 1997*, DOE/EIA-0219(97) (Washington, DC, April 1999). **Projections:** EIA, World Energy Projection System (2000).

On the production side, a new LNG facility in Trinidad and Tobago came on stream in 1999, loading the first Atlantic Basin export cargoes in April [31]. Starting in 2000, the U.S. LNG company Cabot will take 1.8 million metric tons of LNG per year from the facility, some of which will be delivered to Puerto Rico. A new receiving terminal now under construction in Puerto Rico will enable increased use of gas in power generation. BP Amoco and Repsol are promoting plans to add two more LNG trains in Trinidad and Tobago, with Spain lined up as a buyer for at least some of the increased output and Cabot also expected to take more gas [32].

Much of the gas market growth in South America involves Brazil, a large country with large projected gas demand. In the *IEO2000* reference case, gas use in Brazil grows from 0.2 trillion cubic feet in 1997 to 2.5 trillion cubic feet in 2020. A new Bolivia-Brazil pipeline began operating in July 1999 after years of negotiation. Initially, the line was expected to begin carrying about 78 million cubic feet per day to Brazil, rising to 200 million cubic feet per day by the end of 1999 and then to 318 million cubic feet per day in 2000, when a take-or-pay contract begins. By 2006, volumes could exceed 1 billion cubic feet per day, worth about $400 million per year. Import volumes at the end of 1999 were much lower, however, ranging only around 22 to 53 million cubic feet per day due to a significant rise in Brazilian gas prices. Prices are linked via formulae to fuel oil prices (which were rising) and were also affected by a 40-percent devaluation of the Brazilian real. Pipeline operators are estimating the associated financial loss at about $500,000 per month [33].

A second line to Brazil is now under consideration following significant gas discoveries in Bolivia, but construction is not anticipated before 2001. In addition, an Enron project involving a 390-mile pipeline from Bolivia to Cuiaba in Brazil has received approval from the U.S. Overseas Private Investment Corporation (OPIC) for a $200 million credit, despite objections from environmentalists. Protestors opposed the pipeline route through what may be the world's largest intact dry forest (called Chiquitana). Backers of the project have emphasized planned environmental mitigation measures (including a partial rerouting to avoid the most environmentally sensitive areas and a pledge of $20 million over 15 years for conservation efforts) as well as benefits of new gas supplies replacing diesel use and local firewood demand. A 490-megawatt Cuiaba power plant and the pipeline together amount to a $570 million effort involving Shell and Bolivian firm Transredes (50 percent owned by Enron and Shell) [34].

In addition, a pipeline to Brazil from Argentina is under construction to enable the first imports of Argentine gas. The 272-mile Transportadora de Gas del Mercosur, SA (TGM) pipeline from Parana in Entre Rios, Argentina, to Uruguaiana, Brazil, will connect to Argentina's existing domestic line, Transportadora de Gas del Norte (TGN). CMS Energy holds equity in both pipelines. TGM's other owners include Canada's TransCanada Pipelines, Argentina's CGC and Techint, and Malaysia's Petronas. The pipeline is scheduled for completion in 2000 and will initially transport about 100 million cubic feet per day, with capacity expandable up to 425 million cubic feet per day [35].

To address higher gas prices in Brazil and encourage gas-related investment, particularly in the power sector, the government announced that it may control prices in some regions. In southern, southeastern, and central western states as well as the Brasilia Federal District, power projects signing 20-year contracts could have a price ceiling equivalent to $2.26 per million Btu (in 1999 dollars). In Brazil's northeast, where gas will initially be domestically supplied, there is a proposed price ceiling for the first 5 years of $1.94 per million Btu and $2.26 for the remainder of the 20-year period. As many as 23 power plants with a combined capacity of 7,400 megawatts could benefit from the price control policy. The move is welcome by foreign investors and will facilitate project funding, because it fixes long-term contracts in dollars rather than the more volatile Brazilian real [36].

Elsewhere in South America, the GasAtacama pipeline began operating in mid-1999, sending gas from Argentina to Chile. It is the second pipeline linking the two countries, after TransCanada's GasAndes pipeline in central Chile, which started operation in 1997. The

$400 million GasAtacama line was built by CMS Energy and Chilean generator Endesa in a project that includes a $350 million, 370-megawatt power plant. A parallel and rival line, Norandino, backed by Belgian Tractabel and Southern Energy, began operation at the end of 1999. The 585-mile GasAtacama has a capacity of 300 million cubic feet per day but will operate significantly below that at first. Because the project's associated long-term electricity sales contract with Chilean Emel and its distribution companies (via the Nopel power plant) does not begin until 2002, it will sell power on the spot market at first. A pipeline extension is being planned to reach other power plants [37].

A pipeline has also been proposed for transporting Colombian gas into Central America. Backed by local firm Promigas (operated by Enron) and supported by state oil company Ecopetrol, such a project could supply gas to Panama, Costa Rica, and Nicaragua. Conceived to find markets for Colombia's abundant known and potential reserves, the line could face a possible competing or complementing project from Mexico [38]. Venezuela has also expressed an interest in exporting gas to Central America and/or forming a strategic alliance with Ecopetrol [39].

In Peru, the government repeatedly postponed bidding deadlines in 1999 for the sale of the $2 billion Camisea natural gas project. Although the delays purportedly were to give interested parties more time, they followed the resignation of energy and mines minister Daniel Hokama, as well as reports of disagreement within the government's Camisea committee. Potential investors have complained that the terms of the project are not sufficiently attractive and that gas prices should be set by the market, while the government has maintained that set prices are needed as incentive to potential power producers [40].

Venezuela's election of populist president Hugo Chavez has led to turmoil and power struggles for Petroleos de Venezuela (PDVSA), one of Latin America's biggest companies. The president of PDVSA, Roberto Mandini, has resigned, to be replaced by Hector Ciavaldini, who is said to back reforms promoted by Venezuelan President Chavez including a government effort to control the oil industry. PDVSA has been described as the economic backbone of Venezuela, the largest producer and consumer of natural gas in South and Central America [41]. Reorganization at PDVSA includes the establishment of a Natural Gas Division, reflecting a government preference to emphasize gas resource development. Although PDVSA will no longer pursue plans with Shell for an offshore Cristobal Colon LNG project because of poor economics, it is still looking for potential LNG or pipeline projects to commercialize gas reserves. Several domestic pipeline projects will also be prioritized [42].

Asia

Gas market activity in Asia during 1999 reflected ongoing recovery in the region from the Asian financial and economic crisis. Demand growth recovered, most notably in South Korea and Thailand, and new projects moved forward. In the *IEO2000* reference case, the growth rate for natural gas use through 2020 in the whole of Asia (both industrialized and developing) averages 5.6 percent per year, increasing consumption to 31.5 trillion cubic feet from 8.9 trillion cubic feet in 1997. The projected demand growth in developing Asia is much higher than in the industrialized countries of the region (Figure 53).

Industrialized Asia

For the countries of industrialized Asia, natural gas consumption is expected to rise from 3.2 trillion cubic feet in 1997 to 5.2 trillion cubic feet in 2020. Gas use in Japan, the world's largest consumer of LNG, accounts for 70 percent of the incremental increase over the period, increasing from 2.3 to 3.7 trillion cubic feet in the *IEO2000* reference case. Although in the short run Japanese demand growth is affected by economic recession there, LNG imports to Japan are expected in the long run to grow slowly from their large base. The September 1999 occurrence of Japan's worst nuclear accident in history could ultimately affect gas use, which is consumed primarily in the electricity sector, competing with nuclear power.

As not only the largest but also the first LNG importer in Asia, Japan has several major contracts that end over the next 5 to 10 years. Whether or not Japanese utilities seek

Figure 53. Natural Gas Consumption in Asia by Region, 1980-2020

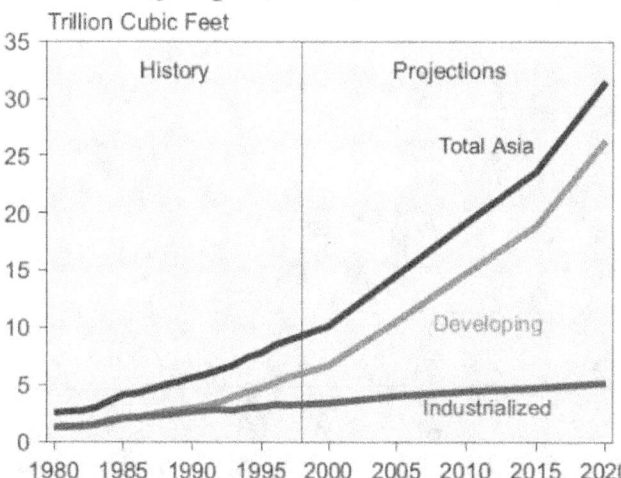

Sources: **History:** Energy Information Administration (EIA), Office of Energy Markets and End Use, International Statistics Database and *International Energy Annual 1997*, DOE/EIA-0219(97) (Washington, DC, April 1999). **Projections:** EIA, World Energy Projection System (2000).

to renew those contracts and/or secure LNG from other suppliers could have a major impact on the gas industry in the region. In September 1999, Tokyo Electric (Tepco) and Tohuku reached an initial agreement with Indonesia's Pertamina to extend their purchase of Arun LNG for another 5 years from the 2005 conclusion of their current contract. Volumes would drop from 3.5 million metric tons per year in 1999 to 1 million metric tons per year (reflecting Japan's uncertain demand outlook and desire to diversify suppliers). In view of increasing competition among suppliers, the agreement is good news for Pertamina, although a firm contract must still be finalized [43]. Tokyo Electric and Tokyo Gas also have contracts with Malaysia that will expire before 2005.

Australia is not a large gas consumer (fifth largest in the Asian region in 1997), but it is an important LNG exporter using its sizable resources located in the northwest, far from domestic demand centers. In 1999, Texaco and Chevron announced the discovery of a new field holding several trillion cubic feet of gas near the existing Gorgon reserves offshore Western Australia (Figure 54).

Figure 54. Australia's Natural Gas Pipelines

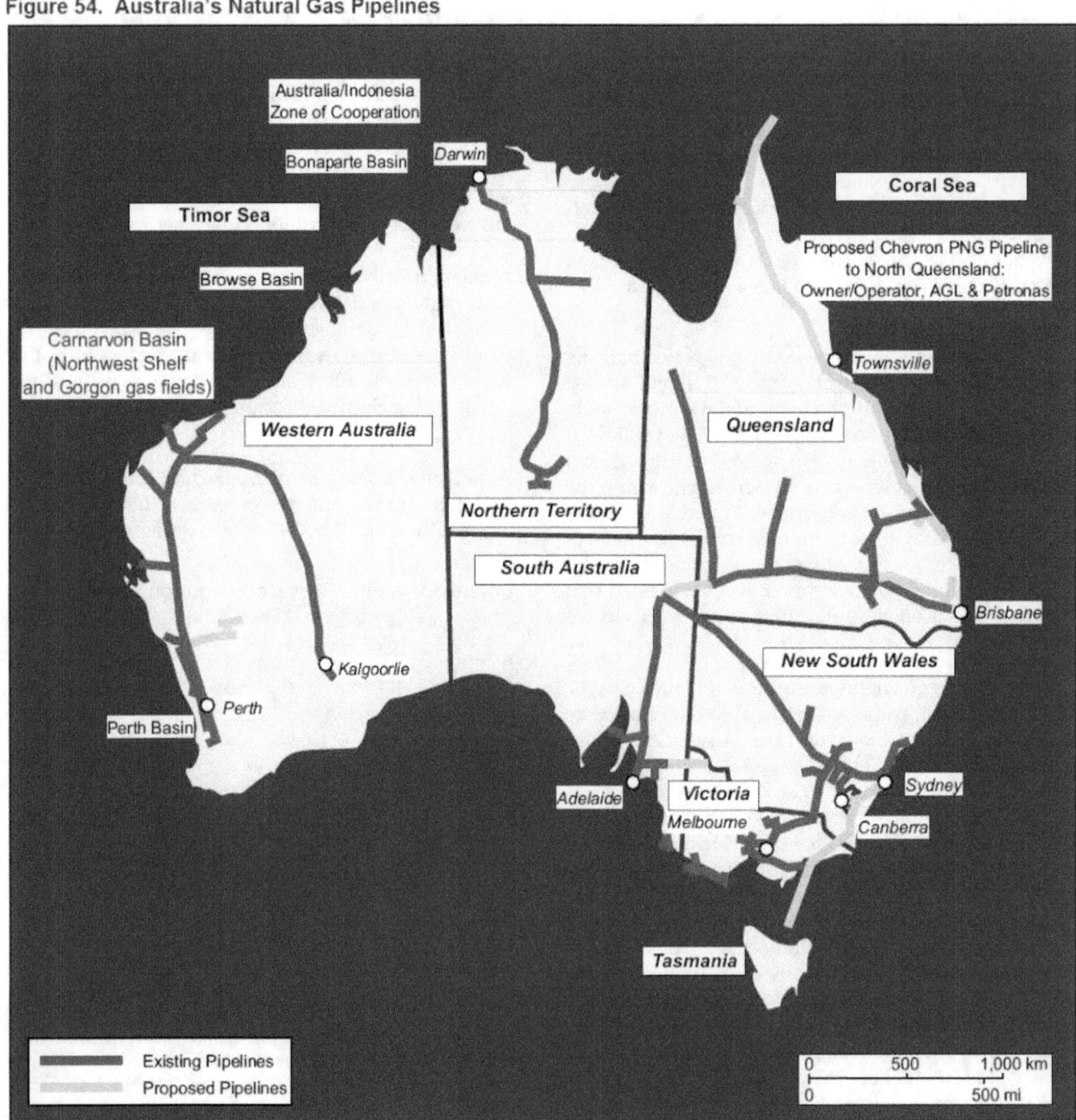

Source: International Energy Agency, *Natural Gas Information 1998* (Paris, France, 1999), p. VI.3.

Although the Gorgon and nearby fields contain proven, probable, and possible reserves of 21.5 trillion cubic feet—enough to more than double Australia's current LNG exports—no buyer has yet signed a final sales contract, delaying development plans [44]. Australia is promoting its LNG in China (a future importer) and Taiwan, and a number of high-level government and private-sector meetings were held in 1999.

At the same time, there are also plans for Australia to become a gas importer, via a pipeline planned from Papua New Guinea (PNG) to northeastern (Queensland) Australia. In July 1999, the $3.5 billion dollar project involving Chevron received a boost with the finalization of sales contracts between upstream producers and Australian buyers. The Australian side of the pipeline will be built and owned by a joint venture between Australian Gas Light Company and Malaysia's Petronas. On the PNG side, owners of the gas assets are expected to set up an entity for pipeline construction, funding, and ownership [45].

Developing Asia

Developing Asia includes the two most populous countries in the world, China and India—emerging giants that are geographically large, have large economies, and are expected to have a tremendous impact on gas use in the region (see box on page 57). At the same time, China and India accounted for only about 15 percent of gas use in the region in 1997, and both use a much smaller proportion of gas in their total energy mix than the region as a whole.

For now, much of the gas market growth and significant infrastructure projects are taking place elsewhere in the region, largely, in Southeast Asia. At the beginning of 1999, Singapore agreed to import gas via pipeline from Indonesia's Natuna West field. A contract involving SembCorp Industries, Ltd. (a government-linked company in Singapore) calls for deliveries of 325 million cubic feet of gas per day over a 22-year period and links the price primarily to Singapore spot prices for high-sulfur fuel oil. The project has raised controversy in Indonesia, including allegations of corruption because of linkages to Mohamad "Bob" Hasan, a businessman and close friend of former president Suharto. A pipeline construction contract worth $335 million was awarded to McDermott, and Hasan holds about 18 percent equity in McDermott Indonesia. In October, however, two independent audits reported that the contract was awarded fairly. The pipeline is under construction, with deliveries still scheduled to start in 2001 [46].

A second deal between Singapore and Indonesia, worth about $7 billion and leading to further pipeline trade, was also finalized in September 1999. The 20-year contract set to begin in 2002 calls for state-owned Singapore Power to buy from Pertamina initially 150 million cubic feet per day of gas, with volumes increasing up to 350 million cubic feet per day by 2008. Gas could be used increasingly in Singapore for power generation and petrochemicals [47].

Despite its political and economic upheaval, Indonesia is completing an eighth train at its Bontang LNG plant. Train H is set to begin exporting gas to Taiwan and South Korea by the end of 1999 or early 2000. Already the world's largest LNG exporter, Indonesia would also like to proceed with construction of a grassroots LNG plant, "Tangguh," in Wiriagar, Irian Jaya, using reserves discovered by British Gas and ARCO (which will merge with BP Amoco). To do so, Pertamina must find buyers for the LNG, which it is actively marketing to China in competition with Australian producers, among others [48]. In its domestic market, the Indonesian government will reduce natural gas prices from around $2.50 per million Btu (U.S. dollars) to about $1.50 per million Btu, in response to a dramatic plunge in gas use over the past 2 years, during which domestic contracts were priced in dollars and the Indonesian rupiah depreciated sharply against the U.S. dollar [49].

Elsewhere in Southeast Asia, production and delivery of contracted gas from the Yadana field in Myanmar (formerly Burma) was delayed until 2000 because of the postponed completion of Thailand's Ratchaburi power plant from its originally scheduled July 1998 startup. The Petroleum Authority of Thailand (PTT), which has a take-or-pay contract with the Yadana consortium, recently reached settlement to pay for some of the gas not taken. It was at first claiming "force majeure" following paralyzing events of the Asian financial and economic crisis [50]. Early estimates from PTT are that it will transmit 10 percent more gas in 1999 than the previous year, suggesting some recovery from the Asian crisis in the gas sector [51].

Completion of the Ratchaburi plant and recovery of Thai gas demand growth will also be important, because Malaysia and Thailand have reached agreement over the development of gas resources in their joint development area (JDA), with infrastructure construction to begin as early as 2000. Initially, project equity is split with 50 percent held by Petronas and 50 percent by PTT. Thailand may offer equity to foreign investors. In September 1999, Thai cabinet ministers approved a PTT investment plan for the project, which includes a pipeline and gas separation plant [52]. Malaysia could take all of the gas during the initial years of production if Thai demand cannot at first absorb the supply [53]. The first phase of the project (with targeted completion by 2002) calls for a 220-mile off/onshore pipeline from the JDA through the southern Thai town of Songkhla to the Malaysian province of Kedah. A second phase would involve a pipeline connecting JDA gas fields to the PTT grid in the Gulf of Thailand. Thai environmentalists quickly objected to the approval of investment plans before public hearings could be convened [54].

In 1998, Petronas of Malaysia signed eight oil and gas production sharing contracts (PSCs), the highest number in 10 years, reflecting the 1997 introduction of new revenue-over-cost formulae in PSCs [55]. Petronas also reached an agreement with Metropolis (wholly owned by Enron) to supply 2.6 million metric tons of LNG to India for a 20-year period starting in 2002. If a final contract is signed, it could be the first arrangement for delivery of Asian LNG into India [56].

India moved forward in 1999 with other plans to begin importing gas, which involve a host of projects and potential LNG terminals. In May, Enron's 624-megawatt Dabhol I power plant began operating, and the 1,624-megawatt Dabhol II reached financial closure [57]. Dabhol I is initially burning naphtha but will switch to LNG, and Dabhol II should use LNG at startup. Enron is making shipping arrangements for LNG imports, which are planned to start in 2002 (from Oman) and will eventually send fuel to the Dabhol power plants. Petronet is still pursuing plans for two initial LNG import terminals, and other terminals are also planned. In July, a 25-year agreement was signed with Rasgas for the purchase of 7.5 million metric tons of LNG per year, although a final contract must still be signed. The agreement calls for delivery of 5 million metric tons per year to a terminal at Dahej in Gujarat and 2.5 million metric tons per year to Cochin in Kerala [58]. Deliveries are scheduled to begin July 2003.

Gujarat Pipavav LNG, involving British Gas and Sea King Engineers, obtained agreement from India's National Thermal Power Corporation (NTPC) to take on a 26-percent equity stake in their project. British Gas and Sea King will then jointly hold a 50-percent share and offer the balance to public-sector gas users. As part of the agreement, an international tender will be issued to select an LNG supplier, although British Gas had earlier signed a memorandum of understanding with Yemen LNG. NTPC also continues to pursue an equity stake in Petronet and will eventually seek the best LNG price possible to fuel its power plants [59].

Another Indian LNG project in strong standing involves Total and Tata Electric Companies (TEC) at Trombay near Mumbai (formerly, Bombay). Equity in the project's joint subsidiary, Indigas, will be acquired by Gas Authority of India (GAIL). TEC and GAIL have agreed to initial purchases of 3 million metric tons of LNG imports per year. TEC will use the gas to fuel its Trombay power station, and GAIL will market gas to

industries around Mumbai [60]. Plans for gas imports to India reflect expectations of rising consumption. In the *IEO2000* reference case, gas use in India is projected to grow at an average annual rate of nearly 8 percent per year from 1997 to 2020, increasing sixfold (Figure 55).

China, also moving toward initiating LNG imports, has undertaken a feasibility study of importing LNG in Guangdong. Australia has remained visible as the potential supplier, with visits to southern China by Australia's Foreign Minister. The Chinese import project involves a consortium led by China National Offshore Oil Corporation (CNOOC), which is likely to retain a 36-percent equity share. CNOOC is to invite bids from foreign companies, which will be allowed to acquire a 35-percent share [61]. In addition, Shell has reached a gas and power joint venture agreement with China National Petroleum Company (CNPC) for a $3 billion project to construct pipelines and convert power plants and industrial facilities from coal to gas supplied from the Ordos Basin (Changbei field) [62].

In LNG-importing South Korea, monopoly importer Kogas faces deregulation and privatization. Presenting a restructuring plan in October 1999, the Commerce, Industry and Energy Ministry proposed spinning off and selling importing and wholesale units of Kogas as well as selling off its stakes in storage facilities and the main pipeline network by 2001 [63]. Taiwan, another LNG importer, was hit by a major earthquake in September 1999. Damage to the power grid could cause Taipower to take less LNG from Chinese Petroleum Corporation (CPC) in 2000 for power generation.

Figure 55. Natural Gas Consumption in India, 1990-2020

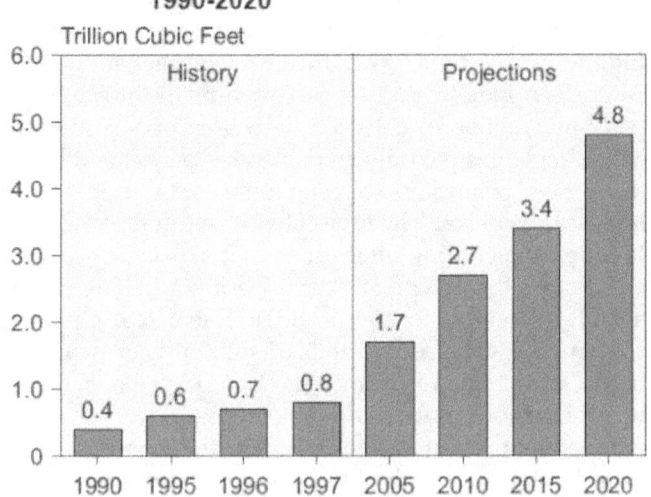

Sources: **History:** Energy Information Administration (EIA), Office of Energy Markets and End Use, International Statistics Database and *International Energy Annual 1997*, DOE/EIA-0219(97) (Washington, DC, April 1999). **Projections:** EIA, World Energy Projection System (2000).

Two Giants Emerge: Natural Gas in China and India

China and India are the two most populous countries in the world, with nearly 40 percent of the world's population. Both are typically labeled "developing," although their economies behave uniquely. They are geographically large and have extensive agricultural sectors that consume large quantities of fertilizer, and both have substantial coal resources that have served as the primary fuel for recent development. Both have some oil production but are net oil importers, and both are actively pursuing gas sector development and imports of liquefied natural gas (LNG).

The differences between the Chinese and Indian national governments are reflected in their pursuit of gas sector development. India, the world's largest (and most complex) democracy, has been through a series of unstable coalition governments in recent years and has held three parliamentary elections in as many years. India's states and localities also have unique, complicated politics that affect government, foreign investment, and gas development. One of India's oldest LNG import projects—led by Enron to fuel its power plants at Dabhol in the state of Maharashtra—has had to renegotiate finalized deals when political power in the state has shifted from one party to another. On one hand, the national government participates in LNG projects through Petronet, a consortium that includes the state oil and gas companies Gas Authority of India, Ltd. (GAIL), Oil and Natural Gas Corporation (ONGC), Indian Oil Corporation (IOC), and Bharat Petroleum Corporation Limited (BPCL). On the other hand, Petronet also competes in the private sector for other LNG projects, as it did when it bid for and lost a project in Tamil Nadu.

China's government, controlled by the Communist Party, is more centralized. Plans to import LNG waited several years for approval by the central government, which is now ready to announce the country's first LNG project. The project is likely to be led by China National Offshore Oil Corporation (CNOOC), with foreign companies allowed to acquire a 35-percent share. Other LNG projects may follow in China but have not yet been approved or announced. In India, by comparison, a number of LNG projects (at least 12 sites have been discussed) are in various stages, some with central government involvement and some without. Front-runners include the Enron Dhabol project and a Petronet project for an LNG terminal at Dahej in the Gujarat state, near India's major gas pipeline (see map).

In both countries, the initiation of gas imports will be affected by the existing infrastructure and industry and will also require new infrastructure and contracts with end users. India has a 1,550-mile natural gas pipeline with a capacity of 1.17 billion cubic feet per day running inland from the coast, and along its path are many gas-hungry users, ranging from fertilizer and petrochemical industries to power producers. Much of India's industrial development is also occurring along its coastlines. India's west coast, in particular, lies near to gas supplies in the Middle East. Lower transportation costs as well as lower overall prices for LNG imports from the Middle East may facilitate the process in India of guaranteeing sufficient end-use demand to absorb the imports.

In China, it is expected that Guangdong Power would take about 70 percent of imported gas from the first LNG import terminal, with the remaining 30 percent supplying town gas to Guangdong cities, including Shenzhen, Guangzhou, Foshan, and Dongguan. Although China, like India, has captive energy users along its developed southeastern seaboard, it does not have a major gas pipeline running inland from the coast. Both countries, in fact, will need more transmission infrastructure for sustained gas sector development. Given the size of the Chinese and Indian economies, pursuit of that development will affect not only new gas supply projects and the regional gas market but also the global gas market, associated markets for competing fuels, and related investment.

Selected LNG Projects in India

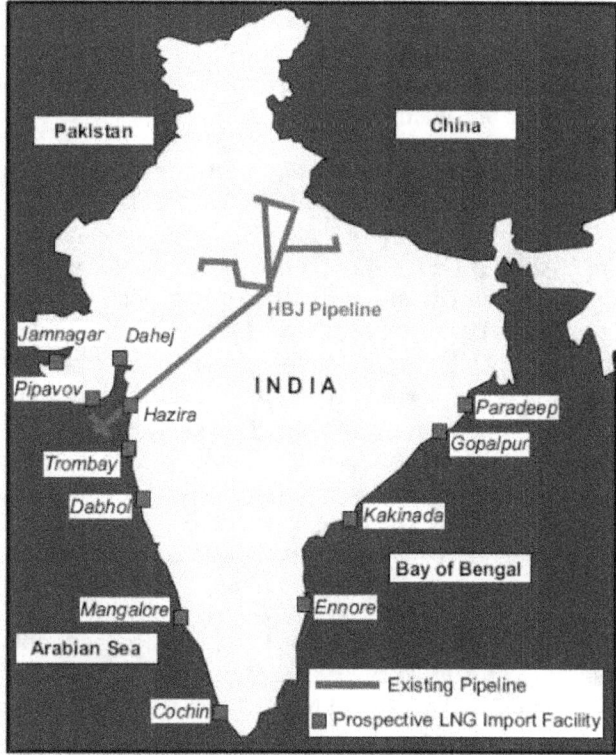

Source: East-West Center, "Center-Stage: Natural Gas in India," *Energy Advisory*, No. 231 (April 1999).

Middle East

As a region, the Middle East has the second largest natural gas reserves after the former Soviet Union. Reserves in the Middle East were estimated at 1,750 trillion cubic feet as of January 1, 2000. Iran, Qatar, and the United Arab Emirates (UAE) have the second, third, and fourth largest reserves in the world, respectively, following Russia. Middle East reserves, which expanded rapidly in the late 1980s and early 1990s, include the super-giant gas structure involving Qatar's North Field and Iran's South Pars. With its large reserves, the Middle East is a strong producer and growing exporter of natural gas, although domestic consumption to date has generally been quite low in the region. The *IEO2000* reference case projects a doubling of Middle East gas consumption between 1997 and 2020, rising from 6.0 to 12.0 trillion cubic feet (Figure 56).

Turkey, one of the fastest growing gas markets, suffered a major earthquake in August 1999 (and another in November), taking at least 16,000 lives and causing infrastructure damage, especially to an oil refinery, but not to gas pipelines. Fierce competition continued among gas projects proposing to supply the Turkish market. Russian Gazprom's Blue Stream project (involving a pipeline from Russia directly to Turkey under the Black Sea) signed a memorandum of understanding with Italy's Eni in February 1999 and then received a series of approvals in Russia. Construction of the pipeline could begin in 2000, according to project participants. The competing Trans-Caspian Pipeline proposal (which would export Turkmen gas to Turkey via the Caspian Sea) also received a boost when commercial agreements were signed in November in Istanbul with President Clinton attending [64].

In an effort to provide gas for domestic development in the south Persian Gulf region, the UAE Offsets Group (UOG) has begun promoting the "Dolphin" gas venture. The UAE and Oman have little gas for local demand, because resources are earmarked for export. Thus, a pipeline has been proposed from Qatar's North Field to Abu Dhabi, Dubai, and Oman—with a proposed future extension on to Pakistan. UOG has signed a memorandum of understanding with Mobil Oil Qatar, Inc., for gas volumes in the range of 300 to 500 million cubic feet per day. Developers say construction could start in 2000, with deliveries by 2003 [65]. Elsewhere in the UAE, BC Gas of Canada began connecting gas consumers to a new domestic gas grid in Sharjah [66].

In Qatar, the Mobil-led RasGas LNG venture loaded its first cargo in August 1999. Korean owned and operated LNG carriers will ship the LNG to Korea under a 25-year contract. A second LNG train under construction is on schedule and expected to start up in the first half of 2000. In addition, further trains can be expected if a RasGas

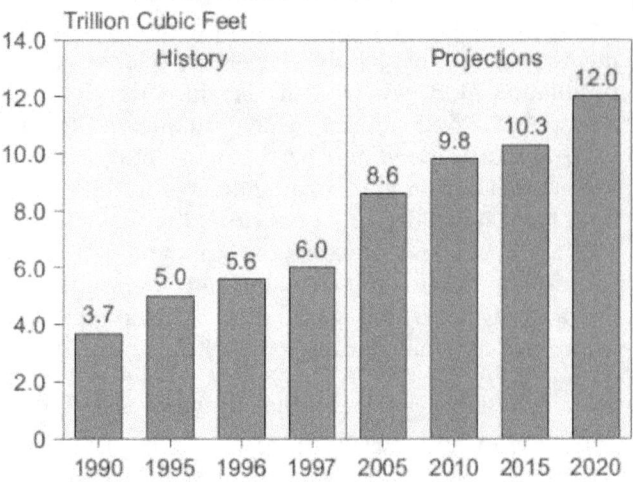

Figure 56. Natural Gas Consumption in the Middle East, 1990-2020

Trillion Cubic Feet — History: 1990: 3.7; 1995: 5.0; 1996: 5.6; 1997: 6.0. Projections: 2005: 8.6; 2010: 9.8; 2015: 10.3; 2020: 12.0

Sources: **History:** Energy Information Administration (EIA), Office of Energy Markets and End Use, International Statistics Database and *International Energy Annual 1997*, DOE/EIA-0219(97) (Washington, DC, April 1999). **Projections:** EIA, World Energy Projection System (2000).

agreement is finalized with India for 7.5 million metric tons per year of LNG [67]. Qatar has also announced intentions to develop a commercial-scale gas-to-liquids (GTL) conversion project that would use its substantial domestic reserves for emerging GTL technologies (see box on pages 59-60).

Elsewhere in the Middle East, Israel has announced that natural gas will power 25 percent of its electricity needs by 2005, but it is still seeking gas supplies. British Gas has discussed exploring for gas in Israel, and Egypt is also a possible source, although talks between the two countries stalled in 1999. Israel is prequalifying bidders for an upcoming tender on development of domestic gas infrastructure, which is due to be issued in the first quarter of 2000 [68]. A pipeline is currently under construction from Egypt to Jordan and on to the Palestinian Authority areas [69], and plans are nearing finalization for a pipeline from Syria to Lebanon that would supply Lebanon with 105 million cubic feet of gas per day. The pipeline would have two sections: a 75-mile stretch from the Syrian city of Homs to Deir Ammar in northern Lebanon and a 90-mile portion continuing to the southern Lebanese town of Zahrani. Conoco and Elf already signed a deal to process gas from the Syrian Deir al-Zur field [70].

Saudi Arabia is also seeking to increase domestic gas use and has undertaken a $4 billion program to develop gas infrastructure and the non-associated gas reserves from the Khuff reservoir. The gas will go to the new Hawiyah gas processing plant (due on stream in 2001) and to power plants in Riyadh. Saudi Arabia's 204 trillion cubic feet of gas reserves are primarily associated gas, which

Gas-to-Liquids Technology: The Current Picture

Much of the world's endowment of identified, recoverable natural gas resources lies in remote locations or in smaller accumulations that make typical approaches for project development, such as delivery via pipeline or LNG tanker, uneconomical. One natural gas marketing option under development would use gas-to-liquids (GTL) technology—generally, the recombination of the carbon and hydrogen atoms in natural gas molecules as synthetic petroleum products, either liquids or petroleum wax. The discussion here focuses primarily on GTL to produce middle distillate products, such as diesel for transportation, because the large volumes of those products consumed worldwide are indicative of the corresponding market potential. Other possible products include methanol and ammonia, but the limited markets for them would not support widespread adoption of the GTL technology.

GTL technology offers a number of advantages as a gas marketing option. Marketing GTL products would avoid costly associated investments by relying on the existing infrastructure for petroleum products, including tankers, terminals, storage facilities, and marketers. GTL technology is expected to be scalable, allowing design optimization and potential application to smaller gas deposits. Also, the technology offers a number of environmental advantages that may enhance the economic attractiveness of GTL projects.

As transportation fuels, GTL products are expected to reduce exhaust emissions from vehicles significantly, which may be reflected in premium prices. Emission reductions realized will depend on such factors as the relative mix of synthetic and petroleum-based fuels in the product consumed, the type and age of vehicles using the fuels, and the specific process by which the synthetic fuel is produced. In one test using "older Pittsburgh transit buses," 100-percent synthetic diesel used in place of No. 2 diesel fuel produced lower levels of nitrogen oxides (by 8 percent), particulate matter (by 31 percent), carbon monoxide (by 49 percent), and hydrocarbons (by 35 percent).[a]

Another potential environmental advantage of GTL technology stems from concern in some countries about the disposition of gas produced in combination with crude oil (called associated-dissolved, or AD, gas). Without local use or infrastructure to ship it to markets, AD gas often is flared or vented into the air, releasing greenhouse gases such as methane and carbon monoxide. A GTL project can use gas that would otherwise be vented or flared as a feedstock.

The commercial success of GTL technology has not yet been fully established, and expected net returns for investment in GTL projects depend on a number of risky factors. The financial benefits depend on the market prices for petroleum products and possible price premiums for the environmental advantages of GTL-produced fuels, the value of byproducts such as heat and water, and potential government subsidies. Unit production costs will reflect the cost of the feedstock gas; the capital cost of the plants; marketability of byproducts such as heat, water, and other chemicals (e.g., excess hydrogen, nitrogen, or carbon dioxide); the availability of infrastructure; and the quality of the local workforce.

The cost of feedstock gas for GTL projects may vary widely, depending on its perceived value and other conditions. In fact, an arguably acceptable price for a project that uses gas that otherwise would be flared can be zero (or even negative) if its use in the project avoids either monetary penalties for violations of environmental regulations or increased costs related to compliance with environmental restrictions. Changes in gas feedstock costs of $0.50 per thousand cubic feet would shift the implied competitive crude oil price by roughly $4 to $5 per barrel.

Capital costs for GTL projects currently tend to be in a range between $20,000 and $30,000 per daily barrel of capacity (compared with refinery costs of $12,000 to $14,000 per daily barrel), and the cost of GTL-produced fuel could vary by approximately $1.50 per barrel with a shift of $5,000 in capital cost.[b] Estimates of the crude oil prices necessary to allow positive economic returns from a GTL project vary widely, with optimistic estimates ranging as low as $14 to $16 per barrel. More typical estimates indicate that expected oil prices would have to average over $20 per barrel on a sustained basis to lead to commitments for large-scale projects.[c]

An EIA cost analysis of a hypothetical GTL project—based on capital costs of $10.48 per barrel ($25,000 per daily barrel over 12 years at 12 percent), operating costs of $5.50 per barrel, and feedstock costs equivalent to $8.92 per barrel of crude oil (including conversion losses of 35 percent)—estimated the cost of GTL fuel at

(continued on page 60)

---

[a]"Mossgas FT Diesel Cuts Emissions in Older City Diesel Buses," *Gas-to-Liquids News* (June 1999), p. 5.
[b]Capital costs are from Howard, Weil, Labouisse, and Friedrichs, Inc., *Fischer-Tropsch Technology* (Houston, TX, December 18, 1998), p. 44. Cost impacts were estimated by EIA's Office of Oil and Gas, based on analysis in Cambridge Energy Research Associates, *New Developments in Gas-to-Liquids Technology: Fundamental Change or Just a Niche Role?* (Cambridge, MA, August 1997).
[c]Cambridge Energy Research Associates, *"Gas-to-Liquids" Two Years Later—Still Just a Niche Opportunity?* (Cambridge, MA, October 1999).

Gas-to-Liquids Technology: The Current Picture (Continued)

almost $25 per barrel. Thus, under conditions that may be considered reasonable, a GTL project with present technology could be cost competitive only if crude oil prices were in the range of $25 per barrel; however, adverse shifts in any of the key cost factors could raise the competitive price significantly. Indeed, uncertainty surrounding both cost factors and world oil prices has tended to limit GTL growth to date.

GTL technology is widely considered to be a gas development alternative that would compete with liquefied natural gas (LNG) projects. While it is true that GTL projects could ultimately reduce the volume of gas available for gas-consuming markets, they are not expected to be detrimental to supplies in the near term, and they may actually enhance worldwide gas supplies by encouraging the development of LNG projects. For very large gas deposits, the two technologies can be applied as complementary development options. Joint development of GTL and LNG projects would allow for shared labor and infrastructure, reducing the costs to both projects—an approach that could benefit gas markets by accelerating the development of some LNG projects. In a number of locations, such as Malaysia, Nigeria, and Qatar, there are such large gas reserves that GTL projects are operating or planned in addition to existing LNG projects.

Only two GTL facilities have operated to produce synthetic petroleum liquids at more than a demonstration level: the Mossgas Plant (South Africa), with output capacity of 23,000 barrels per day, and Shell Bintulu (Malaysia) at 12,500 barrels per day (see map).[d] The Shell plant is being restored after an explosion on December 25, 1997, and it is expected back on line in 2000 with expanded capacity. Other plants are in the planning stages. A joint project of Chevron and Sasol, Ltd. (South Africa) was announced earlier this year for a 30,000 barrel per day plant in Nigeria that would cost $1 billion. It is expected to begin operations in 2003 at costs competitive with crude oil prices of $16 to $18 per barrel.[e] The Nigeria project will benefit from the infrastructure already in place for nearby oil and gas production and export facilities, although it is unclear whether, or to what extent, subsidies or other considerations have helped to lower the estimated costs.

Estimates of worldwide natural gas reserves indicate a prolific resource base (see Table 16 on page 46) of roughly 5,100 trillion cubic feet—the equivalent of more than 900 billion barrels of crude oil—including both volumes currently in production and many more deposits awaiting development.[f] The standard marketing option for gas deposits close to markets is transportation via pipeline. Deposits further removed from markets require either long-distance pipelines or other options to market the gas. For up to 50 percent of global reserves that are estimated to be stranded without local markets and may be jeopardized by logistical difficulties or high costs of development, GTL technology is a promising approach.[g]

GTL projects are not expected to be adopted widely in the near term, and the long-term outlook is subject to considerable uncertainty. Crude oil prices in excess of $20 per barrel on a sustained basis, or enhanced economic returns from further technological advances, may be needed before operators will be motivated sufficiently to invest in large-scale GTL projects.

The economic prospects for GTL technology are sensitive to key cost components that are themselves uncertain. Only limited growth is projected for GTL production in the *IEO2000* reference case, reflecting the expectation that investment in the technology will be constrained by lingering economic risks. The *IEO2000* analysis assumes a threshold world oil price of $27 per barrel before GTL projects could attract substantial investments. At lower prices, production is limited to small demonstration projects.

GTL Projects Worldwide

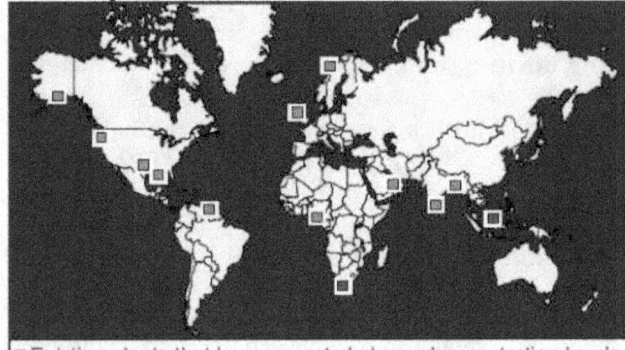

■ Existing plants that have operated above demonstration levels.
■ Test plants or planned capacity.

Source: "Gas-to-Liquids At-a-Glance Reference Guide 1999," *Hart Gas-to-Liquids News*, in association with Syntroleum.

---

[d]"Gas-to-Liquids At-a-Glance Reference Guide 1999," *Hart Gas-to-Liquids News*, in association with Syntroleum.

[e]Assumptions behind this estimated price level include feedstock gas at $0.50 per million Btu (considered the rough equivalent of $5 per barrel of crude oil, or less at strict Btu equivalence), capacity costs of $25,000 per daily barrel, and operating costs of $5 per barrel. Source: "Advanced Technology Puts Sasol in GTL Driver's Seat," *Gas-to-Liquids News* (July 1999), p. 6.

[f]The crude oil equivalence volume was calculated on the basis of an assumed heat content of 1,030 Btu per cubic foot and 5.8 million Btu per barrel.

[g]Howard, Weil, Labouisse, and Friedrichs, Inc., *Fischer-Tropsch Technology* (Houston, TX, December 18, 1998), p. 31.

was flared until the early 1980s. Gas is currently produced from about 10 of more than 80 known fields and is used in fertilizer and petrochemical plants at Jubail and Yanbu [71].

Africa

Gas reserves in Africa account for nearly 8 percent of global stocks. With new additions in Egypt and Nigeria, the region's reserves amount to 394 trillion cubic feet. Roughly 70 percent of Africa's domestic gas consumption and more than 80 percent of its production occurs in Algeria and Egypt. Algeria exports 70 percent of its domestic production via pipeline and LNG tanker. Within Africa, natural gas remains the least utilized fossil fuel. Low growth in Africa's gas consumption (Figure 57) reflects a lack of economic growth in much of the region as a result of political instability, which has been particularly severe in sub-Saharan Africa. In Ghana and the Ivory Coast (in West Africa), economic growth has led to rising demand for energy and associated interest in natural gas development, particularly for electricity generation. Domestic use of natural gas in Africa is heavily for power generation, amounting to 40 percent of the region's gas demand [72].

In the *IEO2000* forecast, Africa's natural gas consumption continues to grow at a relatively slow pace. The average projected growth rate of 1.8 percent annually from 1997 to 2020 is the slowest among the developing regions, including the Middle East, developing Asia, and Central and South America. In the reference case, total gas use in Africa rises from 1.8 trillion cubic feet in 1997 to 2.8 trillion cubic feet in 2020.

In 1999, Nigeria's LNG (NLNG) export facility came on stream amid continued protests by local activists demanding jobs and cash payments. Protests that closed down the Bonny plant in September, just 2 weeks after production began, and again in late September and early October, ended after negotiations with government officials. NLNG was able to repair related damage and make its first delivery to France at the end of October [73]. In early 1999, Shell awarded a construction contract for a third Nigerian LNG train to a consortium involving Technip, Snamprogetti, MW Kellogg, and JGC Corp. More than 70 percent of the LNG from the expansion has already been sold to Spain's Enagas over a 21-year period [74]. Like Qatar, Nigeria has tentative plans for a GTL project (see box on pages 59-60), with South Africa's Sasol as a partner.

Plans to supply Nigerian gas to Ghana, primarily for power generation, also made progress in 1999. The West African Gas Pipeline (WAGP) joint venture agreement was signed in August by six energy companies, including Chevron, Shell, and the oil companies of Nigeria, Ghana, Benin, and Togo. The consortium seeks to build a

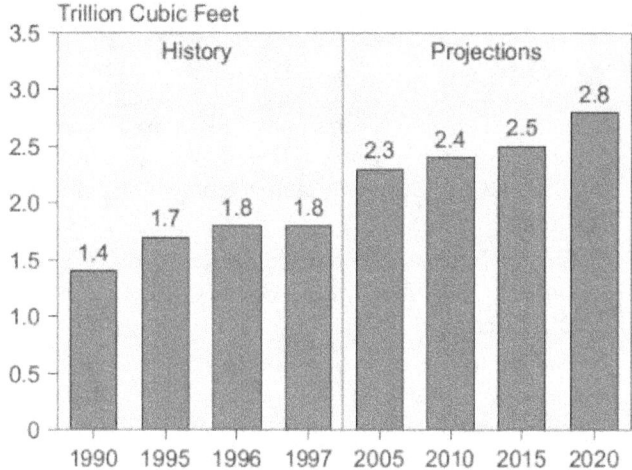

Figure 57. Natural Gas Consumption in Africa, 1990-2020

Sources: **History:** Energy Information Administration (EIA), Office of Energy Markets and End Use, International Statistics Database and *International Energy Annual 1997*, DOE/EIA-0219(97) (Washington, DC, April 1999). **Projections:** EIA, World Energy Projection System (2000).

$400 million pipeline by 2002 that would send initially about 120 million cubic feet per day of now flared Nigerian gas to Ghana, Togo, and Benin [75]. In addition, Ghana has started up the second of two gas-fired power plants using offshore Ivoirian gas [76]. There is also development of domestic resources for power generation in the Tano Fields Development and Power Project, involving a 100- to 140-megawatt power plant fueled by offshore gas [77].

Egypt, too, is moving toward expanding its domestic gas market, in part to switch from oil to gas and save oil for export. The state-owned Egyptian General Petroleum Company (EGPC) has amended agreements with BP Amoco and Eni, allowing production to expand at offshore gas fields (in the Ras el-Barr concession) in order to supply the domestic market. Investments of up to $200 million will be used to expand production by up to 399 million cubic feet per day by 2002. In addition, BP Amoco together with Burlington Resources has also signed a separate gas sales agreement with EGPC for gas off the coast of northern Sinai. Requiring a $150 million investment, production from three fields in the area, which could reach 110 million cubic feet per day by 2004, will also be used for the domestic market [78]. EGPC will also form a joint venture with UK-based British Gas and Italy's Edison International to develop Egypt's largest gas field, Sacarab/Saffron, in the Nile Delta. Production of high-quality gas from the field is expected to start in 2003 and build quickly to deliveries of about 530 million cubic feet per day for a contract period of at least 17 years. The European Investment Bank (EIB) will advance a loan to Egyptian Natural Gas Company (Gasco) for the construction and operation of a pipeline to move gas from the Suez Canal region to a distribution plant north of Cairo, which will reinforce the domestic grid [79].

References

1. British Petroleum Company, *BP Amoco Statistical Review of World Energy 1999* (London, UK, June 1999), web site www.bpamoco.com/worldenergy/naturalgas.

2. Energy Information Administration, *Annual Energy Outlook 2000*, DOE/EIA-0383(2000) (Washington, DC, December 1999).

3. Energy Information Administration, *Country Analysis Briefs: Canada*, web site www.eia.doe.gov/emeu/cabs/canada.html (September 1999).

4. Energy Information Administration, *Country Analysis Briefs: Canada*, web site www.eia.doe.gov/emeu/cabs/canada.html (September 1999).

5. "FERC OKs Southern LNG Terminal, Operations Set To Begin 2002," *Natural Gas Week* (December 20, 1999).

6. "Columbia LNG Moving To Reactivate Cove Point, MD LNG Terminal," *Inside F.E.R.C.'s Gas Market Report* (December 10, 1999), p. 12.

7. U.S. Department of Energy, Office of Fossil Energy, *Natural Gas Imports and Exports, Third Quarter 1999*, DOE/FE-0412 (Washington, DC, December 1999), pp. vi-vii.

8. "Europipe II Ready for Start-Up," *Financial Times: Gas Markets Week*, No. 21 (September 27, 1999), p. 2; and "Europipe II Commissioned 1 October," *Financial Times: Gas Markets Week*, No. 22 (October 4, 1999), p. 12.

9. "New Leg of Yamal-Europe Gas Pipeline Constructed," Itar-Tass News Wire (September 28, 1999).

10. "Ruhrgas Starts on New Gasline," *Financial Times: International Gas Report*, Vol. 383 (October 1, 1999), p. 14.

11. "EU TotalFina/Elf Probe Targets LPG," *Financial Times: International Gas Report*, Vol. 384 (October 15, 1999), p. 3.

12. "Veba-Viag Seal Major Merger," *Financial Times: International Gas Report*, Vol. 383 (October 1, 1999), p. 7.

13. "Italy's Eni Seals $5.5 Billion Libya Gas Deal," *Financial Times: International Gas Report*, Vol. 379-380 (August 6, 1999), pp. 1-2.

14. "Edison-Mobil Italy LNG Plan 'Moves Ahead'," *Financial Times: International Gas Report*, Vol. 382 (September 17, 1999), p. 3.

15. "Gas Demand 'Set To Double'," *Financial Times: International Gas Report*, Vol. 384 (October 15, 1999), p. 15.

16. "Gas/Power Sectors Open Up," *Financial Times: International Gas Report*, Vol. 383 (October 1, 1999), p. 10.

17. British Petroleum Company, *BP Amoco Statistical Review of World Energy 1999* (London, UK, June 1999), web site www.bpamoco.com/worldenergy/naturalgas.

18. "Uzbekistan Threatens To Cut Gas Supplies to Kyrgyzstan," *Slovo Kyrgyzstana* (November 11, 1999).

19. "World Briefing," ITAR-TASS News Agency Release (December 14, 1999).

20. "Belarus Government Compels Domestic Customers To Pay Off Gas Debts," Belapan News Agency (December 9, 1999).

21. Energy Information Administration, *Country Analysis Briefs: Ukraine*, web site www.eia.doe.gov/emeu/cabs/ukraine.html (June 1999).

22. D. Hoffman, "Russia Gets Planes Left in Ukraine; Natural Gas Debt Traded for Strategic Bombers," *International Herald Tribune* (November 2, 1999), p. 10.

23. "Ukraine Pays Off Part of Debt to Turkmenistan," Intelnews Agency (October 12, 1999).

24. "Itera Threatens To Stop Gas Supply to Ukraine," *Russian Oil and Gas Report* (September 29, 1999).

25. "Gazprom To Open Gas Pipeline," *Hart's Offshore Petroleum Newsletter*, Vol. 24, No. 37 (September 21, 1999).

26. "Gazprom To Eventually Reduce Gas Exports Through Ukraine," Interfax News Agency (December 3, 1999).

27. "Turkish Energy Minister Visits Turkmenistan, Discusses Purchase of Gas," Anatolia News Agency (October 9, 1999).

28. "Afghan Taleban Official in Odessa for Talks on Gas Supplies," Intelnews News Agency (September 17, 1999).

29. "Ukraine Hoping To Buy Fuel From Kazakhstan," Intelnews News Agency (September 16, 1999).

30. "Poland Signs Gas Supply Contract With Norway," Gazeta Wyborcza, December 12, 1999, p. 25; Rzeczpospolita, December 12, 1999, p. B1, from Polish News Bulletin.

31. "Atlantic LNG Launched, But Expansion Accord Still Pending," *World Gas Intelligence*, Vol. 10, No. 8 (April 29, 1999), p. 2.

32. "T&T Minister Flags Gas Sector Tax Hike," *Financial Times: International Gas Report*, Vol. 384 (October 15, 1999), p. 2; and "Caribbean: Atlantic LNG Expansion Moves Nearer," *Latin America Monitor*, Vol. 16, No. 9 (September 1999), p. 7.

33. "Bolivia-Brazil Pipeline Opens," *Latin American Energy Alert*, Vol. 6, No. 17 (July 22, 1999), p. 10; and "Bust in Brazilian Gas Purchases May Prompt Bolivia To Allow Flaring," *Latin American Energy Alert*, Vol. 6, No. 23 (October 21, 1999).

34. "OPIC Approves Gas Pipeline Loan," *Latin American Energy Alert*, Vol. 6, No. 16 (July 5, 1999), p. 8.

35. "Ground Broken for TGM Pipe," *Financial Times: Power in Latin America*, No. 50 (August 1999), p. 13; and "Construction of TGM Pipeline From Argentina to Brazil," NewsPage, web site www.newspage.com (August 3, 1999).

36. "Government Fixes Gas Price," *Financial Times: International Gas Report*, Vol. 383 (October 1, 1999), pp. 35-36.

37. "First GasAtacama Gas Flows," *Financial Times: International Gas Report*, Vol. 377 (July 9, 1999), p. 30; "Gas Atacama Opens But Glutted Region May Be Soft Market," *Latin American Energy Alert*, Vol. 6, No. 17, p. 5; "GasAtacama Starts Pumping," *Financial Times: Power in Latin America*, No. 49 (July 1999), p. 13; and "CMS, Endesa Plan To Build Spur Off Gas Atacama Pipeline," *Latin American Energy Alert*, Vol. 6, No. 19 (August 25, 1999), p. 7.

38. "Ecopetrol To Support Plan To Build Gas Pipeline into Central America," *Latin American Energy Alert*, Vol. 6, No. 18 (August 5, 1999), p. 8.

39. "Enron Eyes Colombia Gas for Panama," *Financial Times: International Gas Report*, Vol. 384 (October 15, 1999), p. 5.

40. "Peru Sets Dates for Camisea Sale," *The Oil Daily*, Vol. 49, No. 152 (August 10, 1999), p. 8; and "New Knock for Camisea Hopes," *Financial Times: International Gas Report*, Vol. 383 (October 1, 1999), p. 34.

41. R. Colitt, "New President for PDVSA," *Financial Times London*, U.S. Edition (September 1, 1999), p. 17.

42. "PDVSA Adopts New Management Structure, Fills Vacancies," *Latin American Energy Alert*, Vol. 6, No. 23 (October 21, 1999), p. 5.

43. "Japanese LNG Buyers To Extend Contracts," *The Jakarta Post* (September 10, 1999).

44. "Gas Discovery Off Australia Termed 'Very Significant'," *The Houston Chronicle* (October 3, 1999).

45. S. Wyatt, "Demand Grows for Gas Pipeline," *Financial Times London* (July 30, 1999), p. 30.

46. "Independent Audits Say McDermott Pipe Deal Okay," Reuters (October 11, 1999).

47. "Gas Deal Boosts New Line Plan," *Financial Times: International Gas Report*, Vol. 383 (October 1, 1999), pp. 26-27; *International Herald Tribune* (September 22, 1999); and "Jakarta Signs Singapore Deal But Brushes Reform Bill Aside," *World Gas Intelligence*, Vol. 10, No. 18 (September 30 1999), p. 1.

48. "Japanese LNG Buyers To Extend Contracts," *The Jakarta Post* (September 10, 1999).

49. "Government To Lower Gas Prices for Domestic Market," *The Jakarta Post* (September 28, 1999), p. 1.

50. "New Delay to Yadana Gas Field Supply," *Financial Times: Power in Asia*, Vol. 283-284 (August 9, 1999), pp. 1-2; and "New Thai Delay Hits Burma Gas Output," *Financial Times: International Gas Report*, Vol. 379-380 (August 6, 1999), pp. 2-3.

51. "Thai State Firms Clash Over Gas," *Financial Times International Gas Report*, No. 383 (October 1, 1999), p. 6.

52. "Thailand Endorses Spending on Pipeline To Also Serve Malaysia," *Asian Wall Street Journal* (September 15, 1999), p. 4.

53. *Cedigaz News Report*, No. 36 (September 10, 1999).

54. "Malay-Thai $1 bn Project Set for Lift Off," *Financial Times: International Gas Report*, Vol. 382 (September 17, 1999), p. 1.

55. "Malaysia Signed 8 New PSCs in Past Year," web site www.oilonline.com/news_spotlight_asianoil_041399malay.html (October 1999).

56. "Malaysia LNG Tiga Plant Set To Feed India," *Financial Times International Gas Report*," Vol. 378 (July 23, 1999), pp. 1-2.

57. "Electric Power, Dabhol II: 'Slow and Steady' Wins the Race," *Asian Energy Insights* (Canbrdige, MA: Cambridge Energy Research Associates, July 1999), pp. 9-10.

58. "On the Asia Beat: India," *Financial Times: Power in Asia*, Vol. 283-284 (August 9, 1999), p. 29.

59. "NTPC Takes Stake in Pipavav LNG Project," *Financial Times: International Gas Report*, Vol. 383 (October 1, 1999), p. 3.

60. *Cedigaz News Report*, No. 36 (September 10, 1999).

61. "China Moves to Oz Deal?" *Financial Times: International Gas Report*, No. 379-380 (August 6, 1999), pp. 23-4; and *Lloyd's List London* (October 1, 1999).

62. "Shell Inks $3bn China Gas/Power Deal," *Financial Times: International Gas Report*, Vol. 383 (October 1, 1999), p. 5.

63. "KOGAS To Spin Off, Sell Importing, Wholesale Units," *Korea Herald* (October 1, 1999).

64. T. Gustafson, "The Gas Race For Turkey—Can Blue Stream Win?" (Cambridge, MA: Cambridge Energy Research Associates, June 1999).

65. "Elf, Enron, Mobil 'Front Runners' for Dolphin Gas Venture in UAE," *AFX News* (September 13, 1999).

66. "What's New Around the World," *World Gas Intelligence*, Vol. 10, No. 7 (April 16, 1999), p. 12.

67. "Mobil Announces First RasGas LNG Shipment to Korea," *Business Wire* (August 23, 1999).

68. "El Paso Confirms Interest," *Financial Times: International Gas Report*, Vol. 384 (October 15, 1999), p. 20.

69. "Egypt Warns Israel Against Crossing Border in Exploring Gas," *Al-Sharq al-Awsat* (September 16, 1999).

70. "Syria and Lebanon Finalise Plans for Gas Pipeline," Agence France-Presse (October 26, 1999).

71. Standard & Poor's Platt's, *World Energy Service: Africa/Middle East* (Lexington, MA, 1999), pp. 223-224.

72. Standard & Poor's Platt's, *World Energy Service: Africa/Middle East* (Lexington, MA, 1999), pp. 9-10.

73. "New Setback Hits Nigerian LNG," *Financial Times: International Gas Report*, Vol. 383 (October 1, 1999), p. 23.

74. "Shell Unit Announces Nigeria LNG Train Expansion," *AFX News* (March 15, 1999), p. 1.

75. "Pipeline Projects Reflect Two Kinds of Potential," Africa News Service (September 7, 1999).

76. "What's New Around the World," *World Gas Intelligence*, Vol. 10, No. 3 (February 12, 1999), p. 11.

77. "Offshore Fields Move Ahead," *Financial Times: International Gas Report*, Vol. 384 (October 15, 1999), p. 20.

78. "BP Amoco, Eni in Egypt Gas Deal," *Financial Times London* (September 27, 1999).

79. "EIB Funds Egypt Gasline," *Financial Times: International Gas Report*, Vol. 383 (October 1, 1999), p. 24.

Coal

Although coal use is expected to be displaced by natural gas in some parts of the world, only a slight drop in its share of total energy consumption is projected by 2020. Coal continues to dominate many national fuel markets in developing Asia.

Historically, trends in coal consumption have varied considerably by region. Despite declines in some regions, world coal consumption has increased from 84 quadrillion British thermal units (Btu) in 1985 to 93 quadrillion Btu in 1997. Regions that have seen increases in coal consumption include the United States, Japan, and developing Asia. Declines have occurred in Western Europe, Eastern Europe, and the countries of the former Soviet Union (FSU). In Western Europe, coal consumption declined by 33 percent between 1985 and 1997, displaced in considerable measure by growing use of natural gas and, in France, by nuclear power. The countries of Eastern Europe and the former Soviet Union (EE/FSU) saw an even sharper decline in coal use during the period (a 38-percent decline), primarily the result of reduced economic activity.

Although coal has lost market share to petroleum products, natural gas, and nuclear power, it continues to be a key source of energy, especially for electric power generation. In 1997, coal accounted for 24 percent of the world's primary energy consumption (down from 27 percent in 1985) and 36 percent of the energy consumed worldwide for electricity generation (Figure 58).

In the *International Energy Outlook 2000 (IEO2000)* forecast, coal's share of total energy consumption falls only slightly, from 24 percent in 1997 to 22 percent in 2020. Its historical share is nearly maintained, because large increases in energy use are projected for the developing countries of Asia, where coal continues to dominate many national fuel markets (Figure 59). Together, two of the key countries in the region, China and India, are projected to account for 33 percent of the world's total increase in energy consumption over the forecast period and 97 percent of the world's total increase in coal use (on a Btu basis).

Coal for electricity generation accounts for virtually all the projected growth in coal consumption worldwide. In other sectors where coal is used, such as industrial and residential/commercial, other energy sources—primarily natural gas—are expected to gain market share. One exception is China, where coal continues to be the primary fuel in a rapidly growing industrial sector, in view of the nation's abundant coal reserves and limited access to alternative sources of energy. Consumption of coking coal is projected to decline slightly in most regions of the world as a result of technological

Figure 58. Coal Share of World Energy Consumption by Sector, 1997 and 2020

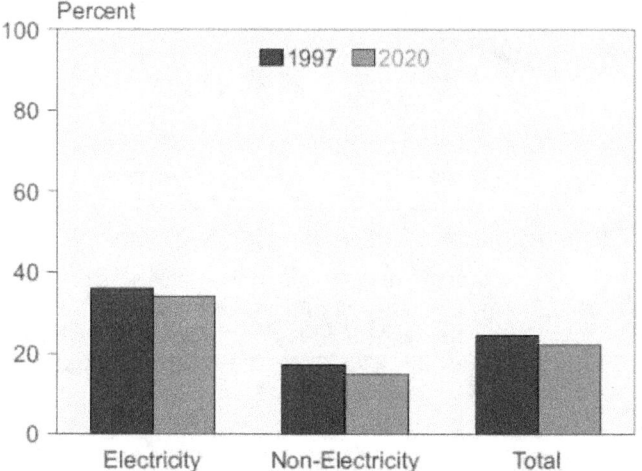

Sources: **1997**: Energy Information Administration (EIA), *International Energy Annual 1997*, DOE/EIA-0219(97) (Washington, DC, April 1999). **Projections**: EIA, World Energy Projection System (2000).

Figure 59. Coal Share of Regional Energy Consumption, 1970-2020

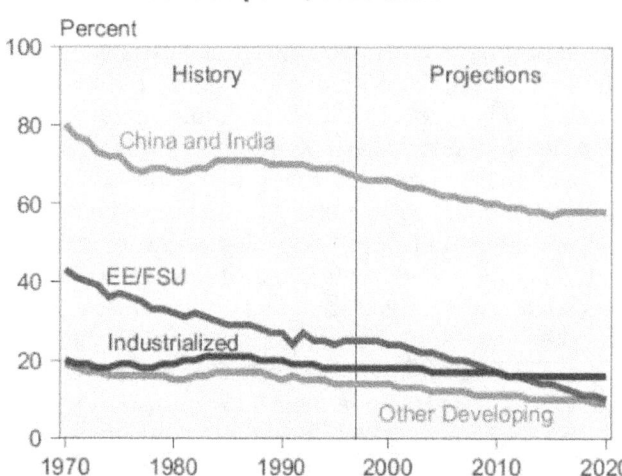

Sources: **History**: Energy Information Administration (EIA), Office of Energy Markets and End Use, International Statistics Database and *International Energy Annual 1997*, DOE/EIA-0219(97) (Washington, DC, April 1999). **Projections**: EIA, World Energy Projection System (2000).

advances in steelmaking, increasing output from electric arc furnaces, and continuing substitution of other materials for steel in end-use applications.

Because the Kyoto Protocol is not currently a legally binding agreement, the *IEO2000* projections do not reflect the commitments made by the signatory countries to reduce or moderate their emissions of greenhouse gases. If their commitments do become legally binding, however, it is likely that the coal outlook for the industrialized countries will differ substantially from the *IEO2000* projections. In *IEO2000*, coal consumption in the industrialized countries is projected to increase by 11 percent over the forecast period, rising from 36.6 quadrillion Btu in 1997 to 40.5 quadrillion Btu in 2020.

In a study completed in October 1998, the Energy Information Administration (EIA) projected that for the United States to meet its Kyoto emissions target, annual U.S. coal consumption would need to be reduced by as little as 18 percent or by as much as 77 percent (on a Btu basis) by 2010, relative to a reference case forecast without the Kyoto carbon emissions constraints [1]. The largest reduction in coal consumption was projected in a case which assumed that the United States would be required to reduce its carbon emissions to 7 percent below the 1990 level through fuel switching, increased penetration of energy-efficient technologies, and reductions in overall energy use. Other cases modeled in the study assumed that the United States would meet its Kyoto emissions target through a combination of actions such as fuel switching, emissions trading, joint implementation, reforestation, and reductions in emissions of other greenhouse gases.

The most significant difference between the *IEO99* and *IEO2000* coal projections, particularly in the short term, is a significant change in international coal trade patterns (see box on pages 76-77). Most of the major coal-exporting countries, including Australia, South Africa, Canada, Indonesia, and Russia, reduced the prices of their export coal considerably in 1999. Both Australia and Indonesia showed a considerable increase in their coal exports for the year, primarily at the expense of U.S. coal exports. Lower prices benefitted coal importers and improved coal's ability to compete with other fuels worldwide. For coal producers in countries such as the United Kingdom and Germany, however, lower prices for coal imports are expected to lead to some reductions in output at domestic mines and an accelerated schedule for mine closures, as domestic

consumers switch from indigenous coal to increasingly less expensive imports.

Highlights of the *IEO2000* projections for coal are as follows:

- World coal consumption is projected to increase by 2.3 billion tons, from 5.3 billion tons in 1997 to 7.6 billion tons in 2020 (Figure 60).[8] World coal consumption in 2020 could be as high as 9.1 billion tons or as low as 5.6 billion tons, based on alternative assumptions about economic growth rates.[9]

- Coal use in developing Asia alone is projected to increase by 2.4 billion tons. China and India, taken together, are projected to account for 33 percent of the total increase in energy consumption worldwide between 1997 and 2020 and 97 percent of the world's total projected increase in coal use, on a Btu basis.

- China is projected to add an estimated 180 gigawatts of new coal-fired generating capacity (600 plants of 300 megawatts each) by 2020 and India approximately 50 gigawatts (167 plants of 300 megawatts each).

- Coal's share of the world's total primary energy consumption is expected to decline from 24 percent in 1997 to 22 percent in 2020. The coal share of energy consumed worldwide for electricity generation also

Figure 60. World Coal Consumption, 1970-2020

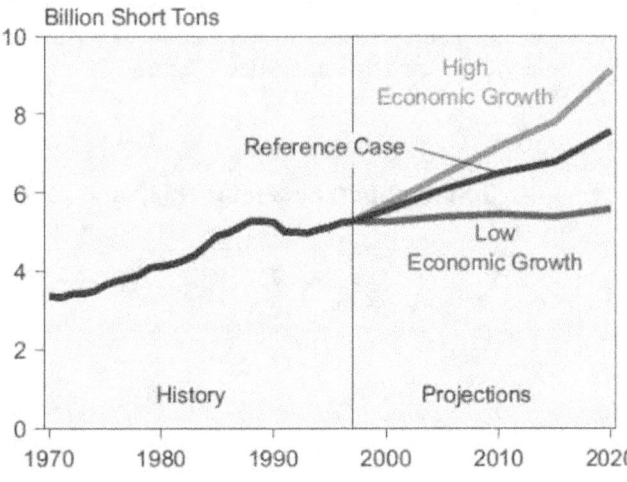

Sources: **History:** Energy Information Administration (EIA), Office of Energy Markets and End Use, International Statistics Database and *International Energy Annual 1997*, DOE/EIA-0219(97) (Washington, DC, April 1999). **Projections:** EIA, World Energy Projection System (2000).

---

[8]Throughout this chapter, tons refers to short tons (2,000 pounds).

[9]In the *IEO2000* reference case, world gross domestic product (GDP) is projected to increase at a rate of 2.8 percent per year between 1997 and 2020. In the low and high economic growth cases, world economic growth rates are assumed to be 1.3 percent lower and 1.2 percent higher, respectively, than in the reference case. By region, the dispersion in economic growth rates across the cases is less symmetrical than for the world as a whole, resulting in slightly asymmetrical variations in the projections of world coal consumption. In the low and high economic growth cases, the expected economic growth rates for China are 3.0 percent lower and 1.5 percent higher, respectively, than in the reference case.

declines, from 36 percent in 1997 to 34 percent in 2020.

- World coal trade is projected to increase from 546 million tons in 1998 to 708 million tons in 2020, accounting for approximately 9 to 10 percent of total world coal consumption over the period. Steam coal (including coal for pulverized coal injection at blast furnaces) accounts for most of the projected increase in world coal trade.

Environmental Issues

In future years, coal will face tough challenges, particularly in the environmental area. Increased concern about the harmful environmental impacts associated with coal use has taken a toll on coal demand throughout industrialized areas. Coal combustion produces several air pollutants that adversely affect ground-level air quality.

One of the most significant pollutants from coal is sulfur dioxide, which has been linked to acid rain. Many of the industrialized countries have implemented policies or regulations to limit sulfur dioxide emissions. Such policies typically require electricity producers to switch to lower sulfur fuels or invest in technologies—primarily, flue gas desulfurization (FGD) equipment—that reduce the amounts of sulfur dioxide emitted.

In the developing countries of Asia, only minor amounts of existing coal-fired capacity currently are equipped with FGD equipment. For example, in China, the world's largest emitter of sulfur dioxide, data for 1995 indicated that only about 3 percent of existing coal-fired generating capacity (less than 4 gigawatts out of a total of 140 gigawatts) had FGD equipment in place [2, 3]. To date, major coal importing countries in the region have typically relied on the use of low-sulfur coal from Australia and Indonesia as a strategy for controlling emissions of sulfur dioxide [4]. In the future, however, greater use of FGD equipment at new coal plants is expected as a result of increased opposition from environmental groups and local residents, stricter adherence to World Bank standards on environmental performance by project developers, and adoption of stricter environmental standards by national governments [5, 6, 7, 8, 9].

In addition to sulfur dioxide, increased restrictions on emissions of nitrogen oxides, particulates, and carbon dioxide are likely, especially in the industrialized countries. Although the potential magnitudes and costs of additional environmental restrictions for coal are uncertain, it seems likely that coal-fired generation worldwide will face steeper environmental cost penalties than will

new gas-fired generating plants. For nuclear and hydropower, which compete with coal for baseload power generation, the future is unclear. Proposals have been put forth in several of the developed countries to phase out nuclear capacity in full or in large measure. In other countries, it has become difficult to site new capacity because of unfavorable public reaction. The siting of new large hydroelectric dams is also becoming more difficult because of increased environmental scrutiny. In addition, suitable sites for new large hydropower projects in the industrialized countries are limited [10].

By far the most significant emerging issue for coal is the potential for a binding international agreement to reduce emissions of carbon dioxide and other greenhouse gases. On a Btu basis, the combustion of coal produces more carbon dioxide than that of natural gas or of most petroleum products [11]. Carbon dioxide emissions per unit of energy obtained from coal are nearly 80 percent higher than from natural gas and approximately 20 percent higher than from residual fuel oil—the petroleum product most widely used for electricity generation. In the *IEO2000* forecast, carbon emissions are projected to rise between 1990 and 2010 in many countries, including increases of 33 percent for the United States, 21 percent for Japan, and 9 percent for Western Europe (Figure 61). On the other hand, carbon emissions for the FSU are projected to be 30 percent lower in 2010, and emissions in Eastern Europe are projected to be 13 percent lower than in 1990. Ratification of the Kyoto Protocol could have a substantial adverse impact on coal, particularly in the United States, which relies heavily on coal to meet its energy needs and could face relatively severe cutbacks in carbon emissions under the Protocol from those currently projected for 2010 (Figures 61 and 62).

In the *IEO2000* forecast, coal continues to be the second largest source of carbon emissions, accounting for 34 percent of the world total in 2020. Oil, at 41 percent in 2020, remains the largest source of carbon emissions, and natural gas, at 25 percent, accounts for the remaining portion. By country, the world's dominant coal consumers—the United States and China—were also the top two contributors to world carbon emissions in 1997, at 24 percent and 13 percent of the world total, respectively (Figure 63). By 2020, however, the U.S. share of world carbon emissions is projected to decline to 20 percent, with China's share increasing to 21 percent. The substantial increase in carbon emissions in China over the period is attributable to expectations of strong economic growth and the country's continuing reliance on coal as its primary source of energy.

Figure 61. Projected Cumulative Growth in World Carbon Emissions by Region, 1990-2010

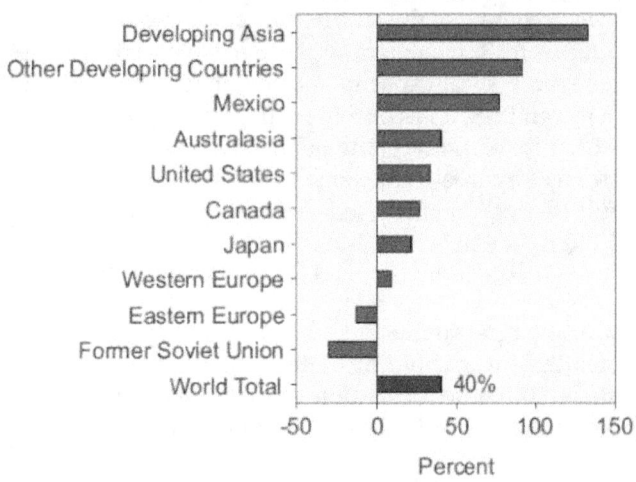

Source: Energy Information Administration, World Energy Projection System (2000).

Figure 62. Coal Share of Total Carbon Emissions by Region, 1997 and 2010

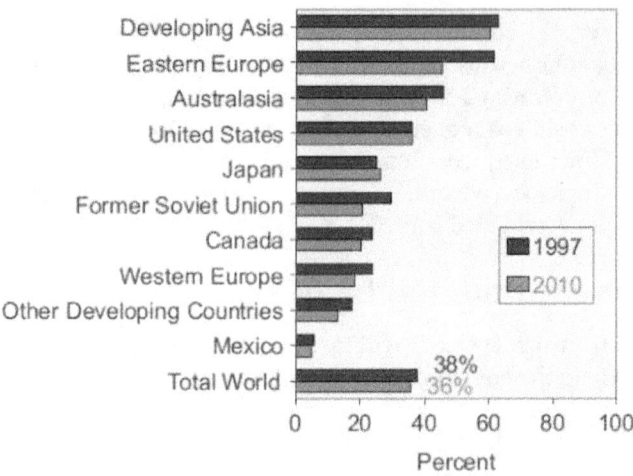

Sources: **1997:** Energy Information Administration (EIA), *International Energy Annual 1997*, DOE/EIA-0219(97) (Washington, DC, April 1999). **2010:** EIA, World Energy Projection System (2000).

Reserves

Total recoverable reserves of coal around the world are estimated at 1,088 billion tons—enough to last approximately 200 years at current production levels (Figure 64).[10] Although coal deposits are widely distributed, 60 percent of the world's recoverable reserves are located in three regions: the United States (25 percent); FSU (23 percent); and China (12 percent). Another four countries—Australia, India, Germany, and South Africa—account for an additional 29 percent. In 1997, these seven regions accounted for 81 percent of total world coal production [12].

Quality and geological characteristics of coal deposits are other important parameters for coal reserves. Coal is a much more heterogeneous source of energy than is oil or natural gas, and its quality varies significantly from one region to the next and even within an individual coal seam. For example, Australia, the United States, and Canada are endowed with substantial reserves of premium coals that can be used to manufacture coke. Together, these three countries supplied 85 percent of the coking coal traded worldwide in 1998 (see Table 18 on page 76).

At the other end of the spectrum are reserves of low-Btu lignite or "brown coal." Coal of this type is not traded to any significant extent in world markets, because of its relatively low heat content (which raises transportation costs on a Btu basis) and other problems related to transport and storage. In 1997, lignite accounted for 18 percent of total world coal production (on a tonnage basis) [13]. The top three producers were Germany (195 million tons), Russia (91 million tons), and the United States

Figure 63. Regional Shares of World Carbon Emissions, 1997 and 2020

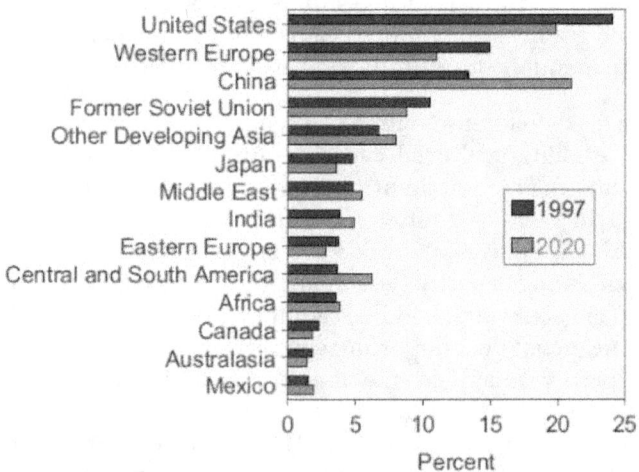

Sources: **1997:** Energy Information Administration (EIA), *International Energy Annual 1997*, DOE/EIA-0219(97) (Washington, DC, April 1999). **2020:** EIA, World Energy Projection System (2000).

(86 million tons), which as a group accounted for 40 percent of the world's total lignite production in 1997. On a Btu basis, lignite deposits show considerable variation. Estimates by the International Energy Agency for coal produced in 1997 show that the average heat content of lignite from major producers in countries of the Organization for Economic Cooperation and Development (OECD) varied from a low of 4.7 million Btu per ton in Greece to a high of 12.3 million Btu per ton in Canada [14].

[10]Recoverable reserves are those quantities of coal which geological and engineering information indicates with reasonable certainty can be extracted in the future under existing economic and operating conditions.

Figure 64. World Recoverable Coal Reserves

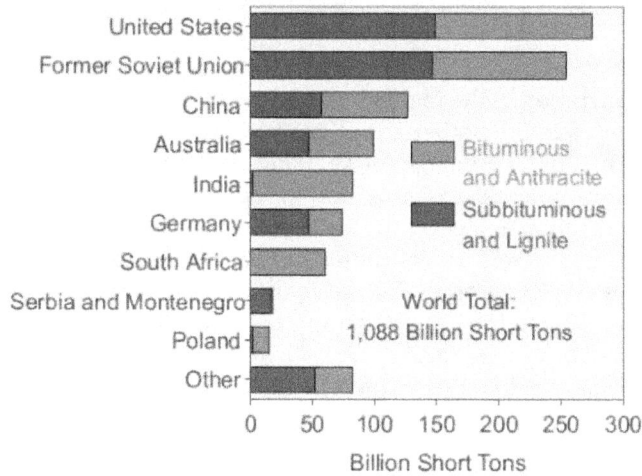

Note: Data represent recoverable coal reserves as of January 1, 1997.

Source: Energy Information Administration, *International Energy Annual 1997*, DOE/EIA-0219(97) (Washington, DC, April 1999), Table 8.2.

Regional Consumption

Asia

The large increases in coal consumption projected for China and India are based on an outlook for strong economic growth (6.3 percent per year in China and 5.4 percent per year in India) and the expectation that much of the increased demand for energy will be met by coal, particularly in the industrial and electricity sectors (Figure 65). The *IEO2000* forecast assumes no significant changes in environmental policies in the two countries. It also assumes that necessary investments in the countries' mines, transportation, industrial facilities, and power plants will be made.

Coal remains the primary source of energy in China's industrial sector, primarily because China has limited reserves of oil and natural gas. In the non-electricity sectors, most of the increase in oil use comes from rising demand for energy for transportation. Growth in the consumption of natural gas comes primarily from increased use for space heating in the residential and commercial sectors. A substantial portion of the increase in China's demand for both natural gas and oil is projected to be met by imports.

In the electricity sector in China, coal use is projected to grow by 4.8 percent a year, from 7.6 quadrillion Btu in 1997 to 22.5 quadrillion Btu in 2020. In comparison, coal consumption by electricity generators in the United

States is projected to rise by 1.3 percent annually, from 18.0 quadrillion Btu in 1997 to 24.0 quadrillion Btu in 2020. One of the key implications of the substantial rise in electricity coal demand in China is that large financial investments in new coal-fired power plants and in the associated transmission and distribution systems will be needed. The projected growth in coal demand implies that China will need approximately 360 gigawatts of coal-fired capacity in 2020.[11] At the beginning of 1997, China had 179 gigawatts of fossil-fuel-fired (coal, oil, and gas) generating capacity [15].

In China, 59 percent of the total increase in coal demand is projected to occur in the non-electricity sectors, for steam and direct heat for industrial applications (primarily in the chemical, cement, and pulp and paper industries) and for the manufacture of coal coke for input to the steelmaking process. Strong growth in steel demand is expected in China as infrastructure and capital equipment markets expand.

In India, projected growth in coal demand occurs primarily in the electricity sector. Between 1997 and 2020, coal use for electricity generation in India is projected to rise by 3.1 percent per year, from 4.3 quadrillion Btu in 1997 to 8.6 quadrillion Btu in 2020, implying that India will need approximately 125 gigawatts of coal-fired capacity in 2020.[12] At the beginning of 1997, India's total fossil-fuel-fired generating capacity amounted to 73 gigawatts [16].

Figure 65. World Coal Consumption by Region, 1980, 1997, 2020

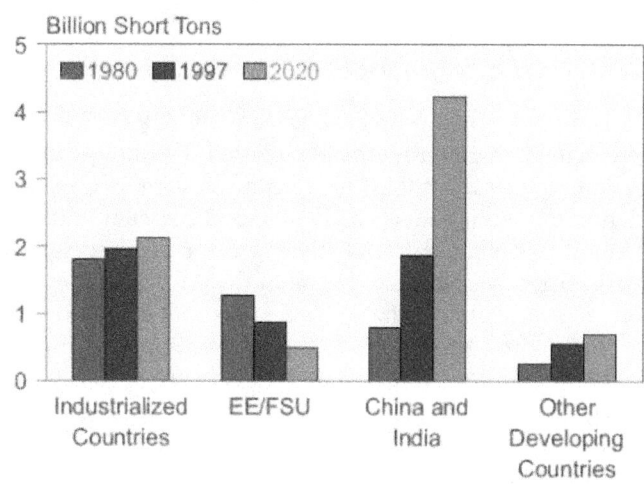

Sources: **History:** Energy Information Administration (EIA), Office of Energy Markets and End Use, International Statistics Database and *International Energy Annual 1997*, DOE/EIA-0219(97) (Washington, DC, April 1999). **Projections:** EIA, World Energy Projection System (2000).

---

[11]Based on a 10-percent improvement in the average heat rate (or conversion efficiency) and a rise in the average capacity factor from approximately 55 percent in 1996 to 65 percent by 2020.

[12]Based on a 10-percent improvement in the average heat rate (or conversion efficiency) and a rise in the average capacity factor from approximately 50 percent in 1996 to 60 percent by 2020.

In the remaining areas of developing Asia, a substantial rise in coal consumption is expected over the forecast period, based on projected strong growth in coal-fired electricity generation in South Korea, Taiwan, and the member countries of the Association of Southeast Asian Nations (primarily, Indonesia, Malaysia, the Philippines, Thailand, and Vietnam). In the electricity sector, coal use in the other developing countries of Asia (including South Korea) is projected to rise by 3.2 percent per year, from 2.5 quadrillion Btu in 1997 to 5.2 quadrillion Btu in 2020.

Most of the new coal-fired capacity in the countries of developing Asia is expected to be built by independent power producers (IPPs). Although much of the expected new capacity was seen as a relatively sure bet a couple of years ago, the financial crisis that rippled through the region in 1997 and 1998, along with other factors, such as environmental concerns and slower economic growth, has led to a reconsideration of a number of the projects planned or under construction.

Currency devaluations in the region have proven to be problematic for IPP projects, primarily because of pressure to price their electricity lower than originally agreed to in long-term contracts with host governments and national utilities. Because most of the costs of IPP projects in the region are based in U.S. dollars, the acceptance of lower prices by project owners would mean lower or negative returns on project investments. On the other hand, electricity sales to end-use consumers in the developing countries of Asia are denominated in local currencies. Thus, with currency devaluations, utilities in developing Asia could pay much higher prices for electricity purchased from IPPs while receiving no additional revenue from sales to end-use consumers.

In Thailand, power purchase agreements (PPAs) for IPP projects have generally been resolved. Unlike in other Asian countries, PPAs in Thailand were originally denominated in bahts (Thailand's local currency) rather than U.S. dollars, directly placing the risks associated with currency exchange rates with the IPP project developers. Realizing that the IPP projects would go bankrupt without some adjustment to the electricity prices stipulated in the PPAs, Thailand revised existing PPAs, effectively transferring a major portion of the currency exchange rate risks from the IPP project developers to the government [17, 18]. In contrast, problems with PPAs continue to persist in Indonesia, where the local currency is still worth only about one-third its value before the financial crisis [19]. Specifically, IPP project owners have been resistant to revisions in the electricity prices stipulated in their PPAs. One strategy currently being considered by Indonesia's state-run utility, Perusahlaan Umunm Listrik Negara (PLN), is to actually purchase IPP projects from the owners directly, thus eliminating the need for renegotiating the electricity

prices stipulated in the PPAs [20]. In 1999, PLN's losses exceeded $1 billion.

Other issues clouding the future for coal-fired electricity generation in the region include increased interest in the use of indigenous natural gas resources for electricity generation (primarily in Thailand and the Philippines), environmental concerns, and privatization of electricity generation assets. South Korea recently divided the generating assets of its state-run utility, Korea Electric Power Corporation (KEPCO), into six separate companies [21], which the government plans to sell to interested buyers by the end of 2000. In 1998, KEPCO's plans for new generating capacity called for the construction of 10 gigawatts of new coal-fired generating capacity between 1997 and 2015 [22].

Despite such uncertainties, coal is generally considered to be one of the most economical choices for power generation in developing Asia, and it should continue to be a strong contender for new electricity generating capacity. In 1999, major coal-fired generating plants came on line in Malaysia, Indonesia, and the Philippines. In addition, many other large coal-fired units remain scheduled for startups during the next several years.

In Japan, coal consumption is projected to increase at a much slower pace than in the other countries of Asia. In the electricity sector, coal use is projected to rise at a rate of 1.4 percent per year, from 1.3 quadrillion Btu in 1997 to 1.8 quadrillion Btu in 2020. The projected increase implies that Japan will need to build less than 10 gigawatts of new coal-fired generating capacity between 1997 and 2020. In contrast, the most recent outlook provided by Japan's Ministry of International Trade and Industry projects a need for 24 gigawatts of new coal-fired capacity between 1997 and 2007 [23]. The *IEO2000* projections, compared with the Ministry's outlook, show slower growth in Japan's overall electricity demand (1.6 percent per year between 1997 and 2010 compared with 1.9 percent per year between 1997 and 2007) and stronger growth in natural-gas-fired electricity generation.

Western Europe

In Western Europe, environmental concerns play an important role in the competition among coal, natural gas, and nuclear power. Recently, other fuels—particularly natural gas—have been gaining economic advantage over coal. Coal consumption in Western Europe has declined by 37 percent over the past 8 years, from 927 million tons in 1989 to 583 million tons in 1997. The decline was smaller on a Btu basis, at 31 percent, reflecting the fact that much of it resulted from reduced consumption of low-Btu lignite in Germany. The decline in coal consumption is expected to continue over the forecast period, but at a slower rate.

Between 1989 and 1997, German lignite production declined by 258 million tons [24]. The sharp decline in German lignite production followed the conversion from lignite-based town gas[13] to natural gas in the eastern states of Germany after reunification in 1990, as well as substitution of natural gas and other fuels for lignite in home heating [25, 26]. A second factor was the collapse of industrial output in the eastern states. Reduced economic activity in eastern Germany contributed to an 8.5-percent decline in total energy consumption in Germany between 1988 and 1994. In the *IEO2000* forecast, further declines in lignite production in Germany are projected to be small in view of the competitiveness of German lignite with other imported fuels and planned investments to refurbish or replace existing lignite-fired plants using best available combustion and pollution control technologies. Over the next few years, Germany's RWE Energie company plans to spend more than $10 billion on new lignite-fired generating capacity [27].

The recent trend in the consumption of hard coal in Western Europe is closely correlated with the trend in the production of hard coal.[14] Following the closure of the last remaining coal mines in Belgium in 1992 and Portugal in 1994, only four member States of the European Union (the United Kingdom, Germany, Spain, and France) continue to produce hard coal [28], and all have seen their output of hard coal decline since 1989. In Germany, Spain, and France, recent agreements between the governments, mining companies, and labor unions on future coal production subsidies indicate that further production declines are forthcoming. In the United Kingdom, production subsidies have been phased out, forcing coal producers into direct competition with North Sea gas and international coal.

Hard coal production in the United Kingdom declined from 111 million tons in 1989 to 54 million tons in 1997 [29]. Most of the decline resulted from privatization in the electricity sector, which led to a rapid increase in gas-fired generation at the expense of coal. Substantial improvements have been made in the country's mining operations in recent years, with average labor productivity rising from less than 1,000 tons per miner-year in 1989 to 3,400 tons per miner-year in 1997 [30].

Despite productivity improvements that have led to a substantial decline in domestic production costs, coal producers in the United Kingdom continue to face an uncertain future [31]. Coal from domestic mines, which was gaining price parity with coal imports during the mid-1990s, is having a difficult time competing with now much lower-priced imports (see box on pages 78-79). In addition to imports, UK coal competes directly with North Sea gas for electricity generation. In mid-1998, the potential negative impacts on the British coal industry and mining jobs resulting from the massive switch to natural gas prompted the issuance of a temporary moratorium by the British government on the construction of new gas-fired generating plants.

In 1998, Britain's energy minister requested an analysis of the nation's power industry to evaluate how the issues of fuel diversity and security of supply should be considered in the approval process for new power projects. The requested study was completed by the Department of Trade and Industry in October 1998 [32]. The report considered issues related not only to the diversity and security of energy supply but also to the design, operation, and structure of the electricity market. A key finding of the review was compelling evidence that the country's wholesale electricity market (the Electricity Pool) was not achieving a competitive economic outcome. For existing coal-fired capacity—typically, large generating plants—consolidation in ownership weakened the incentives of the participants to bid their power into the Pool at competitive prices. In turn, small generators—typically, new gas-fired plants—were bidding their electricity into the Pool at very low prices, thus assuring that their plants would be fully dispatched while receiving the price submitted by the highest bidder—typically, large generators with coal-fired capacity.

In response to the study's findings, the British government initiated a program of reforms in the electricity market intended to create a more competitive environment—one in which existing coal-fired capacity is expected to compete more effectively with generation from new gas-fired plants. Actions taken thus far include a divestiture of several large coal-fired generating plants by British utilities to other generation companies (including the sale of the 4-gigawatt Drax plant in 1999 by National Power to U.S.-based AES Corporation for $3 billion) and steps taken by the government to replace the current Electricity Pool with new trading arrangements for electricity by late 2000 [33, 34]. The new arrangements, which will allow for bilateral trading of electricity between generators, suppliers, traders and customers, are expected to create a more competitive economic market for electricity.

---

[13]"Town gas" (or "coal gas"), a substitute for natural gas, is produced synthetically by the chemical reduction of coal at a coal gasification facility.

[14]Internationally, the term "hard coal" is used to describe anthracite and bituminous coal. In data published by the International Energy Agency, coal of subbituminous rank is classified as hard coal for some countries and as brown coal (with lignite) for others. In data series published by the Energy Information Administration, subbituminous coal production is included in the bituminous category.

Coal subsidies continue to support high-cost production of hard coal in Germany, Spain,[15] and France. For 1998, the European Commission authorized coal industry subsidies of $5,357 million in Germany and $1,297 million in Spain.[16] In each country, the average subsidy per ton of coal produced exceeds the average value of imported coal (Table 17), and all three are currently taking steps to reduce subsidy payments, acknowledging that some losses in coal production are inevitable.

Germany's hard coal production, which is highly subsidized, declined from 88 million tons in 1989 to 56 million tons in 1997 [35]. In March 1997, the federal government, the mining industry, and the unions reached an agreement on the future structure of subsidies to the German hard coal industry. Subsidies to the industry are to be reduced from DM10.5 billion in 1996 to DM5.5 billion by 2005,[17] resulting in an estimated decline in production to 33 million tons [36]. The agreement calls for the closure of 8 to 9 of Germany's 19 hard coal mines, resulting in an estimated decline in employment from 55,000 miners in 1996 to about 36,000 in 2005. In the *IEO2000* reference case, increased imports of coal are expected to compensate for a portion of the expected decline in output from indigenous mines.

In Spain, hard coal production declined from 29 million tons in 1989 to 20 million tons in 1997 [37]. In January 1998, a new restructuring plan for the coal industry was agreed to by the government, labor unions, and coal companies [38]. Under the plan, which has been endorsed by the European Commission, hard coal production will be reduced from 19.8 million tons in 1997 to a maximum level of 16.2 million tons by 2001. Over the same period, coal industry employment will be reduced from 9,800 to 6,500.

In France, production of hard coal declined from 14 million tons in 1989 to 7 million tons in 1997 [39]. A modernization, rationalization, and restructuring plan submitted by the French government to the European Commission at the end of 1994 foresees the closure of all coal mines in France by 2005 [40]. The coal industry restructuring plan was based on a "Coal Agreement" reached between France's state-run coal company, Charbonnages de France, and the coal trade unions. Over the forecast period, consumption of hard coal in Spain and France is expected to decline roughly in accordance with the reductions in indigenous coal production, as other fuels—primarily, natural gas, nuclear, and renewable energy—are expected to compensate for most of the reduction in domestic coal supply.

Coal use in other major coal-consuming countries in Western Europe is projected either to decline or to remain close to current levels. In the Scandinavian countries (Denmark, Finland, Norway, and Sweden), environmental concerns and competition from natural gas are expected to reduce coal use over the forecast period. The government of Denmark has stated that its goal is to eliminate coal-fired generation by 2030 [41]. In 1997, 65 percent of Denmark's electric generation was supplied by coal-fired plants [42].

Italy's coal consumption is projected to remain relatively constant in the *IEO2000* forecast. Currently, investments in environmental and coal-handling equipment are underway at two of the country's large multi-fuel-fired generating plants (3,200 megawatts of generating capacity) to better accommodate the use of coal [43]. On the other hand, a carbon tax introduced by the Italian government in December 1998 is expected to increase the cost of coal-fired generation relative to other fuels in future years [44]. The carbon tax on coal (also applied to petroleum coke and orimulsion) consumed for electricity generation, set at 907 lire per ton ($0.50 per ton) in 1999, is targeted for a gradual increase to 37,957 lire per ton by 2005 (approximately $21 per ton, based on the currency exchange rate for 1999). The law states that the progressive increases in taxes after 2000 will be implemented consistently with the progress of fiscal harmonization processes within the European Union.

Of the major coal-consuming countries in Western Europe, the Netherlands imposes the most substantial carbon-related tax on coal. In 1999, the Netherlands general fuel tax on coal was set at 21.65 guilders per ton ($10.47 per ton) [45]. Over the forecast period, coal-fired generation at existing plants is projected to be displaced gradually by gas-fired generation and by increased imports of electricity [46, 47]. All but one of the country's coal plants are equipped to burn both coal and natural gas.

Partly offsetting the declines in coal consumption elsewhere in Europe is a projected increase in consumption of indigenous lignite for electricity generation in Greece. Under an agreement reached by the countries of the European Union in June 1998, Greece committed to capping its emissions of greenhouse gases by 2010 at 25 percent above their 1990 level [48]—much less severe

[15]In Spain, subsidies support the production of both hard coal and subbituminous coal.

[16]In local currencies, coal subsidies in 1998 were DM9.4 billion in Germany and Pta193.8 billion in Spain. Coal industry subsidies for France have not been approved by the European Commission for 1997 or 1998. The Commission authorized $863 million dollars (FF4.4 billion) in state aid for France's hard coal industry for 1996.

[17]In U.S. dollars, Germany's approved coal industry subsidies for 1996 were $6.9 billion. Planned subsidies for 2005 are approximately $3.0 billion (based on currency exchange rate for 1999). Source: U.S. Federal Reserve Bank, "Foreign Exchange Rates (Monthly)," web site www.bog.frb.fed.us.

Table 17. Western European Coal Industry Subsidies, Production, and Import Prices, 1998

Country	Coal Industry Subsidies (Million 1998 Dollars)	Hard Coal Production (Million Tons)	Average Subsidy per Ton of Coal Produced (1998 Dollars)	Average Price per Ton of Coal Imported (1998 Dollars)
Germany.	5,357	45.5	118	37
Spain.	1,297	18.0	72	36
France	—	5.8	—	41

Sources: **Coal Production Subsidies:** European Commission, *Directory of Community Legislation in Force, Section 12.20.10—Promotion of the Coal Industry,* web site www.europa.eu.int/eur-lex/en/lif (accessed December 10, 1999); and U.S. Federal Reserve Bank, "Foreign Exchange Rates (Annual)," web site www.bogfrb.fed.us (January 3, 2000). **Production:** Energy Information Administration, *International Energy Annual 1998,* DOE/EIA-0219(98) (Washington, DC, January 2000). **Average Price of Coal Imports:** International Energy Agency, *Coal Information 1998* (Paris, France, July 1999).

than the emissions target for the European Union as a whole, which must reduce its emissions to 8 percent below 1990 levels by 2010 [49].

Eastern Europe and the Former Soviet Union

In the EE/FSU countries, the process of economic reform continues as the transition to a market-oriented economy replaces centrally planned economic systems. The dislocations associated with institutional changes in the region have contributed substantially to declines in both coal production and consumption. Coal consumption in the EE/FSU region has fallen by 570 million tons since 1988, to 877 million tons in 1997 [50]. In the future, total energy consumption in the EE/FSU is expected to rise, primarily as the result of increasing production and consumption of natural gas. In the *IEO2000* reference case, coal's share of total EE/FSU energy consumption declines from 25 percent in 1997 to 10 percent in 2020, and the natural gas share increases from 42 percent in 1997 to 54 percent in 2020.

The three main coal-producing countries of the FSU—Russia, Ukraine, and Kazakhstan—are facing similar problems. All three countries have developed national programs for restructuring and privatizing their coal industries, but they have been struggling with related technical and social problems. Of the three, Kazakhstan has shown the most rapid progress. Many of Kazakhstan's high-cost underground coal mines have been closed, and its more competitive surface mines have been purchased and are now operated by international energy companies [51, 52].

In Russia and Ukraine, efforts have been aimed primarily at shutting down inefficient mines and transferring associated support activities—such as housing, kindergartens, and health and recreation facilities—to local municipalities. The closure of inefficient mines in both countries has been slow, however, leading to delays in the scheduled disbursement of money, via loans, from the World Bank. In both countries, coal-mining regions continue to wield considerable political clout, putting

pressure on the leadership through strikes and their ability to influence election results. In 1998, a World Bank report noted that the Ukrainian Coal Ministry's desire to maintain operational authority over mines and to slow or halt the shrinking of the industry proved stronger than the vision of a reformed, economically healthy mining sector [53]. To date, the World Bank has provided $1,050 million in loan assistance to the Russian coal industry and $150 million to Ukraine [54, 55, 56, 57]. The Bank plans to disburse an additional $250 million and $150 million to the Russian and Ukrainian coal industries, respectively, when specific conditions of progress are met. As a supplement to loans from the World Bank, the Export-Import Bank of Japan is committed to supplying additional loans of $400 million to the Russian coal industry [58, 59].

The transfer of support activities from mining associations to local municipalities has also been problematic. Most of the planned transfers in Russia and Ukraine have already occurred, but the municipalities do not have sufficient funding [60]. Thus, the quality of health care and other services in mining communities has deteriorated considerably. Even efficient mines in Russia and Ukraine are not without problems. Payment arrears of large customers have been making it nearly impossible for mines to pay workers and purchase needed mining supplies and equipment. During 1999, the Russian government was exploring countertrade arrangements with Germany by which Russia could receive mining equipment and other solid goods in exchange for coal and/or electricity.

On a positive note, coal exports through a new Russian port, Ust-Luga, located on the Gulf of Finland are set to begin in late 2000 [61]. Initially, the port's coal terminal will have an annual throughput capacity of approximately 1 million tons [62]. An additional construction phase will eventually increase its export capacity to 9 million tons.

Poland is the key coal producer and consumer in Eastern Europe. In 1997, coal consumption in Poland totaled 195 million tons, 46 percent of Eastern Europe's total coal consumption for the year [63]. Poland's hard coal industry produced 151 million tons in 1997, and lignite producers contributed an additional 70 million tons [64]. In other Eastern European countries, coal consumption is dominated by the use of low-Btu subbituminous coal and lignite produced from local reserves. In 1997, the region's other important coal-consuming countries were the Czech Republic (16 percent of the region's total coal use), Romania (11 percent), Serbia and Montenegro (10 percent), Bulgaria (7 percent), and Hungary (4 percent). Eastern Europe relies heavily on local production, with seaborne imports of coal to the region totaling only 6 million tons in 1997 [65].

At present, Poland's hard coal industry is operating at a loss [66]. Over the past several years, a number of coal industry restructuring plans have been put forth for the purpose of transforming Poland's hard coal industry to a position of positive earnings, eliminating the need for government subsidies. The most recent plan was announced by Poland's Ministry of the Economy in March 1998. It calls for the closure of 24 of the country's 50 unprofitable mines over the next 4 years, reducing the total number of mines in Poland from 65 in 1998 to 41 by 2002. In addition, the restructuring plan aims to reduce the number of miners by one-half, from 245,000 in 1998 to 128,000 by 2002 [67, 68]. The government hopes to achieve most of the planned reduction in force through normal retirements and voluntary separations. All miners leaving the industry before retirement age (either voluntarily or involuntarily) under the restructuring program will receive financial compensation packages and assistance in either moving to a new job or establishing a business.

The Polish government projects that sales of hard coal from domestic mines will decline from 100 million tons in 1998 to 88 million tons by 2010 and to 77 million tons by 2020. The World Bank has indicated its willingness to loan the Polish government up to $1 billion over a 3-year period to help cover the costs of the restructuring program, including economic assistance for miners leaving the industry [69]. In June 1999, the World Bank approved a $300 million Hard Coal Sector Adjustment Loan in support of the Polish government's restructuring program [70].

North America

In North America, coal consumption is concentrated in the United States, which, at 1,028 million tons, accounted for 93 percent of the regional total in 1997. By 2020, U.S. coal consumption is projected to rise to 1,279 million tons. With its substantial supplies of coal reserves, the United States has come to rely heavily on coal for electricity generation and continues to do so over the forecast. Coal provided 53 percent of total U.S.

electricity generation in 1997 and is projected to provide 49 percent in 2020 [71]. To a large extent, EIA's projections of declines in both minemouth coal prices and coal transportation rates are the basis for the expectation that coal will continue to compete as a fuel for U.S. power generation. In Canada and Mexico (the other countries of North America), coal consumption is projected to rise from 74 million tons in 1997 to 85 million tons in 2020.

Canada's increased use of coal in the *IEO2000* forecast results primarily from the expected retirement of some of the country's older nuclear units after 2010, and the subsequent need to replace that generation [72]. Between 2010 and 2020, Canada's nuclear generation is projected to decline by 22 percent. In addition, a temporary decrease in Canada's nuclear generation results in higher coal consumption early in the forecast, as increases in both coal- and oil-fired generation compensate for a portion of the generation lost. During the summer of 1997, Ontario Hydro shut down 7 of its 19 nuclear reactors for major overhauls after the discovery of widespread safety and performance problems. Of the 7 units shut down, 4 are located at the utility's Pickering station and 3 at its Bruce station [73]. Their combined generation capacity is 4.3 gigawatts.

As in other parts of the world, natural gas is expected to be the fuel of choice for most new generating capacity in Mexico. In 1997, Mexico consumed 12 million tons of coal. Two coal-fired generating plants, operated by the state-owned utility Comision Federal de Electricidad (CFE), consume approximately 10 million tons of coal annually [74], most of which originates from domestic mines.

Currently, CFE is in the process of switching its dual-fired Petacalco plant, located on Mexico's Pacific coast, from oil to coal. The plant has burned fuel oil since its startup in 1995, but CFE plans to switch most, if not all, of the plant's six generating units to coal. The utility estimates that the 2.1-gigawatt plant will require more than 5 million tons of imported coal annually. A coal import facility adjacent to the plant, with an annual throughput capacity of more than 9 million tons, will serve both the power plant and a nearby integrated steel mill [75]. Although initial coal burn at the plant is not expected until mid-2000, CFE received some initial supplies of coal in 1999 from producers in Australia and Russia [76].

Africa

In Africa, coal production and consumption are concentrated almost entirely in South Africa. In 1997, South Africa produced 243 million tons of coal, 70 percent of which was routed to domestic markets and the remainder to exports [77]. Ranked third in the world in coal exports since the mid-1980s (behind Australia and the United States), South Africa moved into the number two coal-exporting position in 1999, when its exports exceeded those from the United States. South Africa also holds the distinction of being the world's largest producer of coal-based synthetic liquid fuels. In 1997, almost one-fifth of the coal consumed in South Africa was used to produce coal-based synthetic oil, which in turn accounted for approximately 30 percent of all liquid fuels consumed in South Africa during the year [78, 79].

For Africa as a whole, coal consumption is projected to increase by 29 million tons between 1997 and 2020, primarily to meet increased demand for electricity. Contributing to the increase in electricity demand is South Africa's commitment to an aggressive electrification program, which aims to increase the percentage of households connected to the electricity grid from 44 percent at the end of 1995 to 75 percent by 2000 [80, 81]. There are also substantial opportunities for trade in electricity and natural gas between South Africa and neighboring countries. New power transmission lines have been completed or are planned to facilitate flows of electricity between South Africa, Mozambique, Zimbabwe, Swaziland, and Namibia [82]. Such international connections could open new markets for underutilized or idle coal-fired power plants in South Africa.

Elsewhere in Africa, the completion of four additional coal-fired units at Morocco's Jorf Lasfar plant near Casablanca should increase coal consumption there from about 2 million tons in 1997 to more than 5 million tons [83, 84]. When all units are completed, the plant is expected to account for approximately one-third of Morocco's total power generation.

Central and South America

Coal has not been an important source of energy in Central and South America, accounting for less than 6 percent of the region's total energy consumption since 1970. In the electricity sector, hydroelectric power currently meets much of the region's electricity demand. Over the forecast period, both hydropower and natural gas are projected to fuel much of the projected increase in electricity generation.

In 1997, Brazil accounted for one-half of South America's total coal demand (on a Btu basis), with Colombia, Chile, and Argentina accounting for much of the remaining portion. In Brazil, the steel industry accounts for more than 75 percent of the country's total coal consumption, relying on imports of coking coal to produce coke for use in its blast furnaces [85]. In the forecast, increased use of coal for steelmaking (both coking coal and coal for pulverized coal injection) accounts for much of the projected increase in Brazil's coal consumption [86]. New power projects and coke-making facilities in Colombia and Peru account for most of the remaining growth in coal consumption projected for South America [87, 88].

In Central America, petroleum products and hydropower are the key sources of primary energy consumption (accounting for 70 percent and 24 percent of the total, respectively, in 1997) [89]. The only coal consumption in the region is a small quantity used in Panama for industrial purposes [90]. Coal use in the region is set to increase somewhat in 2000, however, with the completion of a 120-megawatt coal-fired generating plant in Guatemala [91, 92]. The plant, built by a consortium of U.S. and Guatemalan companies and located on the Pacific coast, is the first coal-fired power plant in Central America.

Middle East

Turkey accounts for most of the coal consumed in the Middle East. In 1997, a total of 70 million tons of coal was consumed in Turkey, most of it low-Btu, locally produced lignite (approximately 7.4 million Btu per ton) [93, 94]. Over the forecast period, Turkey's coal consumption (both lignite and hard coal) increases by 15 million tons, primarily to fuel additional coal-fired generating capacity.

Israel and Iran accounted for most of the remaining 11 million tons of coal consumed in the Middle East in 1997 [95]. Over the forecast, Israel's coal consumption is projected to rise by approximately 3 million tons with the completion of two new coal-fired generating units at Israel Electric Corporation's Rutenberg plant between 2000 and 2005 [96, 97]. Currently, coal accounts for approximately 75 percent of the country's total electricity generation [98]. In Iran, approximately 1 million tons of coal consumption has been met historically by indigenous suppliers [99]. In addition, Iran's National Steel Corporation imports approximately 0.5 million tons of coking coal annually [100, 101].

Trade

Overview

The amount of coal traded in international markets is small in comparison with total world consumption. In 1998, world imports of coal amounted to 546 million tons (Table 18 and Figure 66), representing 10 percent of total consumption. By 2020, coal imports are projected to rise to 708 million tons, accounting for a 9-percent share of world coal consumption. Although coal trade has made up a relatively constant share of world coal consumption over time and should continue to do so in future years, the geographical composition of trade is shifting.

In recent years, international coal trade has been characterized by relatively stable demand for coal imports in Western Europe and expanding demand in Asia (Figure 67). Rising production costs in the indigenous coal industries in Western Europe, combined with

Table 18. World Coal Flows by Importing and Exporting Regions, Reference Case, 1998, 2010, and 2020
(Million Short Tons)

| | Importers | | | | | | | | | | | |
| | Steam[a] | | | | Coking[b] | | | | Total | | | |
Exporters	Europe[c]	Asia	America	Total[d]	Europe[c]	Asia	America	Total[d]	Europe[c]	Asia	America	Total[d]
	1998											
Australia	11.2	77.3	2.5	92.2	17.9	63.2	4.9	91.6	29.1	140.4	7.4	183.8
United States	8.4	5.0	17.6	31.0	28.6	6.7	11.8	47.1	37.0	11.7	29.4	78.0
South Africa.	45.9	14.8	0.4	66.3	1.2	4.2	1.9	7.6	47.1	19.0	2.3	73.9
Former Soviet Union .	8.9	2.5	0.0	11.4	1.7	3.0	0.0	4.7	10.6	5.5	0.0	16.1
Poland	14.4	0.0	0.0	14.4	6.0	0.0	0.0	6.0	20.4	0.0	0.0	20.4
Canada	0.7	5.2	0.6	6.4	6.7	21.9	2.7	31.2	7.4	27.0	3.2	37.7
China	2.4	25.7	0.0	30.2	0.1	7.7	0.0	5.3	2.5	33.4	0.0	35.5
South America[e]	25.5	0.0	9.9	38.9	0.3	0.3	0.2	0.8	25.8	0.3	10.1	39.7
Indonesia[f]	2.5	39.9	2.0	55.0	0.8	4.6	0.2	6.0	3.4	44.5	2.2	61.0
Total.	119.8	170.3	33.0	345.6	63.5	111.6	21.6	200.4	183.3	281.9	54.6	546.0
	2010											
Australia	10.0	132.2	1.3	143.5	32.9	82.0	8.5	123.4	42.9	214.1	9.8	266.9
United States	15.0	4.6	8.9	28.5	17.7	1.3	16.4	35.5	32.7	6.0	25.3	64.0
South Africa.	45.1	35.5	2.4	83.0	1.1	5.9	0.0	7.1	46.2	41.4	2.4	90.1
Former Soviet Union .	12.1	2.8	0.0	14.9	1.5	0.5	0.0	2.1	13.7	3.3	0.0	17.0
Poland	8.0	0.0	0.0	8.0	3.6	0.0	0.0	3.6	11.7	0.0	0.0	11.7
Canada	5.1	6.0	0.0	11.0	4.6	22.0	2.8	29.3	9.6	28.0	2.8	40.4
China	0.0	41.0	0.0	41.0	0.0	6.1	0.0	6.1	0.0	47.1	0.0	47.1
South America[e]	33.6	0.0	26.6	60.2	0.0	0.0	0.0	0.0	33.6	0.0	26.6	60.2
Indonesia[f]	2.3	62.2	0.0	64.5	0.9	4.0	0.0	5.0	3.2	66.2	0.0	69.4
Total.	131.2	284.2	39.3	454.7	62.5	121.9	27.7	212.0	193.6	406.1	67.0	666.7
	2020											
Australia	8.2	146.6	0.9	155.7	35.8	86.3	10.6	132.8	44.0	232.9	11.5	288.5
United States	7.7	5.1	9.9	22.7	14.4	1.5	19.0	34.9	22.1	6.6	28.9	57.7
South Africa.	46.3	40.3	2.4	89.0	0.9	5.7	0.0	6.6	47.2	46.0	2.4	95.6
Former Soviet Union .	12.1	3.9	0.0	16.0	1.5	2.8	0.0	4.3	13.7	6.6	0.0	20.3
Poland	5.5	0.0	0.0	5.5	3.4	0.0	0.0	3.4	8.9	0.0	0.0	8.9
Canada	5.1	3.3	0.0	8.4	4.3	20.4	3.6	28.3	9.3	23.8	3.6	36.7
China	0.0	46.5	0.0	46.5	0.0	6.6	0.0	6.6	0.0	53.1	0.0	53.1
South America[e]	38.1	0.0	29.8	67.9	0.0	0.0	0.0	0.0	38.1	0.0	29.8	67.9
Indonesia[f]	0.0	74.8	0.0	74.8	0.9	4.1	0.0	5.0	0.9	78.9	0.0	79.7
Total.	123.0	320.6	42.9	486.4	61.2	127.4	33.3	221.9	184.1	448.0	76.2	708.3

[a]Reported data are consistent with data published by the International Energy Agency (IEA). The standard IEA definition for "steam coal" includes coal used for pulverized coal injection (PCI) at steel mills; however, some PCI coal is reported by the IEA as "coking coal."

[b]Includes primarily coal consumed to produce coal coke. According to the IEA, a minor exception for 1997 trade data is the classification of 9.6 million tons of coal imported to Japan for PCI at blast furnaces as coking coal. Similarly, the IEA reports that some exports of coal from Australia, South Africa, Indonesia, and Colombia to be used for PCI at steel mills is classified as coking coal, consistent with data reported by importing countries and industry terminology and practice.

[c]Coal flows to Europe include shipments to the Middle East and Africa.

[d]For 1998, total world coal flows include a balancing item used by the International Energy Agency to reconcile discrepancies between reported exports and imports. The 1998 balancing items by coal type were 22.5 million tons (steam coal), 3.7 million tons (coking coal), and 26.2 million tons (total).

[e]Coal exports from South America are projected to originate from mines in Colombia and Venezuela.

[f]For 1998, coal exports from Indonesia include shipments from other countries not modeled for the forecast period. The 1998 non-Indonesian exports by coal type were 8.4 million tons (steam coal), 1.0 million tons (coking coal), and 9.3 million tons (total).

Notes: Data exclude non-seaborne shipments of coal to Europe and Asia. Totals may not equal sum of components due to independent rounding. The sum of the columns may not equal the total, because the total includes a balancing item between importers' and exporters' data.

Sources: **1998:** International Energy Agency, Coal Information 1998 (Paris, France, July 1999); Energy Information Administration, Quarterly Coal Report, October-December 1998, DOE/EIA-0121(98/4Q) (Washington, DC, July 1999). **Projections:** Energy Information Administration, Annual Energy Outlook 2000, DOE/EIA-0383(2000) (Washington, DC, December 1999), National Energy Modeling System run AEO2K.D100199A.

Figure 66. World Coal Trade, 1985, 1998, and 2020

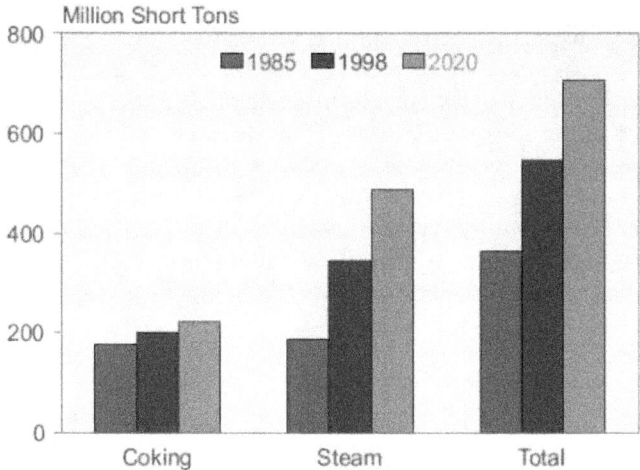

Sources: **1985**: Energy Information Administration (EIA), *Annual Prospects for World Coal Trade 1987*, DOE/EIA-0363(87) (Washington, DC, May 1987). **1998**: International Energy Agency, Coal Information 1998 (Paris, France, July 1999); Energy Information Administration, Quarterly Coal Report, October-December 1998, DOE/EIA-0121(98/4Q) (Washington, DC, July 1999). **2020**: Energy Information Administration, Annual Energy Outlook 2000, DOE/EIA-0383(2000) (Washington, DC, December 1999), National Energy Modeling System run AEO2K.D100199A.

Figure 67. Production and Imports of Hard Coal by Region, 1985, 1990, and 1998

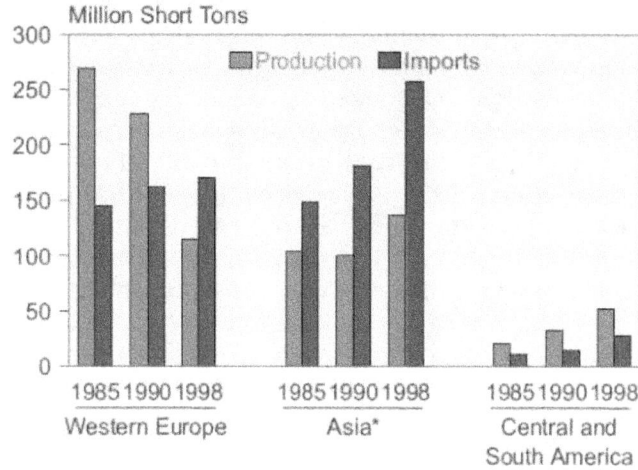

*Data for Asia exclude China, India, and Australasia.
Note: Production and imports include data for anthracite, bituminous, and subbituminous coal.
Sources: Energy Information Administration, Office of Energy Markets and End Use, International Statistics Database.

continuing pressure to reduce industry subsidies, have led to substantial declines in production there, creating the potential for significant increases in coal imports; however, slow economic growth in recent years and increased electricity generation from natural gas, nuclear, and hydropower have curtailed the growth in coal imports. Conversely, growth in coal demand in Japan, South Korea, and Taiwan in recent years has contributed to a substantial rise in Asian coal imports.

International coal markets have gone through some significant changes over the past couple of years. Most noteworthy has been a decline in the price of traded coal (see box on pages 78-79). Price cuts have benefitted electricity generators and steel producers in coal-importing countries and improved the competitiveness of coal-fired generation, but they have also, in most cases, reduced the overall revenues received by coal producers in major exporting countries. U.S. coal producers have been hit particularly hard by the decline in international coal prices, given that coal prices are negotiated internationally in U.S. dollars, and the dollar has increased in value against the currencies of other major coal-exporting countries. In 1999, U.S. coal exports fell sharply, even as coal shipments from most other exporting countries continued to increase. South Africa displaced the United States as the world's number two coal-exporting country in 1999, a position that the United States had held since 1984 [*102*].

Asia

Despite recent setbacks, Asia's demand for imported coal remains poised for additional increases over the forecast period, based on strong growth in electricity demand in the region. Continuing the recent historical trend, Japan, South Korea, and Taiwan are projected to account for much of the regional growth in coal imports over the forecast period.

Japan continues to be the world's leading importer of coal and is projected to account for 25 percent of total world imports in 2020 [*103*], slightly less than its 1998 share of 26 percent [*104*]. In 1998, Japan produced 4 million tons of coal for domestic consumption and imported 140 million tons. The closure of Japan's Miike mine in March 1997 left the country with two remaining underground coal mines and several small surface mines. Production at the two underground mines is expected to end when the government eliminates industry subsidies in 2001, leaving virtually all of Japan's coal requirements to be met by imports [*105, 106*].

As the leading importer of coal, Japan has been influential in the international coal market. Historically, contract negotiations between Japan's steel mills and coking coal suppliers in Australia and Canada established a benchmark price for coal that was used later in the year as the basis for setting contract prices for steam coal used at Japanese utilities [*107*]. Other Asian markets also tended to follow the Japanese price in settling contracts.

World Coal Exports: Prices Decline Sharply in 1999

Fierce price competition prevailed in world coal markets in 1999, substantially affecting trade patterns and the revenues obtained from exports. Australia and Indonesia saw major increases in their coal exports in 1999, while the United States saw a major reduction in its exports for the year, dropping to a level not seen since the mid-1970s.[a] Although both South African and Canadian producers priced their coal exports at very competitive prices in 1999, they did not see substantial increases in shipments over 1998. South African exporters consistently priced their cape size cargoes of steam coal at or below $18 per ton (FOB port of exit) but still were having a difficult time competing with shipments of Russian and Polish coal to Europe.[b] Russian exporters, benefitting from a sharp decline in the ruble, were able to offer coal at a considerable discount from previous years. Canada, which relies heavily on exports of coking coal to Asian steel producers, faced a slight reduction in world coking coal demand in 1999 and strong competition from Australian producers.

A number of factors led to the 1999 drop in world coal prices, including favorable exchange rates for key exporters;[c] productivity improvements; increases in

coal export capacity; aggressive price negotiations on the part of coal importers; and the acceptance of a wider range of coals (in terms of coking quality parameters) for the manufacture of coke for steelmaking. The figures below show FOB port-of-exit prices for steam and coking coal by quarter, as published by the International Energy Agency, in constant 1998 dollars. The figures illustrate a significant divergence in U.S. coal export prices from those of Australia and Canada since about the first quarter of 1998.

Broken Hill Properties, a major international coal producer, pointed to an approximately 25-percent increase in productivity at its Australian operations since the end of 1996 and a weaker Australian dollar as major factors in its ability to cut coking coal prices by approximately $8 per ton in 1999.[d] A little more than half of the price cut was attributed to improvements in productivity that lowered the overall costs of production.

Since 1980, the United States has been gradually closing the price gap with other major exporters, essentially reaching price parity in the past few years. The
(continued on page 78)

Steam Coal Export Prices by Quarter, 1998-1999

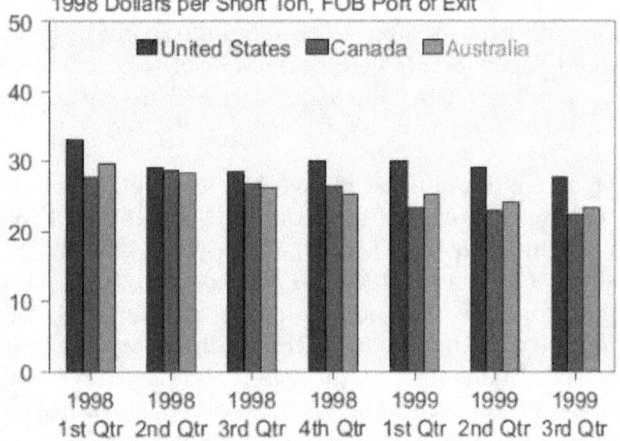

Sources: **Nominal Prices in U.S. Dollars:** International Energy Agency. **GDP Deflators:** U.S. Department of Commerce, Bureau of Economic Analysis.

Coking Coal Export Prices by Quarter, 1998-1999

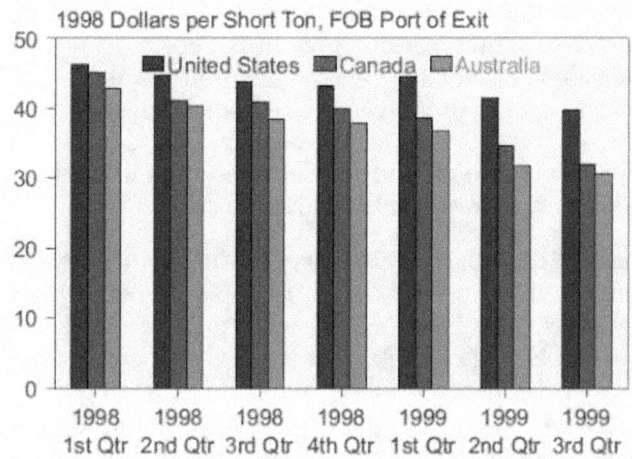

Sources: **Nominal Prices in U.S. Dollars:** International Energy Agency. **GDP Deflators:** U.S. Department of Commerce, Bureau of Economic Analysis.

[a]Energy Information Administration, *Annual Energy Review 1998*, DOE/EIA-0384(98) (Washington, DC, July 1999), Table 7.1.

[b]"International Market Report," *King's International Coal Trade*, No. 1240 (August 23, 1999), pp. 2-3; "International Market Report," *King's International Coal Trade*, No. 1239 (August 16, 1999), p. 1; and "Production Cuts Loom in South Africa," *Financial Times: International Coal Report*, No. 479 (July 26, 1999), p. 5.

[c]Between May 1996 and August 1998, the Australian dollar lost 26 percent of its value compared with the U.S. dollar. Similarly, between January 1996 and August 1998, the South African rand lost 42 percent of its value. More recently, between August 1998 and April 1999, the Russian ruble lost 73 percent of its value compared with the U.S. dollar. Sources: U.S. Federal Reserve Bank, "Foreign Exchange Rates (Monthly)," web site www.bog.frb.fed.us; and International Monetary Fund, IMF Statistics Department, *International Financial Statistics Yearbook* (Washington, DC, various issues).

[d]N. Bristow, BHP Coal, "Metallurgical Coal, Opportunities for the Australian Industry Following the $9 Price Cut," paper presented at the Australian Bureau of Agricultural and Resource Economics's Outlook 99 Conference (Canberra, Australia, March 17-18, 1999), web site www.abare.gov.au.

World Coal Exports: Prices Decline Sharply in 1999 (Continued)

capability of U.S. coal producers to reduce their production costs, and hence prices, consistently at rates equal to or exceeding those of other exporting countries has been a key to the ability of the United States to maintain a strong presence in the world coal market. Despite considerable growth in coal export capacity in most of the other major exporting countries, the United States has held steady as the world's number two coal exporter since 1984, when Australia first displaced the United States as the world leader.

The future, while not entirely clear, does not seem to bode well for U.S. coal exporters. The CEO of Arch Coal, a major producer and exporter of U.S. coal, recently indicated his belief that the United States will not continue as a major exporter in the future.[e] He expects that investments in coal export mines will continue to decline, noting that U.S. coal export terminals currently are looking at new products and at reconfiguring their facilities to handle coal imports.

In 1999, U.S. coal exports fell sharply while coal shipments from most other exporting countries continued to increase, and South Africa displaced the United

States as the world's number two coal-exporting country. Discouraged by the low export prices, some U.S. coal producers idled export capacity in 1999, while others diverted some of their potential exports (both steam and coking coals) to the domestic steam coal market.

In the *IEO2000* forecast, U.S. coal exports are projected to decline slightly over the forecast period. This is in contrast to the growth in coal exports projected for most of the other major exporting countries. As a result, the U.S. share of world coal trade is projected to decline from 14 percent in 1998 to 8 percent by 2020. Australia's share of world coal trade increases from 34 percent in 1998 to 41 percent by 2020.

On the demand side, major coal importing countries also were affected by the sharp decline in international coal prices in 1999. While electricity generators and coke plant operators in coal-importing countries saw lower fuel and raw material input costs as a welcome event, coal producers in countries such as the United Kingdom and Germany saw the decline as yet another setback in their struggle to maintain production levels at their few remaining operations.[f]

Annual Steam Coal Export Prices, 1980-1998

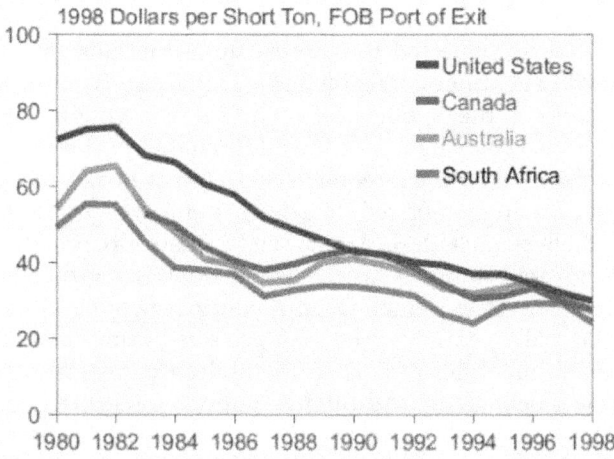

Note: Prices for South Africa's coal exports are the average for all exports of steam and coking coal. Steam coal makes up approximately 90 percent of the total.
Sources: **Nominal Prices in U.S. Dollars:** International Energy Agency and South Africa Minerals Bureau. **GDP Deflators:** U.S. Department of Commerce, Bureau of Economic Analysis.

Annual Coking Coal Export Prices, 1980-1998

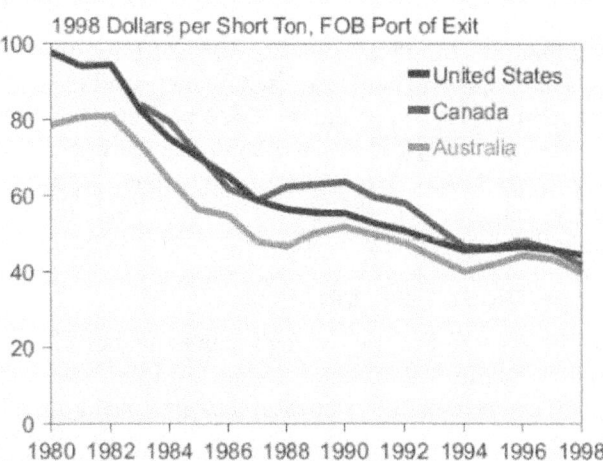

Sources: **Nominal Prices in U.S. Dollars:** International Energy Agency. **GDP Deflators:** U.S. Department of Commerce, Bureau of Economic Analysis.

[e]"International Market Report," *King's International Coal Trade*, No. 1250 (November 1, 1999), pp. 9, 11.
[f]"Coal Boss's Plea As Mine Closes," BBC News, web site news.bbc.co.uk (November 1, 1999).

Japan's influence has declined somewhat over the past several years, however, and the benchmark pricing system that was so influential in setting contract prices for Japan's steel mills was revised substantially in 1996. The revisions reflected a move away from a system which, in

effect, averaged coal prices (with minor adjustments for quality) to a regime with a broad spectrum of prices, where high-quality coking coals received a substantial premium relative to lower quality coals [108]. What seems to be occurring in the Asian coal markets is a shift away from contract purchases to the spot market. Liberalization of the Japanese electricity market is placing increased cost-cutting pressure on utilities, making them less inclined to accept a benchmark price negotiated by any of the other individual utilities. The shift to more competitive coal markets in Asia implies that coal producers in Australia and other exporting countries will be under increased pressure to reduce mining costs in order to maintain current rates of return. It also means that less competitive suppliers, such as the United States, will find it difficult to increase or maintain coal export sales to the region.

China and India, which import relatively small quantities of coal at present, are expected to account for much of the remaining increase in Asian imports. Imports by China and India have the potential to be even higher than the projected amount, but it is assumed in the forecast that domestic coal will be given first priority in meeting the large projected increase (2.3 billion tons) in coal demand.

During the 1980s, Australia became the leading coal exporter in the world, primarily by meeting increased demand for steam coal in Asia. Some growth in exports of coking coal also occurred, however, as countries such as Japan began using some of Australia's semi-soft or weak coking coals in their coke oven blends. As a result, imports of hard coking coals from other countries, including the United States, were displaced. Australia's share of total world coal trade, which increased from 17 percent in 1980 to 34 percent in 1998, is projected to reach 41 percent in 2020 [109]. Australia should continue as the major exporter to Asia, continuing to meet approximately one-half of the region's total coal import demand.

Europe, Middle East, and Africa

Coal imports to Europe (which for the analysis of coal imports includes shipments to the Middle East and Africa) are projected to remain relatively constant over the forecast period. Projected declines in overall imports to the countries of Western Europe are offset by small increases projected for Turkey, Romania, Morocco, and Israel.

In Western Europe, strong environmental lobbies and competition from natural gas are expected gradually to reduce the reliance on steam coal for electricity generation, and further improvements in the steelmaking process will continue to reduce the amount of coal required for steel production. Strict environmental standards are expected to result in the closure of some of Western Europe's older coke batteries, increasing import requirements for coal coke but reducing imports of coking coal.

Projected reductions in indigenous coal production in the United Kingdom, Germany, Spain, and France are not expected to be replaced by equivalent volumes of coal imports. Rather, increased use of natural gas, renewable energy, and nuclear power (primarily in France) is expected to fill much of the gap in energy supply left by the continuing declines in the region's indigenous coal production.

In 1998, the leading suppliers of imported coal to Europe were South Africa (26 percent), the United States (20 percent), Australia (16 percent), and South America (14 percent). Over the forecast period, low-cost coal from South America is projected to meet an increasing share of European coal import demand, displacing some coal from such higher cost suppliers as the United States and Poland.

The Americas

Compared with European and Asian coal markets, imports of coal to North and South America are relatively small, amounting to only 55 million tons in 1998 (Table 18). Canada imported 33 percent of the 1998 total, followed by Brazil (22 percent) and the United States (16 percent) [110]. Almost all (97 percent) of the imports to Brazil were coking coal [111].

Over the IEO2000 forecast period, coal imports to the Americas are projected to increase by 22 million tons, with most of the additional tonnage going to the United States, Brazil, and Mexico. Coal imports to the United States are projected to increase from 9 million tons in 1998 to 20 million tons by 2020 [112]. Coal-fired power plants in the southeastern part of the country are expected to take most of the additional import tonnage projected over the forecast period, primarily as a substitute for higher priced coal from domestic producers. Coal imports to the Brazilian steel industry are projected to rise substantially as the result of strong growth in domestic steel demand and a continuing switch from charcoal to coal coke. Mexico is projected to import additional quantities of coal for both electricity generation and steelmaking. Additional imports of coal to the Americas are projected to be met primarily by producers in Colombia and Venezuela.

Coking Coal

Historically, coking coal has dominated world coal trade, but its share has steadily declined, from 55 percent in 1980 to 37 percent in 1998 [113]. In the forecast, its share of world coal trade continues to shrink, to 31 percent by 2020. In absolute terms, despite a projected decline in imports by the industrialized countries, the total world trade in coking coal is projected to increase slightly over the forecast period as the result of increased demand for steel in the developing countries. Increased imports of coking coal are projected for South Korea, Taiwan, India, Brazil, and Mexico, where expansions in blast-furnace-based steel production are expected.

Factors that contribute to the decline in coking coal imports in the industrialized countries are continuing increases in steel production from electric arc furnaces (which do not use coal coke as an input) and technological improvements at blast furnaces, including greater use of pulverized coal injection equipment and higher average injection rates per ton of hot metal produced. One ton of pulverized coal (categorized as steam coal) used in steel production displaces approximately 1.4 tons of coking coal [114, 115]. In 1997, the direct use of pulverized coal at blast furnaces accounted for 14 percent of the coal consumed for steelmaking in Japan and the European Union [116].

References

1. Energy Information Administration, *Impacts of the Kyoto Protocol on U.S. Energy Markets and Economic Activity*, SR/OIAF/98-03 (Washington, DC, October 1998), Table B1.

2. Energy Information Administration, *International Energy Outlook 1999*, DOE/EIA-0484(99) (Washington, DC, March 1999), pp. 60-61.

3. IEA Coal Research, *Non-OECD Coal-Fired Power Generation—Trends in the 1990s*, CCC/18 (London, UK, June 1999), p. 32.

4. IEA Coal Research, *Southeast Asia—Air Pollution Control and Coal-Fired Power Generation*, IEAPER/39 (London, UK, December 1997), pp. 19-43.

5. IEA Coal Research, *Non-OECD Coal-Fired Power Generation—Trends in the 1990s*, CCC/18 (London, UK, June 1999), p. 57.

6. T.M. Johnson, F. Liu, and R. Newfarmer, *Clear Water, Blue Skies: China's Environment in the New Century*, World Bank No. 14044 (Washington, DC, September 1997), pp. 50-51.

7. IEA Coal Research, *Coal-Fired Independent Power Production in Developing Countries*, CCC/03 (London, UK, April 1998), pp. 13-14.

8. "Greenpeace Hits at Coal Plant," *Financial Times: Power in Asia*, No. 289 (November 1, 1999), pp. 19-20.

9. "Coal Plant Delays To Hit Sri Lanka," *Financial Times: Power in Asia*, No. 282 (July 26, 1999), pp. 15-16.

10. Directorate-General XVII—Energy, European Commission, *Energy in Europe: European Union Energy Outlook to 2020* (Brussels, Belgium , November 1999), pp. 44-45.

11. Energy Information Administration, *Emissions of Greenhouse Gases in the United States 1998*, DOE/EIA-0573(98) (Washington, DC, October 1999), Table B1.

12. Energy Information Administration, *International Energy Annual 1997*, DOE/EIA-0219(97) (Washington, DC, April 1999), Table 2.5.

13. Energy Information Administration, *International Energy Annual 1997*, DOE/EIA-0219(97) (Washington, DC, April 1999), Tables 2.5 and 5.4.

14. International Energy Agency, *Coal Information 1998* (Paris, France, July 1999), pp. II.19-II.21.

15. Energy Information Administration, *International Energy Annual 1997*, DOE/EIA-0219(97) (Washington, DC, April 1999), Table 6.4.

16. Energy Information Administration, *International Energy Annual 1997*, DOE/EIA-0219(97) (Washington, DC, April 1999), Table 6.4.

17. IEA Coal Research, *Coal-Fired Independent Power Production in Developing Countries*, CCC/03 (London, UK, April 1998), p. 32.

18. IEA Coal Research, *Non-OECD Coal-Fired Power Generation—Trends in the 1990s*, CCC/18 (London, UK, June 1999), p. 27.

19. "Asian Currency Watch ($)," *Financial Times: Power in Asia*, No. 291 (November 29, 1999), p. 7.

20. "Tanjung Jati B Buyout Gets Parliamentary Backing," *Financial Times: Power in Asia*, No. 283/284 (August 9, 1999), p. 15.

21. "KEPCO Decides New Structure," *Financial Times: International Coal Report*, No. 485 (October 18, 1999), pp. 2-3.

22. *The Tex Report*, No. 7150 (August 31, 1998).

23. "MITI's New Ten-Year Plan Calls for the Development of 11.3 Mil. kW Nuclear Power To Raise Its Generating Capacity to 56.2 Mil. kW by End of FY2007," *JPET* (May 1998), pp. 2-9.

24. Energy Information Administration, *International Energy Annual 1997*, DOE/EIA-0219(97) (Washington, DC, April 1999), Table 5.4.

25. International Energy Agency, *Coal Information 1998* (Paris, France, July 1999), p. II.201.

26. Directorate-General XVII—Energy, European Commission, *Energy in Europe: European Union Energy Outlook to 2020* (Brussels, Belgium , November 1999), p. 47.

27. "Coal and Lignite Worried by Perk for Gas," *Financial Times: International Coal Report*, No. 487 (November 15, 1999), pp. 15-16.

28. Directorate-General XVII—Energy, European Commission, *The Market for Solid Fuels in the Community in 1996 and the Outlook for 1997* (Brussels, Belgium, June 6, 1997), web site www.europa.eu.int.

29. Energy Information Administration, *International Energy Annual 1997*, DOE/EIA-0219(97) (Washington, DC, April 1999), Tables 5.2 and 5.3.

30. International Energy Agency, *Coal Information 1998* (Paris, France, July 1999), Table 6.5a.

31. Directorate-General XVII—Energy, European Commission, *The Market for Solid Fuels in the Community in 1996 and the Outlook for 1997* (Brussels, Belgium, June 6, 1997), p. 33, web site www.europa.eu.int.

32. UK Department of Trade and Industry, *Conclusions of the Review of Energy Sources for Power Generation, and Government Response to Fourth and Fifth Reports of the Trade and Industry Committee Energy Review White Paper*, CM 4071 (London, UK, October 1998).

33. UK Office of Gas and Electricity Markets, *The New Electricity Trading Arrangements: Ofgem/DTI Conclusions Document* (London, UK, October 1999), web site www.ofgem.gov.uk.

34. UK Office of Gas and Electricity Markets, *The New Electricity Trading Arrangements* (London, UK, July 1999), web site www.ofgem.gov.uk.

35. Energy Information Administration, *International Energy Annual 1997*, DOE/EIA-0219(97) (Washington, DC, April 1999), Tables 5.2 and 5.3.

36. International Energy Agency, *Coal Information 1998* (Paris, France, July 1999), pp. I.210-I.211.

37. Energy Information Administration, *International Energy Annual 1997*, DOE/EIA-0219(97) (Washington, DC, April 1999), Tables 5.2 and 5.3.

38. International Energy Agency, *Coal Information 1998* (Paris, France, July 1999), p. I.210.

39. Energy Information Administration, *International Energy Annual 1997*, DOE/EIA-0219(97) (Washington, DC, April 1999), Tables 5.2 and 5.3.

40. International Energy Agency, *Coal Information 1998* (Paris, France, July 1999), p. I.210.

41. International Energy Agency, *Coal Information 1998* (Paris, France, July 1999), p. I.210.

42. International Energy Agency, *Coal Information 1998* (Paris, France, July 1999), Table 3.3b.

43. International Energy Agency, *Coal Information 1998* (Paris, France, July 1999), p. III.126.

44. International Energy Agency, *Coal Information 1998* (Paris, France, July 1999), p. I.270 and Table 7.2.

45. International Energy Agency, *Coal Information 1998* (Paris, France, July 1999), pp. I.270-I.271 and Table 7.2.

46. "Is the End Nigh for GKE?" and "The Dutch Tax Squeeze," *Financial Times: International Coal Report*, No. 483 (September 20, 1999), pp. 3-5.

47. "International Market Report," *King's International Coal Trade*, No. 1242 (September 6, 1999), pp. 11, 13.

48. Organization for Economic Cooperation and Development, *OECD Economic Outlook* (Paris, France, June 1999), Table V.1.

49. United Nations Framework Convention on Climate Change, *Report of the Conference of Parties on Its Third Session, Held at Kyoto from 1 to 10 December 1997, Kyoto Protocol to the United Nations Framework Convention on Climate Change*, FCCC/CP/1997/L.7/Add.1 (December 10, 1997), web site www.unfccc.de.

50. Energy Information Administration, *International Energy Annual 1997*, DOE/EIA-0219(97) (Washington, DC, April 1999), Table 1.4.

51. J. Chadwick, "CIS Coal," *The Mining Journal* (June 1998).

52. PlanEcon, Inc., *Energy Outlook for Eastern Europe and the Former Soviet Republics* (Washington, DC, October 1999), pp. 13-14.

53. World Bank: Public Policy for the Private Sector, "Coal Industry Restructuring in Ukraine, The Politics of Coal Mining and Budget Crises," Note No. 170 (Washington, DC, December 1998), web site www.worldbank.org.

54. World Bank, *Country Brief: Russian Federation* (Washington, DC, August 1999), web site www.worldbank.org.

55. World Bank, *World Bank Support to the Russian Federation*, web site www.worldbank.org (accessed December 28, 1999).

56. World Bank, "World Bank Supports Russia's Reforms," News Release No. 98/1586/ECA (Washington, DC, December 19, 1997), web site www.worldbank.org.

57. Reuters via NewsEdge Corporation, "Russia Wins World Bank Loan, Eyes IMF Cash," web site www.individual.com (December 28, 1999).

58. Energy Information Administration, *Country Analysis Brief: Russia*, web site www.eia.doe.gov (October 1998).

59. "Market Chatter," *King's International Coal Trade*, No. 1238 (August 9, 1999), p. 14.

60. A. Kudat, *Russia Coal Sector Restructuring*, World Bank Social Assessment Case Study (Washington, DC, February 20, 1998), web site www.worldbank.org.

61. "Exports Through Ust Luga To Start Next August," *Financial Times: International Coal Report*, No. 487 (November 15, 1999), p. 9.

62. "Russia Expands Output and Exports and Goes for Price Hike," *Financial Times: International Coal Report*, No. 493 (Fenruary 21, 2000), pp. 17-19.

63. Energy Information Administration, *International Energy Annual 1997*, DOE/EIA-0219(97) (Washington, DC, April 1999), Table 1.4.

64. Energy Information Administration, *International Energy Annual 1997*, DOE/EIA-0219(97) (Washington, DC, April 1999), Tables 5.2, 5.3, and 5.4.

65. S. Spence and Young Consultancy and Research, Ltd., *Coal Trade Matrix 1997* (London, UK, October 19, 1998).

66. PlanEcon Energy Service, "New Program for Restructuring the Polish Coal-Mining Industry: Already Dead on Arrival," FAX Alert No. 8 (April 28, 1998).

67. Associated Press, "Parliament Chamber Approves Coal Mining Restructuring Plan," web site cnnfn.news-real.com (November 26, 1998).

68. Reuters via NewsEdge Corporation, "Poland To Speed Up Mining Lay-Offs," web site www.individual.com (December 22, 1999).

69. World Bank, *Country Brief: Poland* (Washington, DC, August 1999), web site www.worldbank.org.

70. World Bank, "World Bank Supports Restructuring of Hard Coal Industry in Poland," News Release No. 99/2234/ECA (Washington, DC, June 10, 1999), web site www.worldbank.org.

71. Energy Information Administration, *Annual Energy Outlook 2000*, DOE/EIA-0383(2000) (Washington, DC, December 1999), Table A8.

72. Natural Resources Canada, *Canada's Energy Outlook: 1996-2020* (April 1997), pp. 48-50, web site www.nrcan.gc.ca.

73. Natural Resources Canada, *Canada's Emissions Outlook: An "Events-Based" Update for 2010*, Working Paper (October 1998), pp. 3-4, web site www.nrcan.gc.ca/es/ceo/ceo-2010.pdf.

74. F.G. Jáuregui D., Fuels Purchasing Manager, Comisión Federal de Electricidad, "Steam Coal Requirements of CFE," paper presented at the Western Coal Council's Pacific Coal Forum (Park City, UT, June 23-26, 1997).

75. "International Market Report," *King's International Coal Trade*, No. 1194 (September 11, 1998), pp. 2-3.

76. "CFE Takes Second Russian Cargo," *Financial Times: International Coal Report*, No. 489 (December 13, 1999), p. 8.

77. Energy Information Administration, *International Energy Annual 1997*, DOE/EIA-0219(97) (Washington, DC, April 1999), Tables 1.4 and 2.5.

78. International Energy Agency, *Coal Information 1998* (Paris, France, July 1999).

79. Energy Information Administration, *Country Analysis Brief: South Africa*, web site www.eia.doe.gov (February 1999).

80. Energy Information Administration, *Country Analysis Brief: South Africa*, web site www.eia.doe.gov (February 1999).

81. R.W. Lynch, C. Pinkney, L. Feld, E. Kreill, and A.W. Lockwood, "Opportunities for the Power Industry in South Africa," paper presented at the 58th Annual American Power Conference (Chicago, IL, April 11, 1996), web site www.fe.doe.gov.

82. Energy Information Administration, *Country Analysis Brief: South Africa*, web site www.eia.doe.gov (February 1999).

83. "Jorf Prepares 3rd LT Tender," *Financial Times: International Coal Report*, No. 483 (September 20, 1999), p. 13.

84. "CMS Generation Finalizing Moroccan Power Project," *King's International Coal Trade* (March 29, 1996), p. 13.

85. International Energy Agency, *Coal Information 1998* (Paris, France, July 1999), p. III.13.

86. IEA Coal Research, *Coal Prospects in Latin America to 2010*, IEAPER/23 (London, UK, March 1996).

87. "Market Chatter," *King's International Coal Trade*, No. 1226 (April 23, 1999), p. 15.

88. "Tractebel To Tender for Peru," *Financial Times: International Coal Report*, No. 478 (July 12, 1999), p. 12.

89. Energy Information Administration, *Regional Indicators: Central America*, web site www.eia.doe.gov (June 1999), Table 2.

90. IEA Coal Research, *Coal Prospects in Latin America to 2010*, IEAPER/23 (London, UK, March 1996), p. 65.

91. Energy Information Administration, *Country Analysis Brief: Guatemala*, web site www.eia.doe.gov (March 1999).

92. TECO Power Services, "TECO Power Services Dedicates Guatemala's Largest Generating Unit," News Release (Tampa, FL, December 14, 1999), web site www.tecoenergy.com.

93. Energy Information Administration, *International Energy Annual 1997*, DOE/EIA-0219(97) (Washington, DC, April 1999), Tables 2.5 and 5.4.

94. International Energy Agency, *Coal Information 1998* (Paris, France, July 1999), p. II.21.

95. Energy Information Administration, *International Energy Annual 1997*, DOE/EIA-0219(97) (Washington, DC, April 1999), Table 1.4.

96. International Energy Agency, *Coal Information 1998* (Paris, France, July 1999), pp. III.126.

97. "Market Chatter," *King's International Coal Trade*, No. 1240 (August 23, 1999), p. 16.

98. IEA Coal Research, *Non-OECD Coal-Fired Power Generation—Trends in the 1990s*, CCC/18 (London, UK, June 1999), p. 21.

99. Energy Information Administration, *International Energy Annual 1997*, DOE/EIA-0219(97) (Washington, DC, April 1999), Table 2.5.

100. Energy Information Administration, Office of Energy Markets and End Use, International Statistics Database.

101. "International Market Report," *King's International Coal Trade*, No. 1217 (February 19, 1999), p. 11.

102. SSY Consultancy and Research, Ltd., "Seaborne Coal Trade," *Monthly Shipping Review* (January 10, 2000), p. 5.

103. Energy Information Administration, AEO2000 National Energy Modeling System run AEO2K. D100199A (Washington, DC, October 1999).

104. International Energy Agency, *Coal Information 1998* (Paris, France, July 1999), Table 4.2.

105. International Energy Agency, *Coal Information 1998* (Paris, France, July 1999), p. I.189.

106. "Panel To Urge Technology Transfer at Two Coal Mines," Kyodo News Service/Associated Press, web site www.industrywatch.com (July 1, 1999).

107. International Energy Agency, *International Coal Trade: The Evolution of a Global Market* (Paris, France, January 1998).

108. B. Jacques, "High Turnover, Low Returns," *Financial Times* (July 8, 1996), p. 1.

109. Energy Information Administration, *Annual Prospects for World Coal Trade 1991*, DOE/EIA-0363(91) (Washington, DC, June 1991), Table 1.

110. International Energy Agency, *Coal Information 1998* (Paris, France, July 1999), p. I.132.

111. International Energy Agency, *Coal Information 1998* (Paris, France, July 1999), p. III.13.

112. Energy Information Administration, AEO2000 National Energy Modeling System run AEO2K. D100199A (Washington, DC, October 1999).

113. Energy Information Administration, *Annual Prospects for World Coal Trade 1987*, DOE/EIA-0363(87) (Washington, DC, May 1987), Tables A2 and A3.

114. Energy Information Administration, *Coal Data: A Reference*, DOE/EIA-0064(93) (Washington, DC, February 1995), pp. 33-35.

115. World Coal Institute, "Coal in the Steel Industry," web site www.wci-coal.com (1998).

116. International Energy Agency, *Coal Information 1998* (Paris, France, July 1999), Tables 3.9 and 3.10.

Nuclear Power

In the IEO2000 reference case, nuclear power represents a declining share of the world's total electricity consumption from 1997 through 2020. Plant retirements are expected to produce net reductions in nuclear capacity in most of the industrialized nations.

In 1998, a total of 2,291 billion kilowatthours of electricity was generated by nuclear power worldwide, providing 16 percent of the world's total generation[1]. Among the countries with operating nuclear power plants, national dependence on nuclear energy for electricity varies greatly (Figure 68). Nine countries met at least 40 percent of total electricity demand with generation from nuclear reactors.

The prospects for nuclear power to maintain a significant share of worldwide electricity generation are uncertain, despite projected growth of 2.5 percent per year in total electricity demand through 2020. Over the long term, only the developing nations are projected to have continuous growth in nuclear power capacity through 2020 (Figure 69). Most regions have some increase in capacity through 2010, followed by declines. Countries that are operating older reactors and have other, more economical options for new generating capacity are expected to let their nuclear capacity fade as current nuclear units are retired.

In the *IEO2000* reference case, worldwide nuclear capacity is projected to increase from 349 gigawatts in 1998 to 368 gigawatts in 2010, then begin to decline, falling to 303 gigawatts in 2020. Aggressive plans to expand nuclear capacity, mainly in the Far East, lead to the near-term increase, but plant retirements in the United States and other countries, exceeding total new additions worldwide, produce a decline later in the forecast. Developing Asian countries are projected to add 30 gigawatts of new nuclear capacity by 2020, whereas the industrialized nations overall lose 64 gigawatts. Nuclear consumption in the reference case remains flat through 2010 before starting to fall, but it represents a declining share of electricity consumption throughout the forecast (Figure 70).

Figure 68. Nuclear Shares of National Electricity Generation, 1998

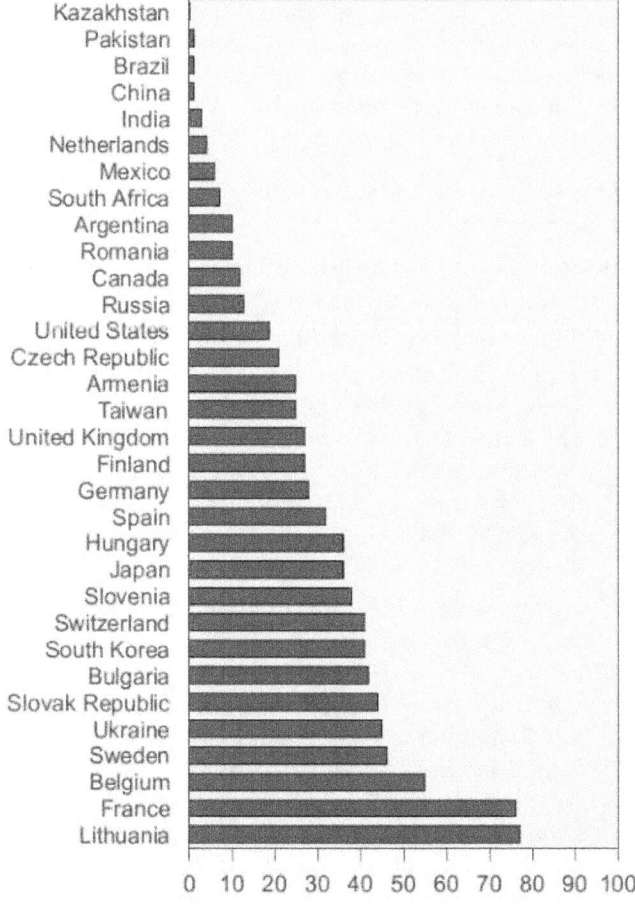

Source: International Atomic Energy Agency, *Nuclear Power Reactors in the World 1998* (Vienna, Austria, April 1999).

Figure 69. World Nuclear Capacity by Region, 1998-2020

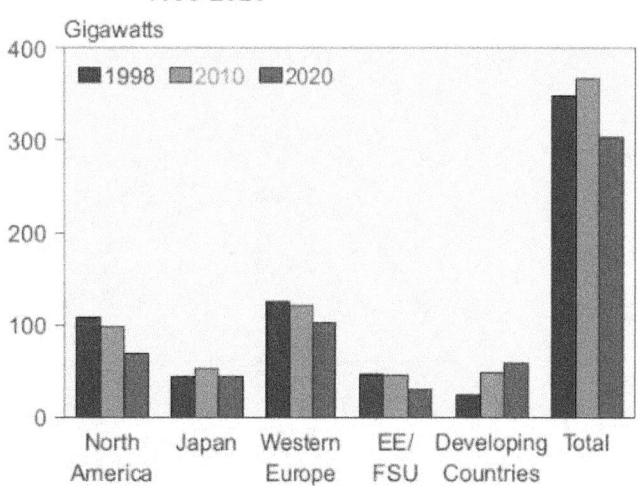

Sources: **History:** International Atomic Energy Agency. *Nuclear Power Reactors in the World 1998* (Vienna, Austria, April 1999). **Projections:** Based on detailed assessments of country-specific nuclear power programs.

Figure 70. World Nuclear and Total Electricity Consumption, 1970-2020

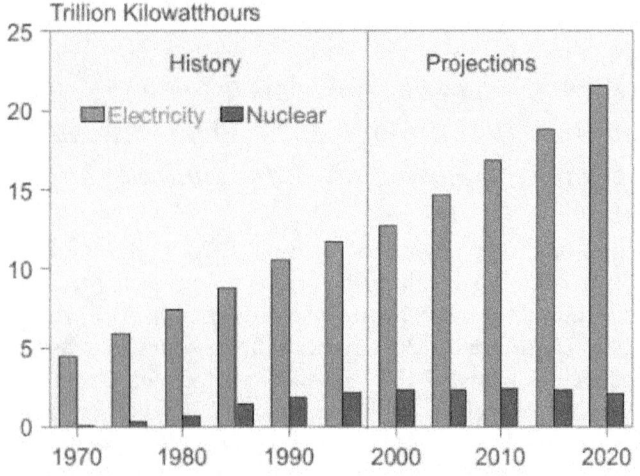

Sources: **History:** International Atomic Energy Agency. *Nuclear Power Reactors in the World 1998* (Vienna, Austria, April 1999). **Projections:** Based on detailed assessments of country-specific nuclear power programs.

Three nuclear capacity scenarios were developed for *IEO2000*, to provide a range of outcomes reflecting the uncertainty surrounding future investment in nuclear technology (Figure 71 and Table 19). The reference case reflects a continuation of present trends; the low and high cases present more pessimistic and optimistic views of the future of the nuclear power industry. For the United States, the reference case assumes that nuclear plants will operate as long as their operation is more economical than building new capacity, with nuclear generation costs increasing over time as aging plants become more expensive to maintain and operate. In the United States, 13 units are expected to be retired

Figure 71. World Nuclear Capacity in Three Cases, 1970-2020

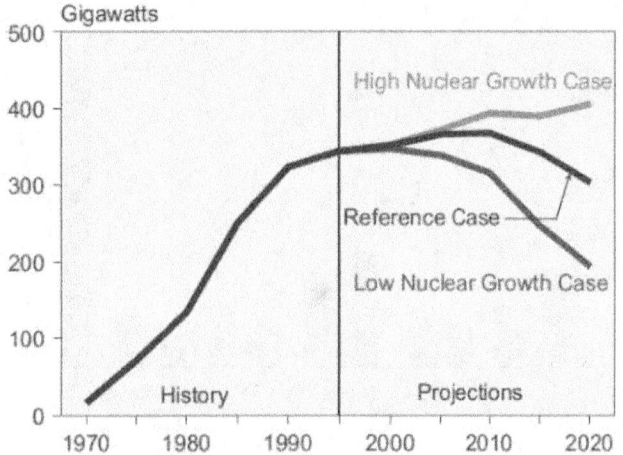

Sources: **History:** International Atomic Energy Agency. *Nuclear Power Reactors in the World 1998* (Vienna, Austria, April 1999). **Projections:** Based on detailed assessments of country-specific nuclear power programs.

before their operating licenses expire, and 11 units are expected to obtain license renewals that will extend their operation. For foreign nuclear projections, the reference case takes into account announced schedules for completion of units under construction and any announced retirement dates. Also considered are political environments, national energy plans, construction management experience, and financial conditions. Complete country-by-country listings of the nuclear capacity projections for the reference, low, and high nuclear growth cases are provided in Appendix A.

The low nuclear growth case projects a more significant decline in nuclear capacity orders, as well as additional retirements of existing units. In the United States, reactors are assumed to face higher aging-related expenses, leading to more early retirements. The forecast for worldwide capacity in 2020 is 193 gigawatts, a 45-percent decline from current capacity.

The high growth case reflects a slight revival for the nuclear power industry, with net capacity growth of 2.6 gigawatts per year over the forecast period. In the United States, the high growth case assumes limited aging effects, with more reactors operating after license renewals. The high growth projections for other countries are based on assumptions that construction times for new units will be shorter, and that provisions will be made to extend the operating lives of existing units beyond current retirement dates.

Some key developments affecting the nuclear power industry in 1999 include:

- **Safety issues moved to the forefront in Asia after several leaks at nuclear power plants and an accident at a fuel reprocessing plant in Japan.**

  - In July, a cracked pipe leaked 51 tons of coolant water from the Tsuruga 2 nuclear plant in Japan. The incident was classified as Level 1 (anomaly) on the International Nuclear Event Scale (INES), but the plant remains down while inspections take place, and plants of similar design are also being inspected [2].

  - A more serious accident occurred on September 30 at a uranium reprocessing facility in Tokaimura, Japan, where 69 people received significant radiation exposure. The accident occurred when a self-sustaining nuclear chain reaction was started, caused by workers violating procedures. The two workers involved in the operation were most severely affected, receiving large doses of radiation [3]. One died on December 21, after having been in the hospital in critical condition since the accident. This Tokaimura criticality event is ranked as the third most serious accident—behind Three Mile Island and Chernobyl—in the history of the nuclear electricity supply industry.

Table 19. Historical and Projected Operable Nuclear Capacities by Region, 1996-2020
(Net Gigawatts)

Region	1997[a]	1998[b]	2000	2005	2010	2015	2020
Reference Case							
Industrialized	**283.6**	**278.6**	**279.9**	**277.4**	**272.7**	**247.2**	**214.1**
United States	99.0	97.1	97.5	93.4	84.1	67.4	57.0
Other North America	13.3	11.6	11.6	13.7	13.7	13.7	11.4
Japan	43.9	43.7	43.7	44.5	53.5	49.3	43.8
France	62.9	61.7	63.1	62.9	62.9	62.9	60.0
United Kingdom	13.0	13.0	13.0	13.0	11.9	11.6	10.6
Other Western Europe	51.5	51.6	51.1	50.0	46.7	42.4	31.4
EE/FSU	**46.3**	**46.6**	**45.3**	**52.0**	**46.3**	**41.1**	**30.4**
Eastern Europe	9.8	10.2	10.6	13.1	11.4	10.6	9.6
Russia	19.8	19.8	19.8	22.7	19.8	17.0	11.9
Ukraine	13.8	13.8	12.1	13.1	13.1	13.1	8.6
Other FSU	2.8	2.8	2.7	3.1	1.9	0.4	0.4
Developing	**22.0**	**23.7**	**26.2**	**36.8**	**48.8**	**54.7**	**58.7**
China	2.2	2.2	2.2	6.6	10.5	12.9	17.6
South Korea	9.8	11.4	13.0	14.9	16.8	16.2	16.2
Other	10.1	10.1	11.0	15.3	21.5	25.5	24.9
Total World	**351.9**	**348.9**	**351.4**	**366.1**	**367.8**	**343.0**	**303.3**
Low Growth Case							
Industrialized	**283.6**	**278.6**	**277.4**	**260.2**	**234.0**	**179.9**	**143.1**
United States	99.0	97.1	96.7	85.4	72.5	53.5	43.7
Other North America	13.3	11.6	11.6	11.6	11.6	7.6	3.3
Japan	43.9	43.7	43.7	43.8	39.6	30.2	28.1
France	62.9	61.7	62.9	62.9	62.9	55.7	51.1
United Kingdom	13.0	13.0	13.0	12.3	11.4	8.1	6.6
Other Western Europe	51.5	51.6	49.6	44.2	36.0	24.9	10.3
EE/FSU	**46.3**	**46.6**	**44.0**	**43.5**	**40.7**	**27.2**	**14.3**
Eastern Europe	9.8	10.2	10.2	10.8	10.6	8.3	5.8
Russia	19.8	19.8	19.8	19.2	17.4	10.4	4.7
Ukraine	13.8	13.8	11.2	11.9	12.7	8.6	3.8
Other FSU	2.8	2.8	2.7	1.6	0.0	0.0	0.0
Developing	**22.0**	**23.7**	**25.8**	**34.9**	**41.5**	**39.4**	**35.8**
China	2.2	2.2	2.2	6.6	8.6	8.6	8.6
South Korea	9.8	11.4	13.0	14.9	14.3	15.0	13.2
Other	10.1	10.1	10.7	13.4	18.6	15.8	14.0
Total World	**351.9**	**348.9**	**347.2**	**338.6**	**316.2**	**246.6**	**193.2**
High Growth Case							
Industrialized	**283.6**	**278.6**	**279.9**	**280.4**	**286.6**	**273.7**	**273.3**
United States	99.0	97.1	97.5	95.1	90.2	79.7	71.1
Other North America	13.3	11.6	11.6	13.7	15.4	15.4	15.4
Japan	43.9	43.7	43.7	44.5	54.2	55.1	61.0
France	62.9	61.7	63.1	63.1	62.9	62.9	67.2
United Kingdom	13.0	13.0	13.0	13.0	12.6	12.1	11.9
Other Western Europe	51.5	51.6	51.1	51.1	51.4	48.6	46.7
EE/FSU	**46.3**	**46.6**	**45.3**	**52.0**	**51.7**	**50.6**	**49.3**
Eastern Europe	9.8	10.2	10.6	13.1	12.3	11.4	11.4
Russia	19.8	19.8	19.8	22.7	23.3	21.9	20.6
Ukraine	13.8	13.8	12.1	13.1	13.1	13.7	14.3
Other FSU	2.8	2.8	2.7	3.1	3.1	3.5	3.0
Developing	**22.0**	**23.7**	**27.4**	**39.8**	**55.1**	**65.7**	**83.2**
China	2.2	2.2	2.2	6.6	10.5	12.9	20.7
South Korea	9.8	11.4	13.0	15.8	16.8	17.7	21.9
Other	10.1	10.1	12.2	17.3	27.9	35.0	40.6
Total World	**351.9**	**348.9**	**352.6**	**372.1**	**393.5**	**390.0**	**405.8**

[a]Status as of December 31, 1997.
[b]Status as of December 31, 1998. Data are preliminary and may not match other EIA sources.
Notes: EE/FSU = Eastern Europe/Former Soviet Union. Totals may not equal sum of components due to independent rounding.
Sources: **United States:** Energy Information Administration, *Annual Energy Outlook 2000*, DOE/EIA-0383(2000) (Washington, DC, December 1999). **Foreign:** Based on detailed assessments of country-specific nuclear power programs.

- Just one week later, a minor event at the Wolsong 3 reactor in South Korea drew attention. A small amount of heavy water was spilled during a normal maintenance outage, due to a damaged seal on a pump being readied for maintenance [4]. The incident was ranked as below Level 1, but it still raised concerns after the events in Japan.

- Finally, Chinese nuclear officials recently admitted that an accident 1 year ago at the country's first domestically designed and built reactor, Qinshan 1, left it off line for the year as repairs were made [5]. Officials say no one was hurt and there was no leak of radiation.

These reports are likely to cause further public concern about the aggressive plans for nuclear capacity expansion in the Far East.

• **Competition in the U.S. electric industry led to sales of existing nuclear plants.** The first sale of an existing U.S. nuclear plant was completed in 1999 when Boston Edison's parent company sold the Pilgrim nuclear plant to Entergy [6]. The 670-megawatt boiling water reactor (BWR), which has operated for 27 years, was sold for $81 million. Entergy has also entered discussions with New York Power Authority regarding the purchase of the Fitzpatrick and Indian Point 3 nuclear plants [7]. AmerGen, a U.S.-British joint venture, is close to completing its purchase of Three Mile Island 1 and has several other purchases under way, including Clinton—which had been out of service for more than 2 years but is now operable—and Oyster Creek 1, one of the highest cost plants in the United States[8]. Plant sales could lead to a consolidation of the U.S. nuclear electricity industry, with a few large companies owning and operating a large number of plants. As a result, better management could lower costs and make nuclear plants more competitive in the deregulated electricity industry. At the same time, the potential market power of large nuclear generating companies could raise concerns about the pricing of nuclear electricity.

• **Repairs were required at France's newest nuclear reactors.** In mid-1998 a leak in the residual heat removal (RHR) system was discovered at Civaux 1, just 6 months after the reactor had been connected to the French power grid. The plant was the third of the new N4 design to be connected to the grid; Civaux 2, the fourth and final unit, had been scheduled to be online by the end of 1998. Instead, the discovery of further fatigue-induced cracking in the RHR system at Civaux 1 led to the redesign and replacement of sections of the pipework on all four N4 reactors. The first two of the series—Chooz B1 and Chooz B2—had

been restarted by mid-1999, and Civaux 1 and 2 were connected to the grid just before the year's end. All four reactors will operate under a special regime designed to minimize thermal stresses, and further inspection will be required after one full cycle of operation [9, 10].

• **Nuclear issues dominated accession talks in the European Union.** The European Commission began negotiations with six candidate member countries in 1998 and more recently proposed to open negotiations with six additional countries in 2000 [11]. Enlargement of the European Union is meant to further the integration of the continent by peaceful means, extending a zone of stability and prosperity to new members. The safety of older, Soviet-designed nuclear reactors in some candidate countries—in particular, Bulgaria, Lithuania, and the Slovak Republic—dominated the discussions, however, and the Commission made further negotiations contingent on commitments to close the reactors. Lithuania agreed to close Ignalina 1 by 2005 and is expected to close Ignalina unit 2 by 2009. The Slovak government proposed to close the first two units at Bohunice in 2006 and 2008. Most recently, Bulgaria agreed to shut down four Soviet-built reactors at Kozloduy between 2003 and 2006. The countries will receive financial aid from the European Commission and the European Bank for Reconstruction and Development to go toward the decommissioning efforts [12].

• **Debate continued in Germany over phaseout schedules for nuclear plants.** During its first year in power, the new coalition of Social Democrats and Greens made little progress in its promise to begin to shut down Germany's 19 nuclear reactors. Discussions with the utilities to negotiate a limit on the plants' lifetimes failed to produce an agreement [13]. The primary disagreement was over what should be the appropriate lifetime for the reactors and when the first plants would be shut down. The Greens are politically committed to shutting at least one or two reactors during the current government, which means by 2002. For this to happen, the utilities must agree to retire reactors after a 30-year lifetime, but they are refusing to negotiate anything less than a 35-year lifetime. A further complication is the transport of spent nuclear fuel, which has been blocked since 1998, requiring all spent fuel from reactors to be stored on site. Several reactors may have to be shut down indefinitely in 2000, when their storage pools reach full capacity. Resolution of the spent fuel transportation issue is also part of the debate between government and the utilities [14].

Regional Overview

Developing Asia

Countries in developing Asia currently operating nuclear power plants include China, South Korea, Taiwan, India, and Pakistan. All expect some growth in the future. At the end of 1998, these five countries had 20.3 gigawatts of nuclear capacity on line. By 2020, nuclear capacity in the region is projected to be between 32.8 and 70.8 gigawatts, including at least one nuclear unit in North Korea, and—in the high growth case—new programs in Indonesia, Thailand, the Philippines, and Vietnam.

The Asian economic crisis of the late 1990s led to financing concerns, lower electricity demand, and delays in some new orders for nuclear power plants in Asia's developing countries. Recovery is expected for the region's national economies, with an eventual return to baseline projections for long-term economic and energy growth. South Korea, currently the largest operator of nuclear power in the region with 14 operable units totaling 11.4 gigawatts of generating capacity, is projected to have between 13.2 and 21.9 gigawatts on line by 2020. China is projected to have at least 8.6 gigawatts of nuclear capacity operating by 2020—four times the current capacity; and in the high nuclear growth case, China's nuclear capacity is projected to reach almost ten times its current level by 2020.

Two new units began commercial operation in South Korea during 1998. Wolsong 3, a 650-megawatt pressurized heavy water reactor (PHWR), was completed in 1998, and the fourth and final unit at the Wolsong site is near completion. The Wolsong units were built and designed by Atomic Energy of Canada Limited (AECL), but all the other units under construction are of the designated Korean standardized design. At the Ulchin site, two new standardized pressurized water reactors (PWRs), units 3 and 4, were completed in 1998, although unit 4 was not commercially operable until 1999. The economic crisis caused a steep drop in electricity demand in South Korea in 1998, which may lead to revised supply decisions for the future. The new president of the Korea Electric Power Corporation, the second in just over a year, is expected to postpone a decision on new nuclear orders as he focuses first on the restructuring of Korea's electricity sector [15].

China also has ambitious plans to build additional nuclear power plants to meet rapid growth in electricity demand. The next two units at the Qinshan site, 600-megawatt PWRs of a Chinese design, are under construction. Two additional 700-megawatt PHWR units supplied by Atomic Energy of Canada Limited are planned for the site. Construction has also started on two French-designed PWRs at Lingao. In addition,

Russia and China have signed a contract for two 1,000-megawatt units based on a modernized Russian design [16], which will be built in Jiangsu Province on China's northeastern seaboard. Previous announcements of even further nuclear expansion in the near future are not likely to be fulfilled. China's tenth 5-year plan, due in early 2000, is expected to reflect a slowdown for nuclear construction. In May, officials announced that no new nuclear orders will be placed in the next 3 years, because demand growth has not met expectations [17].

In India, construction of Kaiga 2, a 220-megawatt PHWR, was completed in 1999, and the unit was connected to the grid by the end of the year. The project was delayed after the collapse of part of the concrete containment facility at unit 1 in 1994, while the unit was under construction. Designs were altered for both units, and Kaiga 1 is expected to be on line in 2000. Two other PHWRs of the same design, Rajasthan 3 and 4, were also delayed for the accident investigation and are expected to be completed in 2000 [18]. In Pakistan, a Chinese designed PWR, the nation's second nuclear power unit, is expected to be operable in early 2000.

In Taiwan, two 1,350-megawatt advanced boiling water reactors (ABWRs) are under construction at the Lungmen power station. Currently, there are no plans for any further nuclear investment in Taiwan, and there is strong public opposition to nuclear power, as well as concern about a site for nuclear waste storage and disposal. Most officials expect that, given its size, Taiwan will not build further nuclear capacity after Lungmen.

Other Developing Countries

Other developing countries that currently operate nuclear power plants are Argentina, Brazil, and South Africa. Countries with the potential to have nuclear programs in place by 2020 include Cuba, Iran, Egypt, and Turkey. Argentina's two nuclear units provided 10 percent of the country's electricity in 1998, and Brazil's one nuclear unit supplied 1 percent of its total electricity generation. South Africa has two nuclear units currently operable, which provided 7 percent of the country's electricity generation in 1997. Argentina has one unit under construction, and Brazil has two units in the construction pipeline. Given the uncertainties in the region, there is a wide range of possible operable capacity (3.0 to 12.4 gigawatts) by 2020.

Brazil began final testing of Angra 2 and expects full operation in 2000. Construction on Angra 2 began in 1976 but was halted from 1989 to 1996 by technical, financial, and political problems. A third unit was also under construction, but no decision has been made about its completion [19]. The expected privatization of Brazil's electric power industry could make the project uneconomical.

Most of the other developing countries do not have the capital for large nuclear programs and are likely to need financial and technical assistance to undertake nuclear power construction. Russia has agreed to complete two units for Iran, at the Bushehr site, where construction was started in the 1970s but halted after Iran's 1979 revolution [20]. Russia has also formed a joint venture with Cuba to complete the Juragua plant, where construction was halted in 1992 after the breakup of the Soviet Union, when Russia said it was unable to fund the completion [21].

The South African utility Eskom has developed a half-scale model of a new nuclear design. The model achieved criticality during 1999, and plans are underway to build a full-scale prototype over the next few years [22]. The new Pebble Bed Modular Reactor (PBMR) is a high-temperature gas-cooled reactor with fuel elements in the form of graphite "pebbles"—a design that is described as inherently safe. The full-scale 115-megawatt reactor is designed to provide power to remote areas, and to create a potential export business for Eskom.

Industrialized Asia

In the industrialized countries of Asia, only Japan has a well-established nuclear program, with 53 units totaling 43.7 gigawatts of operable capacity at the end of 1998. Japan's nuclear share of electricity in 1998 was 36 percent. In March of 1998, Japan's first nuclear unit—Tokai 1, a 159-megawatt gas-cooled reactor (GCR)—was permanently shut down. Because of its unique design (as the only GCR operating in Japan) and small size, the unit was no longer considered economical to continue operating.

Japan has ambitious plans for further nuclear expansion, mainly to help achieve energy independence and to limit emissions of greenhouse gases. However, the uncertainties surrounding long-term capital markets in Asia, as well as public opposition to nuclear power in Japan, will affect new construction decisions. Public support for the industry is faltering in the wake of several incidents at nuclear reactors and a significant accident at a uranium reprocessing facility, where two workers received large doses of radiation, resulting in one death [23]. The accident, which occurred in Tokaimura on September 30, 1999, could contribute to the public's lack of trust in the management of nuclear facilities in general.

In the *IEO2000* reference case, Japan's nuclear capacity is projected to increase by nearly 10 gigawatts to 53.5 gigawatts by 2010, then decline between 2010 and 2020 to 43.8 gigawatts. The capacity forecast for 2020 ranges from 28.1 gigawatts to 61.0 gigawatts. Japan's current expansion plan includes nine units in the construction

pipeline at the end of 1998. The reference case assumes their completion by 2010, but after 2010 retirements exceed new construction. The low nuclear growth case assumes that most of the current construction plans will be deferred, with no new construction completed by 2020.

Western Europe

Western Europe relies heavily on nuclear power to meet electricity demand. In 1998, nuclear generation from Western European countries produced 36 percent of worldwide nuclear generation. In France and Belgium, 76 and 55 percent, respectively, of the national demand for electricity was supplied from nuclear power plants. The overall trend in Western Europe, however, is away from nuclear power builds. Most of the countries in the region have frozen all nuclear construction plans. In the reference case, only France is projected to have new nuclear capacity on line between 1998 and 2020, and Switzerland sees a small increase in capacity as a result of upgrades to existing units.

France permanently shut down its biggest fast breeder reactor, Superphenix, in 1998 [24]. The reactor had had technical problems and generated electricity only sporadically since coming on line in the mid-1980s. The government agreed to downgrade the reactor to research for waste processing in 1994 and now has agreed to close it permanently, using only an older and smaller breeder reactor for research.

The discovery of cracked piping in one of France's newest N4 design nuclear units, Civaux 1, led to service outages for all four of the new N4 reactors (Chooz B1 and B2 and Civaux 1 and 2) during parts of 1998 and 1999. Repairs were made after a redesign of the affected section of piping, and Chooz B1 and B2 were on line by mid-1999. The newer units at Civaux were also connected to the grid by the end of the year. Civaux 2, which was still under construction when the initial cracks were found, became operable for the first time. Future plans in France are uncertain. The current prime minister promised that no decision would be made on a new generation of nuclear units without a large scientific and democratic debate. A decision on new orders is not expected for 4 or 5 years [25].

Finland continues to pursue the prospect of building its fifth nuclear reactor. The utility Teollisuuden Voima Oy (TVO), which operates two BWRs in Finland, will prepare the official application, which will present both PWR and BWR alternatives. Fortum, operator of the other two units in Finland, will provide technical assistance [26].

In Switzerland, a variety of organizations worked together to get enough signatures to ensure future votes on two antinuclear issues [27]. The first seeks to decommission all five of the country's reactors after 30 years of operation. The second proposes that the current moratorium on new nuclear construction, which expires in 2000, be extended for another 10 years. If the petitions are accepted, they will be put to a national vote within the next 2 years.

After a year in power, the new German government has made little progress toward the promised phased shutdown of the country's 19 nuclear stations, which currently provide almost 30 percent of its electricity. Negotiations with the utilities that own the plants have not been successful in setting a schedule for the reactor retirements. In Sweden, almost 20 years after a national vote to phase out nuclear power, an agreement was finally reached to shut down the first unit, Barsebaeck 1, at the end of 1999 [28]. The owner, private utility Sydkraft, fought to keep it open but eventually agreed to a compensation plan in return for the early loss of unit 1.

North America

The United States, Canada and Mexico all have nuclear power programs, but most of the nuclear capacity in the region is in the United States. In 1998, the nuclear share of electricity in the United States was 19 percent. In Canada the nuclear share of electricity generation was 12 percent, and Mexico's two units supplied 6 percent of its electricity. Total nuclear capacity in the region is expected to decline over the forecast in all cases, as existing units age and are removed from service. With no new orders being placed, nuclear plant retirements by 2020 cut U.S. nuclear capacity by 41 percent from the 1998 total. In Canada, several older units currently off line are assumed to return to service during the forecast; however, they are projected to be retired permanently by 2020. The *IEO2000* projections reflect nuclear capacity retirements of between 22.3 and 61.8 gigawatts in the region.

Ontario Hydro (OH), the operating utility for the majority of Canada's nuclear units, shut down seven of its oldest units—three at the Pickering A site (where a fourth unit has been dormant since 1996) and four units at Bruce A. The units may be refurbished and brought back on line eventually.

In the United States, Commonwealth Edison permanently closed its two Zion units, and the Millstone 1 unit was retired in Connecticut, all during 1998. No units were shut down during 1999, and several reactors that had been down for extended maintenance outages were returned to service, including the other two units at the Millstone plant site. In Illinois, the Clinton plant—which owners were considering shutting permanently—was brought back on line and sold to AmerGen Energy

Company. AmerGen has deals underway to purchase at least five plants in the United States [29].

With several companies aggressively pursuing the purchase of nuclear plants, owners of older units may now consider selling rather than shutting down early if concerns about future costs and competitiveness arise. The sale of a nuclear unit in the United States is a complicated process, requiring approval from public utility commissions and the U.S. Nuclear Regulatory Commission as well as favorable rulings on tax issues from the Internal Revenue Service. As more plants go through the process, however, it should become more straightforward.

The companies now trying to buy U.S. nuclear plants have substantial experience managing other reactors. They expect to improve operations and lower costs to make the plants good investments in a competitive environment. They may also be gambling on the prospects for future rewards for generation that does not produce greenhouse gases. The sellers of the plants relieve themselves of the risks and uncertainties surrounding future operating costs as well as the burden of ultimately decommissioning the plants.

Eastern Europe/Former Soviet Union

In the EE/FSU region during 1998, 70 nuclear units (46.6 gigawatts of generating capacity) produced 241.9 billion kilowatthours of electricity; almost 75 percent of the electricity in the region from nuclear plants was generated in the FSU. Reliance on nuclear power varies in the region, with Lithuania supplying 77 percent of its electricity from nuclear power, Russia 13 percent, and Kazakhstan less than 1 percent. Several countries in the region have ambitious plans for additional nuclear capacity, but a number of challenges are likely to limit new nuclear builds. With the potential for future projects uncertain, the region's nuclear capacity is projected to decline 16.2 gigawatts by 2020. The low and high nuclear growth cases project a loss of 32.3 gigawatts and a gain of 2.7 gigawatts by 2020, respectively.

Construction of Mochovce 1 in the Slovak Republic was completed in 1998, and the second unit was in final testing stages in late 1999 [30]. The original Soviet design was upgraded to modern safety standards with the help of Western European companies. Kazakhstan's one unit, a small fast breeder reactor, was shut down in April 1999, and is now being prepared for decommissioning [31]. Officials state that the closure decision is for economic rather than technical reasons. The republic is developing plans for a future nuclear power program, probably based on Russia's new small PWR design. Kazakhstan is also developing its large reserves of uranium.

Many of the countries in the EE/FSU region are negotiating to become part of the European Union (EU). As a result of concerns about maintaining nuclear safety within the EU, Bulgaria, Lithuania, and the Slovak Republic have agreed in their negotiations to shutdown schedules for specific units. The Austrian government, with a long history of opposing nuclear power throughout Europe, has been pushing aggressively to require nuclear shutdowns in return for continued negotiations for accession to the EU. Austria is still fighting for units in the Czech Republic, Hungary, and Slovenia to be either modernized or shut down permanently.

References

1. International Atomic Energy Agency, *Nuclear Power Reactors in the World 1998* (Vienna, Austria, April 1999).

2. "JAPC Provides Tsuruga-2 Leak Findings to the NSC," *Nuclear News*, Vol. 42, No. 11 (October 1999), p. 42.

3. "Nuclear Safety and the Human Factor," *Power in Asia* (October 1999), pp. 4-6.

4. "Heavy-Water Spill at Wolsong-3 Ranked As Anomaly," *Nuclear News*, Vol. 42, No. 12 (November 1999), p. 54.

5. "Chinese Nuclear Accident Revealed," BBC News, Online Network (July 5, 1999), web site http://news.bbc.co.uk.

6. P. Howe, "Pilgrim Deal to Entergy Completed: $81m Sale of Nuclear Plant Is the First in US History," *Boston Globe* (July 14, 1999), p. E1.

7. "NYPA Discussing Nuclear Plant Sale, Representing 1,900 MW, With Entergy," *Electric Utility Week* (November 8, 1999), p. 12.

8. British Energy Company, press releases dated June 24, 1999, October 18, 1999, December 16, 1999, and December 21, 1999, web site www.british-energy.co.uk/media/fr_press_releases.html.

9. A. MacLachlan, "EdF Discovers More RHR Cracking at Civaux, Plans New N4 Repairs," *Nucleonics Week*, Vol. 39, No. 50 (December 10, 1998), p. 1.

10. "Late News," *Nuclear News*, Vol. 42, No. 5 (April 1999), p. 59.

11. "EU Bullish on Accession Despite Nuclear Obstacles," *Financial Times Energy*, No. 97 (October 1999), p. 1.

12. A. MacLachlan and A. Saines, "EU Candidate Countries To Get EC Aid in Shutting Old Reactors," *Nucleonics Week*, Vol. 40, No. 46 (November 18, 1999), p. 11.

13. M. Hibbs, "German Utilities Won't Negotiate Without Three Major Concessions," *Nucleonics Week*, Vol. 40, No. 31 (August 5, 1999), p. 5.

14. M. Hibbs, "Six Reactors, Their Pools Full, May Close in 2000, BMU Reports," *Nucleonics Week*, Vol. 40, No. 38 (September 23, 1999), pp. 5-7.

15. M. Hibbs, "New KEPCO CEO Expected To Delay New Reactor Orders Until 2000," *Nucleonics Week*, Vol. 40, No. 26 (July 1, 1999), pp. 1-2.

16. "China To Get Two Russian Reactors," *Power in Asia* (January 1998), p. 26.

17. M. Hibbs, "China Confirms Zhu Will Slow Nuclear Buildup in Tenth Plan," *Nucleonics Week*, Vol. 40, No. 19 (May 13, 1999), p. 1.

18. N. Patri, "Kaiga-2 Finally Goes Critical, Marking Project Accident Recovery," *Nucleonics Week*, Vol. 40, No. 40 (October 7, 1999), p. 4.

19. A. Schmid, "Angra-2 Readied for Fuel Load, Brazil May Revive Angra-3 Work," *Nucleonics Week*, Vol. 40, No. 39 (September 30, 1999), pp. 9-10.

20. "Bushehr Completion by 2002, With Russian Assistance," *Nuclear News*, Vol. 42, No. 1 (January 1999), p. 49.

21. "Juragua Reconsidered—Again," *Power in Latin America*, No. 49 (July 1999), pp. 17-18.

22. A. MacLachlan, "Pebble Bed Model Goes Critical, Moving Eskom Closer to Decision," *Nucleonics Week*, Vol. 40, No. 29 (July 22, 1999), p. 1.

23. "Nuclear Safety and the Human Factor," *Power in Asia* (October 1999), pp. 4-6.

24. "France To Close Nuclear Reactor Superphenix," Itar-Tass News Wire (February 3, 1998).

25. A. MacLachlan, "Jospin Pledges Extensive Debate on Nuclear's Future in France," *Nucleonics Week*, Vol. 40, No. 35 (September 2, 1999), pp. 6-7.

26. A. Sains, "TVO Is Chosen To Lead Project To Build Fifth Finnish Reactor," *Nucleonics Week*, Vol. 40, No. 35 (September 2, 1999), pp. 1-2.

27. L. Pilarski, "New Swiss Anti-Nuclear Referenda Get Enough Signatures To Get Vote," *Nucleonics Week*, Vol. 40, No. 35 (September 2, 1999), pp. 8-9.

28. A. Sains, "Sydkraft Shuts Barsebaeck-1 as Sweden Agrees to Payment," *Nucleonics Week*, Vol. 40, No. 48 (December 2, 1999), p. 1.

29. British Energy Company, press releases dated June 24, 1999, October 18, 1999, December 16, 1999, and December 21, 1999, web site www.british-energy.co.uk/media/fr_press_releases.html.

30. "Slovakia: Mochovce-2 Fuel Load Begins," *Nucleonics Week*, Vol. 40, No. 40 (October 7, 1999), p. 18.

31. "Kazakhstan: BN-350 Fast Reactor To Be Decommissioned," *Nuclear News*, Vol. 42, No. 9 (August 1999), p. 97.

Hydroelectricity and Other Renewable Resources

The renewable energy share of total world energy consumption is expected to continue at a level of about 8 percent from 1997 through 2020, despite a projected 54-percent increase in consumption of hydroelectricity and other renewable resources.

The development of renewable energy sources is constrained in the *International Energy Outlook 2000* (*IEO2000*) reference case projections by expectations that fossil fuel prices will remain low and, as a result, renewables will have a difficult time competing. Although energy prices rebounded in 1999 from 1998 lows, it remains unlikely that renewable energy can compete economically over the projection period. Failing a strong worldwide commitment to environmental considerations, such as the limitations and reductions of carbon emissions outlined in the Kyoto Climate Change Protocol, it is difficult to foresee significant widespread increases in renewable energy use. Instead, *IEO2000* projects a moderate 54-percent increase in consumption of hydroelectricity and other renewable resources between 1997 and 2020—just enough for renewable energy sources to maintain an 8-percent share of total world energy consumption through 2020 (Figure 72).

The past year was mixed for the hydroelectric industry. In July 1999, the United States began removal of Maine's 3.5-megawatt Edwards dam—the first dismantling of a dam ever ordered by the Federal Energy Regulatory Commission against an owner's wishes—largely in an effort to restore the environmental damage caused by

the dam over the past 162 years [1]. There are proposals to breach several additional hydroelectric dams throughout the United States; in particular, four dams on the Snake River in Washington have been slated for destruction. Those dams, however, provide substantially more infrastructure support (in terms of electricity, navigational waters, and farm irrigation) to local residents than did the Edwards dam (which provided less than 1 percent of Maine's electricity generation), and plans for their dismantling have met opposition from some residents and their Congressional representatives.

Large-scale hydroelectric projects are still being planned and constructed in the developing countries, particularly in developing Asia, where the fastest development of hydroelectricity is projected to occur. The controversial Three Gorges Dam project in China has continued apace despite international protest from environmental groups and a temporary suspension of work after the devastating flooding on the Yangtze river in the summer of 1998. If construction remains on schedule, the project should be completed in 2009. The development of large-scale hydroelectricity helps sustain the growth of renewables in the *IEO2000* forecast.

In terms of other renewable resources, wind power has enjoyed fast-paced development in recent years, mostly in the industrialized world, with Germany, the United States, Spain, and Denmark emerging as the fastest growing wind markets worldwide in 1999. In April 1999, the American Wind Energy Association (AWEA) and the European Wind Energy Association jointly announced that the world's total installed wind capacity had exceeded 10,000 megawatts [2].

Wind energy projects enjoyed a resurgence in the United States in 1999 after several years of lackluster growth. Developers rushed to install wind energy facilities before the threatened elimination of Federal production tax credits for wind power, which expired in June 1999, although legislation was recently passed to extend the provision until December 31, 2001 [3]. Under this provision wind power producers are allowed to claim a tax credit of $0.017 per kilowatthour of electricity produced [4]. Wind developers accelerated installations to qualify for the 10-year period of tax incentives, and U.S. wind capacity surged in 1998 and 1999 with more than $1 billion worth of new generating equipment—representing

Figure 72. World Hydroelectric and Other Renewable Energy and Total Energy Consumption, 1995-2020

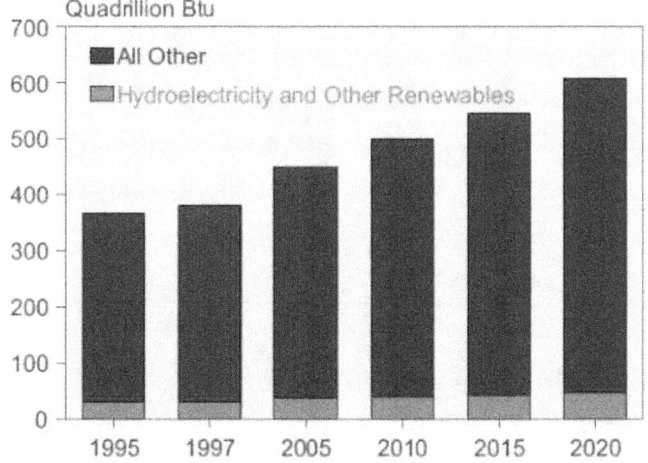

Quadrillion Btu

Legend:
- All Other
- Hydroelectricity and Other Renewables

Sources: **1995, 1997:** Energy Information Administration (EIA), *International Energy Annual 1997*, DOE/EIA-0219(97) (Washington, DC, April 1999). **Projections:** EIA, World Energy Projection System (2000).

some 1,073 megawatts in new or repowered wind capacity—installed between June 1998 and June 1999 [5].

The *IEO2000* reference case projections for hydroelectricity and other renewable energy sources include only on-grid renewables. Although noncommercial fuels from plant and animal sources are an important source of energy, particularly in the developing world, comprehensive data on the use of noncommercial fuels are not available and, as a result, cannot be included in the projections. Similarly, dispersed renewables (renewable energy consumed on the site of its production, such as solar panels used for water heating) are not included in the projections because there are few extensive sources of international data on their use.

Key developments in hydroelectricity and other renewable resources in 1999 include:

- **Wind Energy Had Its Best Year Ever Worldwide.** The AWEA released preliminary estimates of more than 3,600 megawatt of newly installed electricity generating capacity in 1999, bringing the world total to around 13,400 megawatts[6]. The total increase is the largest addition to global wind capacity ever in a single year, a 36-percent increase from 1998. Germany, the United States, and Spain alone accounted for more than 40 percent of the total increase in capacity.

- **The 1999 Drought in Many Latin American Countries May Deter Future Growth of Hydroelectricity.** Countries historically dependent on hydroelectricity for their electricity supplies were hit by a drought that some analysts depicted as the worst of the century. The lack of water in the Mexican state of Sinaloa, for example, left reservoirs filled to only 13 percent of capacity, and electricity had to be imported from the United States to accommodate demand[7]. In Chile, where hydroelectricity typically provides 80 percent of the country's electricity supplies, only 15 percent was supplied by hydropower in 1999[8]. As a result, Chile is expected to pursue more aggressive development of natural-gas-fired electric power.

- **China Continues To Pursue Large-Scale Hydroelectric Projects.** Work on China's 18,200 megawatt Three Gorges Dam Project continued in 1999, amid criticisms that the planned relocation of people at the site of the reservoir to be created for the dam was not progressing smoothly. The country also announced several additional plans to develop hydroelectric resources along the Yangtze and Yellow Rivers, among others. There are also plans to increase the level of other renewable resources in China. The World Bank approved its largest-ever renewable energy loan and its first renewable energy loan for China, $100 million plus a $35 million grant from the

Global Environment Facility. Funds are to be used to develop wind power in Inner Mongolia, Hebei, Fujian, and Shanghai, as well as to develop solar power in isolated rural areas in the northwest of the country.

Regional Activity

North America

Hydroelectricity and other renewables contribute a substantial portion of North America's electricity supply. In 1997, hydroelectricity accounted for 13 percent, 56 percent, and 27 percent of total installed electricity capacity in the United States, Canada, and Mexico, respectively [9]. Geothermal and other renewable resources, on the other hand, accounted for only 2 percent of the installed capacity of both the United States and Mexico, and only a trace amount in Canada. Over the forecast period, *IEO2000* projects that North American consumption of hydroelectricity and other renewable resources will grow from 11.2 quadrillion Btu to 14.4 quadrillion Btu, an increase of almost 30 percent (Figure 73). Much of the growth is expected for renewables other than hydroelectricity, which has already been extensively developed in the United States and Canada. Environmental considerations are likely to cause controversy for any new development plans in either country.

United States

In the U.S. projections for grid-connected renewable generation, hydroelectricity declines slightly over the projection period, from 3.7 quadrillion Btu in 1997 to 3.1 quadrillion Btu in 2020, as increasing environmental

Figure 73. Consumption of Hydroelectricity and Other Renewable Resources in North America, 1980-2020

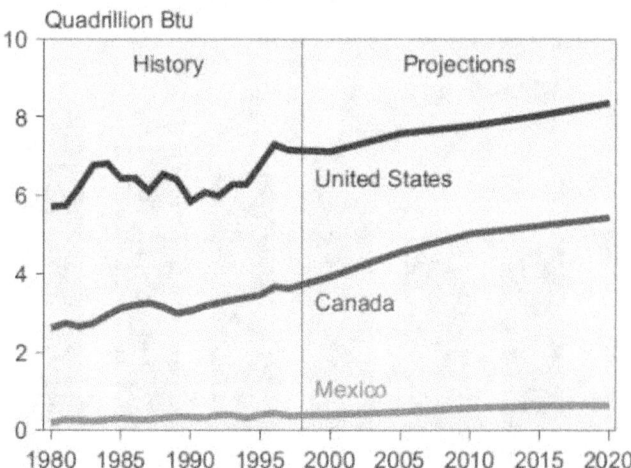

Sources: **History:** Energy Information Administration (EIA), Office of Energy Markets and End Use, International Statistics Database and *International Energy Annual 1997*, DOE/EIA-0219(97) (Washington, DC, April 1999). **Projections:** EIA, World Energy Projection System (2000).

and other competing needs reduce average U.S. hydroelectric productivity [10]. Other renewable energy sources are expected to increase in the United States, from 0.8 quadrillion Btu in 1997 to 1.7 quadrillion Btu in 2020. Most of the growth is attributed to biomass, municipal solid waste (MSW), and wind power. Higher costs disadvantage renewables relative to fossil fuel technologies over the forecast period.

In the United States, the current news surrounding renewable energy is decidedly mixed. Hydroelectricity has become more controversial in recent years, with fears about damaging the environment with dams that have decimated fish populations. In 1999, the United States began removing the 3.5-megawatt Edwards dam on Maine's Kennebec River [11]. The dam was to be completely removed by November 1999, restoring access to 17 miles of Kennebec River fish habitat for the first time in 160 years. The Federal Energy Regulatory Commission issued an order for the involuntary removal of the dam, but eventually a settlement was reached that provided for license transfer and voluntary removal by the State by the end of 1999.

The utility Pacificorp recently announced that it would remove the Condit dam on Washington State's White Salmon River by 2006, rather than invest $30 million that would be required to make the dam more friendly to fish populations [12]. The 125-foot 9.6-megawatt dam is the tallest ever slated for demolition in the United States. Additionally, congressional negotiators agreed to a $12 million appropriation for the removal of the 12-megawatt Glines Canyon dam and the 12-megawatt Elwha dam on the Elwha River on Washington's Olympic Peninsula in October 1999, to allow restoration of salmon populations [13].

The breaching of other, larger hydroelectric dams is also under consideration in the United States. Most notable are four dams on Washington State's lower Snake River, which provide about 4 percent of the Northwest's electricity generation and were constructed to make Lewiston, Idaho, the West Coast's furthest inland seaport at some 500 miles from the Pacific Ocean [14]. Environmental groups and fishing advocates argue that restoring the river to its original form is the only way to save the indigenous salmon from extinction, but residents have come to depend on the hydroelectric system not only for the electricity it provides but also for navigational uses, and they argue that restoring the river might not save the salmon populations. The Columbia and Snake River system, located in Oregon, Washington, and Idaho, transported 43 percent of all U.S. wheat exports in 1997 [15]. The U.S. Army Corps of Engineers is preparing a draft study on alternatives for improving salmon survival, one of which will include the possibility of breaching the four dams. The final report is scheduled for release in May 2000 [16].

Other developments in 1999 may increase the penetration of other renewable energy sources in U.S. electricity supplies. On April 15, 1999, the Clinton Administration submitted the Comprehensive Electricity Competition Act (CECA) to Congress [17]. The proposed legislation includes a renewable portfolio standard (RPS) to stimulate renewables by establishing a minimum annual share of national electricity generation (or sales) that must come from renewable facilities. Under the proposed system, owners or operators of qualifying renewable facilities would receive credits for each kilowatthour generated. The credits could be held for future use (banked) or sold to others to ensure that their mix of power (portfolio) contained the required share of renewables.

The CECA has been submitted in both the House (H.R. 1828, Mr. Bliley, May 17, 1999) and the Senate (S. 1047, Mr. Murkowski, May 13, 1999) and referred to the appropriate committees. Full implementation of an RPS could, according to the Energy Information Administration's *Annual Energy Outlook 2000*, increase renewable generation, particularly in the cases of biomass and wind [18]. The legislation has not yet been enacted, however, and so the reference case projections for the United States do not include its RPS or other provisions.

Wind energy made dramatic gains in the United States between 1998 and 1999, for the most part because wind energy producers rushed to install wind facilities before the expiration of the Federal production tax credit for installed wind turbines. The provision expired in June 1999 but eventually (in November 1999) was extended until December 31, 2001 [19]. The U.S. wind energy industry installed about 892 megawatts of new projects and 181 megawatts of re-powering projects between June 1998 and June 1999, adding more than $1 billion worth of new generating equipment [20]. The added wind capacity was more than double the previous annual record for the United States, 1985, when about 400 megawatts were installed.

More than half the new U.S. wind projects in 1999 were installed in Minnesota and Iowa (Figure 74), where State laws mandate some level of renewable generating capacity. Minnesota—where a 1994 State law requires the State's largest utility to install 425 megawatts of wind power by 2002 in return for the right to store nuclear waste from its power plants within the State—installed 247 megawatts of new wind capacity. Likewise, Iowa's 1983 law requiring utilities to obtain 2 percent of their total electricity from renewables accounted for most of the State's 240 megawatts of additional wind capacity. Since June 1999, Iowa has inaugurated the world's largest wind power generating facility to date in Storm Lake, Iowa, where 257 wind turbines went into operation in September with a combined

Figure 74. Grid-Connected Wind Power Plants in the United States as of December 31, 1998

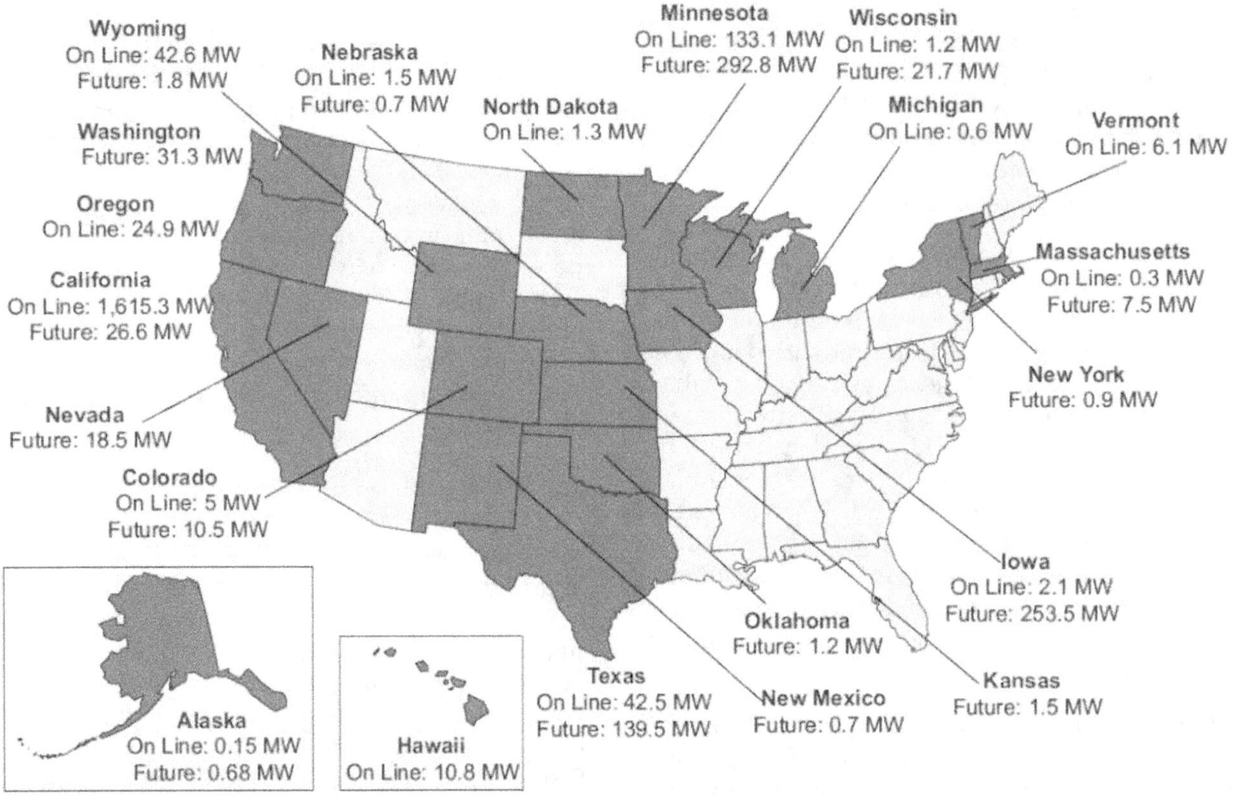

Wyoming
On Line: 42.6 MW
Future: 1.8 MW

Nebraska
On Line: 1.5 MW
Future: 0.7 MW

Minnesota
On Line: 133.1 MW
Future: 292.8 MW

Wisconsin
On Line: 1.2 MW
Future: 21.7 MW

Washington
Future: 31.3 MW

North Dakota
On Line: 1.3 MW

Michigan
On Line: 0.6 MW

Vermont
On Line: 6.1 MW

Oregon
On Line: 24.9 MW

California
On Line: 1,615.3 MW
Future: 26.6 MW

Massachusetts
On Line: 0.3 MW
Future: 7.5 MW

New York
Future: 0.9 MW

Nevada
Future: 18.5 MW

Colorado
On Line: 5 MW
Future: 10.5 MW

Iowa
On Line: 2.1 MW
Future: 253.5 MW

Oklahoma
Future: 1.2 MW

Alaska
On Line: 0.15 MW
Future: 0.68 MW

Hawaii
On Line: 10.8 MW

Texas
On Line: 42.5 MW
Future: 139.5 MW

New Mexico
Future: 0.7 MW

Kansas
Future: 1.5 MW

Source: National Renewable Energy Laboratory and International Energy Agency, *IEA Wind Energy Annual Report 1998* (Golden, CO, April 1999), p. 157.

capacity of almost 193 megawatts [21]. Other States adding new wind projects include Texas (146 megawatts), California (117 megawatts), Wyoming (73 megawatts), Oregon (25 megawatts), Wisconsin (23 megawatts), and Colorado (16 megawatts) [22].

Canada

In Canada, the overwhelming share of installed renewable energy sources are in the form of hydroelectricity. Fifty-six percent of the country's electricity capacity is derived from hydropower [23]. In contrast, less than 0.5 percent of Canada's electricity capacity is from geothermal, wind, solar, and other renewable energy sources. In the *IEO2000* reference case, consumption of hydroelectricity and other renewable resources increases by 50 percent between 1997 and 2020, from 3.6 to 5.4 quadrillion Btu.

There are few plans to expand large-scale hydroelectric facilities in Canada, but several small to mid-sized hydroelectric projects are still being pursued. In 1999, Hydro Quebec completed the $7.4 million Chute Bell hydroelectric project [24]. The project, located on the Riviere Rouge in Canton de Grenville, north of Montreal, was built on the site of a previous dam that was installed there in 1915. Hydro Quebec acquired the original project in 1964 but closed the powerhouse 20 years later because of the high costs of running the facility.

Chute Bell began operating in April 1999, with an installed generating capacity of 9.9 megawatts, which will be used to generate 66,000 megawatthours of electricity per year.

The developers of the Churchill River Power Project recently scaled back plans for a 3,200-megawatt hydropower plant by agreeing to cancel a 1,000-megawatt addition planned for the existing Churchill Falls Project [25]. They also eliminated plans to divert Quebec's St. Jean River to provide water more directly to the existing 5,428-megawatt Churchill Falls powerhouse on Newfoundland's Labrador mainland. Elimination of the new powerhouse at Churchill Falls and the St. Jean diversion should reduce the cost of the project by $333 million to $6.4 billion, with little loss in productive capability. The newly configured project, which includes a $1.6 billion, 25-mile underwater transmission link from Labrador to Newfoundland Island, a 2,264-megawatt powerhouse at Gull Island, and diversion of water from Quebec's Romaine River, would produce 17 million megawatthours of electricity annually.

The Churchill Falls Power Project was, according to the developers, scaled back from its original plan primarily for economic reasons. The project has also been delayed as a result of some friction between the project and the native Innu tribe, which in September 1999 accepted an agreement with the developers to allow the tribe to participate in the environmental review of the proposed project [26]. Government officials in Newfoundland and Quebec believe the agreement will mean a 6- or 7-month delay in a formal agreement to develop the project, which originally was to have been signed in the summer of 1999.

There are also some small to mid-size projects currently under construction or in the planning stages in Canada. Hydro Quebec plans to construct the 231-megawatt Grand-Mere plant, which is scheduled for completion in December 2003 [27]. Work on the project is not scheduled to begin until January 2002. Upon completion, the plant will replace an existing 159-megawatt hydroelectric plant that was built in 1916 on Quebec's Saint Maurice River.

A 440-megawatt hydroelectric station is planned for the Toulnustouc River on the north shore of the St. Lawrence River [28]. Hydro Quebec and the Betsiamites Montagnais Band Council have formulated a $469 million project that would include construction of the Toulnustouc River hydropower project, as well as a plan for the partial diversion of the Portneuf, Sault-aux-Cochons, and Manouane rivers to increase electricity generation at the existing 936-megawatt Bersimis 1 and 798-megawatt Bersimis 2 hydroelectric stations.

Three small hydroelectric projects have been proposed for development in southern British Columbia at an estimated cost of $68 million [29]. The three projects include the 15-megawatt Cascade Heritage Power Park Project, the 60-megawatt Kwioek Creek Hydroelectric Project, and the 3.3-megawatt Slollicum Creek Hydroelectric Development. The development plan is to complete the Cascade Heritage and Slollicum Creek projects in 2001 and the Kwioek Creek project in 2002.

British Columbia has authorized construction of a 25-megawatt, run-of-river plant[18] near Revelstoke called the Pingston Creek Project [30]. Canadian Hydro Developers and Great Lakes Power Ltd. will own the project jointly, and government permission to begin the construction has already been granted. The companies have not announced a schedule for completion of the project.

There have been some attempts to increase the use of wind generated electricity in Canada, but the use of wind and other alternative renewables remains low. In 1999, Canada had a total of 83 megawatts of installed wind capacity in seven facilities, both on and off the national power grid [31]. Four of the facilities are in Alberta Province—Cowley Ridge, in Cowley, and Vision Quest Windelectric installations in Pincher Creek, Belly River, and Blue Ridge. Another facility

owned and operated by Ontario Hydro is located in Tiverton, Ontario, and there are two in Quebec—one owned by Hydro Quebec at Mantane and the other, the Le Nordais project, at Cape Chat. The Canadian Wind Energy Association estimates that Canada will have 133 megawatts of wind energy capacity operating by 2000.

The 52-turbine Cowley Ridge wind plant began operation in 1994 and has produced an average of 55 million kilowatthours of electricity annually [32]. All the electricity from the project is sold to TransAlta Utilities Corporation under a long-term contract that expires in 2014. The 18.9-megawatt Cowley Ridge facility is owned by Canadian Hydro, which owns a total of 46.2 megawatts of installed renewable electric capacity—59 percent in run-of-river hydroelectric facilities and 41 percent in wind projects.

In June 1999, Toronto Hydro announced that it would construct two 20-story waterfront windmills by the end of 2000 [33]. The turbines, estimated to be powerful enough to generate electricity for up to 1,000 homes, are to be used to help alleviate the growing air pollution problems in Toronto, where smog and air quality advisories have been escalating in recent years. The project has been estimated to cost about $816,327 (U.S.) for each turbine. The Canadian government has provided $224,490 in funding for one of the windmills, and another $67,000 has been donated by Environment Canada.

Mexico

Renewable energy was responsible for about 7 percent of total energy consumption in Mexico in 1997. Hydroelectricity accounts for most of the renewable energy used in the country, which has about 10,000 megawatts of installed hydroelectric generating capacity but only about 1,000 megawatts of geothermal power and other renewable sources. At present, about 27 percent of the country's electricity generation is derived from hydroelectricity and 2 percent from geothermal [34].

The worst drought in 40 years hit northern and central parts of Mexico in 1999, forcing eight hydroelectric plants in the state of Sinaloa to stop operating in June because of a lack of water [35]. The drought left the state's reservoirs depleted, at only 13 percent of full capacity. Mexico's Comision Federal de Electricidad (CFE) had to arrange for extra power imports from the United States to compensate for the loss of around 1,000 megawatts of hydroelectric generating capacity.

Mexico has modest plans to expand the use of geothermal power and wind power over the next few years, but much of the 13,182 megawatts of new generating capacity thought to be needed before 2006 is expected to be

[18]In a run-of-river system, the force of the river current (rather than falling water as in a dam system) is used to drive the turbine blades to generate electricity.

fueled by natural gas rather than renewable resources [36]. The country currently has only 3.1 megawatts of wind-driven capacity. No specific plans have been established for installing wind projects to feed national grids, but wind resources in the southern part of the Tehuantepec Isthmus—a particularly rich source of wind resources in Mexico—could, by some estimates, support a wind facility of 2,000 megawatts. At the end of 1998 there were four wind projects in operation in Mexico, with several small units serving niche groups (Figure 75).

Central and South America

Many countries of Central and South America remain heavily dependent on hydroelectric resources for electricity generation, but the situation is expected to change over the projection period as countries diversify their electricity sources. In Brazil, the largest energy-consuming nation in the region, 87 percent of the installed electricity capacity is hydropower. For other major economies in the region, hydropower makes up smaller shares of generating capacity—for example, 43 percent in Argentina, 59 percent in Venezuela, and 53 percent in Chile. Hydroelectric and other renewable energy use is projected to increase by only 0.4 percent per year in Brazil over the forecast period but by 1.4 percent in the region as a whole.

Brazil

Brazil has moved forward with plans to develop renewable energy sources other than hydroelectricity. Brazilian energy officials have announced a goal to increase the number of renewable energy installations in the country in order to provide renewable energy electricity to people who are not connected to the national power grid [37], backed by a commitment to invest $25 billion in the effort [38].

Brazil's government first tried to address the imbalance in access to electricity between rural and urban residents in 1994 when it launched its Programa de Desenvolvimento Energetico de Estados e Municipios (Prodeem). Prodeem started as a public sector social assistance program run by the Ministry of Mines and Energy, with the government performing most of the work, purchasing solar panels and other renewable energy equipment, transporting it to remote communities, and demonstrating and installing the units after drawing up agreements with local and state governments. However, lack of resources, a recent currency devaluation, and the wide potential market caused the government to revise the program, switching it to a decentralized market-based strategy, with the execution and design of projects being handed over to local entities, such as municipal and state governments, nongovernment organizations (NGOs) and cooperatives, and the private sector.

Figure 75. Wind Turbine Installations in Mexico

Notes: 1 = Guerrero Negro, one 600-kilowatt Gamesa Eolica turbine, operated by Mexico's Comision Federal de Electricidad (CFE). 2 = Guerrero Negro, one 250-kilowatt Mitsubishi turbine, operated by Exportadora de Sal. 3 = La Venta I, seven 225-kilowatt Vestas turbines, operated by CFE. 4 = Ramos Arispe, one 550-kilowatt Zond turbine, operated by Cementos Apasco.

Source: National Renewable Energy Laboratory and International Energy Agency, *IEA Wind Energy Annual Report 1998* (Golden, CO, April 1999), p. 109.

Prodeem plans to develop 20,000 megawatts of renewable energy capacity for the 20 million rural inhabitants without access to Brazil's national grids over the next two decades. The program is to focus on market studies and field testing of desalinization, refrigeration, ice-making, and food-drying applications. Already, the Inter-American Development Bank (IDB) has approved an $898,950 grant from a Japanese Special Fund that the IDB administers, to provide training and consultancy services for a Brazilian program aimed at providing electricity over a 20-year period to the estimated 20 million Brazilians not currently connected to a grid [39]. The IDB plans to provide a further grant of $2 million from its Multilateral Investment Fund. An estimated $3 million towards the project is also due to come from the Brazilian government and another $3 million from a variety of institutional sources, including USAid and the European Union.

The Brazilian government also has been working to resume the manufacture of ethanol-fueled vehicles [40]. Ethanol-fueled automobiles accounted for some 90 percent of all car sales as recently as 15 years ago, but a failing distribution network and frequent strikes by sugar cane workers reduced the available fuel supplies and raised prices. At one point, Brazil was forced to import ethanol. Production of alcohol-fueled vehicles fell from 700,000 in 1986 to about 1,000 in 1997 [41].

In October 1999, the government ended tax breaks for gasoline-fueled cars for taxi drivers but continued breaks for ethanol vehicles, hoping to create demand for as many as 50,000 vehicles per year. Today, Brazil has 4 million cars that run on neat ethanol, yet fewer than 1 percent of the country's new cars are neat ethanol vehicles, and government incentives for neat ethanol vehicles are being phased out.

The government has been encouraging the automotive industry to increase production of ethanol-fueled cars and in 1999 secured commitments from General Motors and Ford Motor Company to reintroduce production of the vehicles. Other automobile manufacturers are already increasing their production of alcohol-fueled cars because they believe consumers may be enticed into buying them because of the lower cost of alcohol fuels relative to motor gasoline. In Brazil, a gallon of gasoline costs an average of $2 (U.S.), whereas ethanol costs about $0.80 [42]. Indeed, Italy's Fiat and Germany's Volkswagen have announced plans for significant increases in the production of alcohol-fuel cars, with Fiat increasing from 90 alcohol-fueled cars in August 1999 to 1,300 in September and Volkswagen from 800 to 1,200. In addition, General Motors plans to introduce a new alcohol-fueled model in November, after having suspended production of the cars 3 years ago. Ford Motor Company also recently announced plans to relaunch its alcohol-fueled car line early in 2000.

Other Central and South America

The problems of heavy dependence on hydroelectricity were underscored in Chile in 1999, when the worst drought of the century plagued the country and resulted in a year-long national power shortage [43]. Hydroelectricity accounts for more than 80 percent of the electricity consumed in Chile's 5,500-megawatt central grid in a typical year, but in 1999 it supplied only about 15 percent. Gas-fired generation was introduced in Chile in 1995, and most new investment in electric capacity is expected to be in the form of thermal energy. The 1999 drought, a scarcity of water rights near major consumption centers, and environmental opposition to hydroelectric projects have been factors in the push for diversification.

A recent change in Chile's electricity industry regulations introduced compensation for customers affected by blackouts and, as a result, has reduced the attractiveness of new investment in hydroelectric capacity. The legislation eliminates a clause that exempted generators from responsibility when a drought is more severe than the one in 1968, previously the driest year of the twentieth century. For the industry, this is a serious financial matter. Compensation can hurt far more than fines because, for each megawatt that a utility fails to deliver, customers are entitled to receive the difference between the node or contract price, currently just over $20 per megawatthour, and a penalty spot market price that was set by law at $140 per megawatthour during rationing.

Chile's Endesa generator had previously planned to construct the 570-megawatt Ralco hydroelectric dam, but environmental concerns and the impact of the country's new regulation on compensation for lost generation has caused the project's future to be questioned [44]. The generator was recently taken over by Spain's Endesa electricity company. Endesa has so far committed an estimated 40 percent of the total cost of the $500 million project, which is supposed to begin operation early in 2002, but still may not continue to develop Ralco. The dam would flood land occupied by some of Chile's last remaining Pehuenche Indian communities, and most of the affected people have accepted Endesa's resettlement offer; however, a small group has refused to leave its ancestral land. Rather than engaging in a lengthy and damaging court battle, Endesa may decide to increase its gas-fired capacity as an alternative to the Ralco dam.

Peru also has plans to expand its hydroelectric power. Peru Hydro plans to construct the 525-megawatt Cheves hydroelectric project at a cost of $560 million [45]. The project is to be located on the Huaura River near Churin, some 87 miles north of Lima. There are some doubts about whether the project will actually be constructed, however. A new law to promote development of natural gas was published by the government in early June, and one of its clauses extends the current prohibition on issuing licences for new hydropower projects for another 12 months. The legislation was passed in response to fears that gas from the Camisea fields may not be cheap enough to allow thermal plants to compete with low-cost hydroelectricity.

Also in Peru, construction began in 1999 to repair and increase the capacity of the Machu Picchu hydroelectric plant to 140 megawatts [46]. The plant was damaged by a mudslide caused by El Nino. The construction has been estimated to cost $75 million, and the work should be completed by the end of 2000 [47].

In August 1999, commercial operation of the 20-megawatt Tierras Morenas wind farm began in Costa Rica [48]. Energia Global International, Ltd. (EGI), a Bermuda-based developer of electricity generation projects in Central America, unveiled the project in conjunction with Dallas-based International Wind Corporation (IWC) and Aeorgeneracion de Centro America S.A., a Costa Rican wind energy company. Tierras Morenas is located at Lake Arenal in Guanacaste Province, where wind resources are considered among the best in the western hemisphere [49]. The facility will sell electricity to Costa Rica's state-owned utility, Instituto Costarricense de Electricidad, under the terms of Costa Rican Law 7200. The project was funded through a combination of equity and $24.3 million in loans and grant support provided by DANIDA (a development agency of the Danish government), the Central American Bank for Economic Integration, and a consortium of five Costa Rican banks.

Guatemala also expanded its use of renewable energy sources in 1999 by completing its second geothermal plant at Zunil near Quetzaltenango [50]. The $65 million plant was constructed with funds provided by Ormat (an Israeli company), the International Finance Corporation, the Commonwealth Development Corporation, and other local and regional capital. Electricity from the plant will be sold to Guatemala's INDE distributor at an average price of about 5 cents per kilowatthour under a 25-year power purchase agreement. The country's first commercial geothermal plant went into operation late last year at Amatitlan.

Asia

Large-scale hydroelectric projects are still being constructed in developing Asia. China and India, as well as Laos, Malaysia, and Vietnam, all have plans to add large hydroelectric facilities over the next decade or so. The Asian market for other renewables has limited potential for development, but some projects may be initiated in communities where national grids currently cannot serve the residents, particularly in remote rural areas. Developing Asia's use of hydroelectricity and other renewable energy sources grows by 3.7 percent per year in the *IEO2000* reference case projection (Figure 76).

Development of renewable energy sources in industrialized Asia (Japan, Australia, and New Zealand) has not increased dramatically in recent years. Plans for wind projects in Australia have been hampered by complaints and protests from local residents who do not like the

noise or the visual distraction of wind farms. The Australian Federal Cabinet deferred a decision on a government proposal to require that an extra 2 percent of energy needs be provided with renewable energy sources, jeopardizing development of a $300 million to $600 million (U.S.) Victorian-state-based wind energy industry [51].

Japan's largest wind power plant became fully operational in 1999. The Hisai City, Mie Prefecture, site has four 750-kilowatt turbines [52], but the 3-megawatt plant represents only a small fraction of the 211,000 megawatts of total electric power capacity currently installed in the country. Almost all of Japan's current generating capacity is fired with fossil or nuclear fuels. Consumption of renewable energy in industrialized Asia is projected to grow by 1.2 percent per year over the 23-year *IEO2000* forecast period .

China

The Three Gorges Dam project remains the largest and one of the most controversial hydroelectric projects under construction in the world. Construction on the 18,200-megawatt project began in 1993, but it has been in various stages of planning since 1919, when it was first proposed by the Chinese leader Sun Yat Sen [53]. Supporters of the dam argue that it is needed to help control flooding along the Yangtze River, as well as to provide much-needed electricity from a source that does not produce the greenhouse gas emissions associated with generation from fossil fuels. More than 4,000 people died in China as a result of flooding in 1998, which also caused more than $20 billion of damage [54]. In 1999, more than 800 people died in China's floods.

When completed, Three Gorges Dam will extend 1.4 miles across the Yangtze River and will be 607 feet tall. Its 370-mile-long reservoir will allow shipping through the central Yangtze to increase from 10 million to 50 million tons annually. The total cost to complete the 17-year project has been estimated at $24 billion, although there have been estimates that the project has already exceeded the original estimate by some $3 billion [55]. Three Gorges Dam, when fully operational in 2009, will supply electricity to central and eastern China, including 2,000 megawatts to Chongqing province, 12,000 megawatts to central China, and 4,200 megawatts to eastern China [56]. The project will supply electricity to six major cities in addition to Chongqing: Zhengzhou, Wuhan, Nanjing, Shanghai, Nanchong, and Changsha [57]. Project advocates expect the dam to produce as much as 85 billion kilowatthours of electricity per year, or about 9 percent of the 956 billion kilowatthours of electricity consumed in China in 1997 [58].

Opponents of the dam believe the project will harm the indigenous flora and fauna, threatening such species as

Figure 76. Consumption of Hydroelectricity and Other Renewable Resources in Asia, 1997-2020

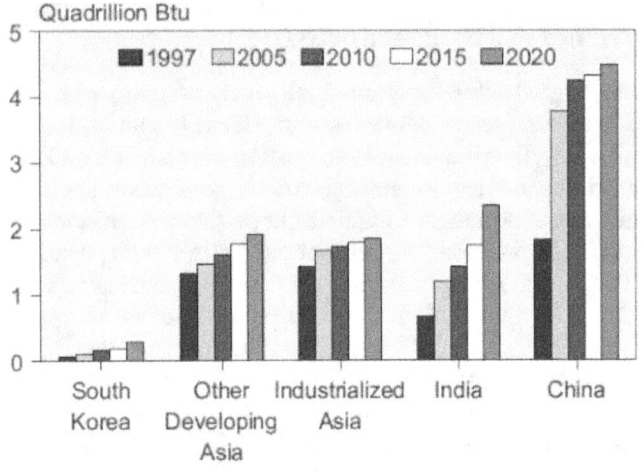

Sources: **1997:** Energy Information Administration (EIA), *International Energy Annual 1997*, DOE/EIA-0219(97) (Washington, DC, April 1999). **Projections:** EIA, World Energy Projection System (2000).

the endangered Yangtze river dolphin and several rare plant and animal species [59]. There are also concerns about the pollution that may be caused by the dam. Critics say that water pollution along the Yangtze will double as the dam traps pollutants from mining operations, factories, and human settlements that used to be washed out to sea by the strong currents of the river. Further, an estimated 1.1 million to 1.9 million people are expected to be resettled before the dam reservoir is completed.

In November 1997, the second phase of the three-phase project began with the successful damming of the Yangtze River [60]. During the second phase, which is scheduled for completion in 2003, the first 700-megawatt generating unit will be installed, and the permanent ship lock will also be completed and ready to receive traffic. In 1999, workers began pouring concrete for the dam. An estimated 27 million cubic meters of concrete will be needed to complete the main structure. Costs for the second phase have run higher than expected, and the estimated cost for construction may double by the time it is completed [61]. There are also concerns about the quality of the construction work. In April 1999 the director of the Three Gorges Construction and Development company, Lu Youmei, announced that he would hire foreign engineers to inspect construction of Three Gorges Dam because of fears that corruption among the Chinese inspectors could compromise the project [62].

In addition to the Three Gorges Dam project, China has several other planned hydropower projects. The Chinese State Development Planning Commission announced its intent to construct two hydroelectric plants on the northern Yangtze—the country's longest river. The combined capacity is expected to approach 18,000 megawatts. The project should take 12 years to complete, with construction beginning as early as 2002 [63, 64]. The $13.2 billion, 12,000-megawatt Xiluodu hydropower plant will be located in central China, 480 miles from the Three Gorges Dam project. Once it has been built, construction will begin on the 6,000-megawatt Xiangjiaba plant on the border between Sichuan and Yunnan provinces in southwest China.

China also approved the establishment of a new power generation company in 1999. The Yellow River Upper Reaches Power Development Company, Ltd., will at first be responsible for developing the 1,500-megawatt Gongboxia power station, the fourth largest on the Yellow River [65], and there are plans for the company to construct two additional hydroelectric stations after that is finished. The Chinese government has announced plans to develop 12 major hydroelectric energy bases in the country. Seven power stations with a combined installed capacity of 5,600 megawatts are either under construction or have been completed on a section of the

river between Longyangxia in Qinghai province and Qingtongxia in the Ningxia Hui autonomous region. Under the country's present development program, a total of 25 power stations with a combined capacity of 15,800 megawatts will eventually be installed in the area. No timetable has yet been established, however.

In 1999, the World Bank approved its largest-ever renewable energy loan and its first renewable energy loan for China, $100 million [66]. The Global Environment Facility is providing an additional $35 million grant. The aims of the project are to develop wind power in four provinces: Inner Mongolia, Hebei, Fujian, and Shanghai, and to develop solar power in isolated rural areas in the northwest of the country, for the first time on a large-scale and competitive basis. Upon completion, China's project will amount to 400 megawatts of wind power (as compared with 700 megawatts in India).

For the wind portion of the project, the World Bank will provide funds to the State Power Corporation of China and provincial and/or municipal companies, which in turn will hire wind farm companies to install a total of 190 megawatts of wind capacity at five sites. The farms will include the initial 100-megawatt installation at Huitingxile in Inner Mongolia in the east side of China, where wind resources are considered vast (100 megawatts on a site with a 1,000-megawatt potential); an installation at Zhangbei in Hebei Province (50 megawatts on a site with 500 megawatts potential); a medium-scale installation at Pingstan in Fujian Province (20 megawatts on a 120-megawatt site); and two small demonstration projects at Chongming (14 megawatts) and Nanhui (6 megawatts), both in Shanghai Province.

The first tenders for the wind projects are scheduled to take place by the end of 1999, and all five projects are scheduled to be completed within 4 years. Each wind farm will be developed on a commercial basis, using power purchase agreements (PPAs) that are meant to pave the way for eventual private sector participation in future wind power projects.

The solar part of the project will use photovoltaic or solar-generated electricity to provide an estimated 1 million people in rural China—in the poor, northwestern regions—with electricity supplies for the first time. Photovoltaic modules will be installed in individual homes, shops, schools, and small businesses to generate electricity during the daytime for use during evening and night hours. In contrast, today, many farmers and nomads who live in Inner Mongolia, for example, use candles, kerosene, and often butter to generate light. Under the terms of the project, about 10 megawatts of photovoltaic capacity—about 300,000 to 400,000 systems—will be supplied to households and institutions in remote areas in six northwestern provinces: Qinghai, Gansu, Inner Mongolia, Xinjiang, Tibet, and Western Sichuan.

The World Bank is funding this project mainly to help China reduce its greenhouse gas emissions, especially carbon emissions, as well as other air pollution (NO_x and SO_2 emissions, in particular). According to the World Bank, annual health and agricultural losses in China associated with coal-caused air pollution equal as much as 6 percent of the nation's gross domestic product. Local awareness of environmental issues is increasing, and a growing number of universities are providing students with courses on the harm of acid rain and greenhouse gases. Another reason for making a commitment to develop renewables is that this form of electricity supply is fundamental for China's rural development, which, in turn, is crucial for the country's development as a whole (especially since much of China is still rural).

In addition to the major project to be financed by the World Bank, the local government in the Ningxia Hui Autonomous Region of northwest China announced that it had adopted a "Sunshine Plan" to increase the penetration of solar energy in a wide range of areas in the region [67]. The government has committed $115,000 for the first phase of the project, which will promote the use of solar energy in both remote mountain and comparatively well-off irrigated areas over a 3- to 5-year period. Ningxia has been developing the use of solar energy since the 1970s, and thousands of solar-powered appliances and dozens of solar-heated houses are in use in many parts of the region today.

India

Hydroelectricity in India is already well established: 22 percent (21,104 megawatts out of 96,803 megawatts) of the country's total installed electricity generating capacity is hydropower. In contrast, only 74 megawatts of India's on-grid capacity is fired by geothermal and other renewable resources. There are plans to increase hydroelectric capacity substantially over the next several years. The Indian government has announced intentions to increase hydroelectric generating capacity by 35,490 megawatts by 2012 [68]. In the *IEO2000* reference case, India's consumption of hydroelectric and other renewable energy increases by 5.6 percent per year over the projection period, a result of expectations that large-scale hydroelectricity will continue to be developed. Twelve large-scale projects have already been approved by the government, all of which are scheduled for completion by 2002 [69].

Construction on the 300-megawatt Chamera II hydroelectric project in Himachal Pradesh in India, scheduled for completion in 2004, moved into its second phase in 1999. The $391 million turnkey project is being constructed for the National Hydroelectric Power Corporation (NHPC) [70]. Chamera II will channel the waters of the Ravi River through a 128-foot-high concrete dam. Its output will be fed into the northern electricity grid and sold to customers in Himachal Pradesh, Jammu and Kashmir, the Punjab, Haryana, Delhi, and Rajasthan. The Himachal Pradesh State Electricity Board, in addition, has invited proposals for several small hydropower projects in the state: Patikri (16 megawatts), Sal Stage 1 (6.5 megawatts), Fozal (6 megawatts), Sainj (5.5 megawatts), Kashang (70 megawatts), Bharmour (45 megawatts), Harsar (60 megawatts), and Kugti (45 megawatts) [71].

The country also has plans for a series of four pumped-storage projects in the Ayodhya hills of West Bengal [72]. The first is the 900-megawatt Purulia project, which currently is facing bureaucratic delays that may postpone the facility's operation until the end of 2005 rather than the 2002 as originally planned. Three other projects would increase the capacity of the total scheme to 3,600 megawatts. The 900 megawatt Turga pumped-storage facility would go into operation 4 years after Purulia, and the two others, linked with the canals of Katlajal and Bandhunala, would be completed after that.

The development of India's wind industry has fallen off sharply since 1995 because of the imposition of new taxes, an economic slowdown, and bureaucratic delays in land allotment and environmental clearances [73]. The government is, however, working on new measures to secure private-sector investment in wind projects. In July 1999, the government proposed several incentives to help boost wind power development, including withdrawal of the minimum tax on renewable energy projects, automatic environmental clearances for units of generation capacity up to 5 megawatts, and softer loans.

Other Developing Asia

Vietnam has announced plans to assemble a feasibility study on the construction of a 3,600-megawatt hydropower plant in Son Law Province [74]. The proposed scheme would involve constructing a dam on the Da River, some 190 miles north of Hanoi [75]. Two proposals have been under consideration: a 3,600-megawatt scheme costing around $3 billion and a smaller 2,400-megawatt scheme estimated at around $2.3 billion. The larger scheme is currently receiving government support, and existing proposals suggest it could become operational between 2007 and 2012. Hydroelectric power currently accounts for around 60 percent of Vietnam's total power generating capacity (around 5,000 megawatts). The dam would also be used to regulate seasonal water flows and would allow better management of flows to the existing 1,900-megawatt HEP scheme at Hoa Binh, further down the Da River. Hoa Binh provides more than one-third of the existing generating capacity in Vietnam.

Although a recent World Bank report on Vietnam's energy needs concluded that the Son La plan would be economically justified, only limited studies have been performed on the environmental and social impacts of the dam on the Son La area. Officials in the state-run media estimate that the 3,600-megawatt project would involve resettlement of 100,000 people. Vietnam hopes to secure approximately 70 percent of the construction costs from international donors.

In 1999, the Malaysian government decided to resurrect plans for the large-scale Bakun hydroelectric project in Sarawak. Plans for the 2,400-megawatt hydropower plant were suspended in September 1997 because of the impact of the Asian economic crisis, compounded by continual contractual disputes between the government and its contractor on the project, Swiss-Swedish ABB [76]. The project, which has been scaled down to 500 megawatts generating capacity, will supply power only to Sabah and Sarawak, not to the mainland as originally planned. There is still opposition to the project, and critics contend that Malaysia would be better served if the government would implement energy conservation measures rather than construct Bakun [77].

Thailand appeared to begin recovering from the 1997-1998 Southeast Asian economic crisis with modest economic growth in 1999, accompanied by increased demand for electricity generation and higher national petroleum consumption (other than for the petrochemical industries) [78]. To that end, in September 1999 Thailand reached an agreement to purchase additional electricity from the newly completed 60-megawatt Nam Luek hydropower station in Laos [79]. The agreement for the new supplies was incorporated into an existing contract between the Electricity Generating Authority of Thailand (Egat) and Electricite du Laos—for the purchase of electricity from the 150-megawatt Nam Ngum hydroelectric plant in Laos—that was up for renewal. Egat began importing electricity from Nam Ngum in 1968. In addition to Nam Ngum, two other Laotian sources of hydroelectric power, the 45-megawatt Xeset and the 195-megawatt Thuen Hinboun hydropower stations, supply electricity to Thailand.

At the end of 1998, India announced plans to develop three hydroelectric projects in Bhutan within the next decade [80]. Output from the plants would be imported by India to supply electricity to several Indian states on the country's eastern grid. The projects include the 900-megawatt Wangchu, the 180-megawatt Bunakha, and the 4,000 megawatt Sankosh projects, along with a 60-megawatt irrigation scheme. India's National Hydroelectric Power Corporation is already constructing the 45-megawatt Kurichu hydroelectric project in Bhutan, and the 1,020-megawatt Tala hydropower plant is being constructed jointly by India and Bhuton.

The government of Pakistan's North-Western Frontier Province called for tenders in 1999 to help develop four mid-scale hydroelectric projects: the 72-megawatt Khan Khwar project and the 35-megawatt Daral Khwar project in the province's Swat district, the 106-megawatt Golen Gol project in the Chitral district, and the 28-megawatt Summar Gah project in the Kohista district [81]. Environmental impact analyses have been prepared for all four, and their construction has been approved under the terms of the analyses. The northern half of Pakistan has an estimated potential 24,000 megawatts of hydroelectric resources for development. The projects are being developed under the government's private independent power project policy, which was established in 1998.

In October 1998, the Export-Import Bank of Japan approved a $302 million loan to help finance the San Roque hydroelectric dam on Agno River in the Cordillera region of northeastern Luzon, the largest island of the Philippines [82]. An additional loan of $400 million is also under consideration. When completed, San Roque will be the tallest dam and the largest private hydroelectric project in Asia, at nearly 660 feet tall and generating 345 megawatts of capacity. Preparation of the construction site began in 1999, and the $937 million project is scheduled for completion by 2004 [83, 84].

Another Asian country that is looking to increase its hydroelectric generating capacity is Nepal. Nepal's potential hydroelectric resource has been estimated to be as high as 83,000 megawatts, although less than 1 percent has been developed [85]. India is a prime potential market, and in 1996 and 1997 the two countries signed agreements defining terms for joint development of hydroelectric power projects in Nepal (the Mahakali Treaty). India and Nepal are developing detailed project plans for construction of the 6,480-megawatt Pancheshwore at a proposed cost of $3 billion. The 144-megawatt, $453 million Kali Gandaki project is already under construction in Nepal, financed in part by the Asian Development Bank and the Japanese government. Other projects expected to be developed either by the Nepalese government or by private developers on a build, own, operate, and transfer basis include the 300-megawatt Upper Karnali run-of-river project, 420 miles from Kathmandu; the 134-megawatt run-of-river Dudh Koshi 1 peakload project, and the 72-megawatt Tamu project to be located in the Taplejung district of eastern Nepal, as well as several additional small hydroelectric projects.

Western Europe

In Western Europe, much of the projected growth in renewable energy use will come from sources other than hydroelectricity, inasmuch as most of the region's hydroelectric resources have already been developed.

Many countries of Western Europe have recently passed legislation to support the development of alternative renewables—mostly in the form of taxes that are to be used to develop renewable resources that would not otherwise be competitive to develop, such as Denmark's Energy 21 program, Germany's Electricity Feed Law, and the United Kingdom's Government Renewable Obligation. Wind-powered generation is the fastest growing among the renewable energy sources in Western Europe, with Germany, Denmark, and Spain alone adding more than 1,400 megawatts of wind capacity in 1998. In the *IEO2000* forecast, renewable energy use is projected to grow by 1.9 percent per year in Western Europe, mostly from renewable sources other than hydroelectricity (Figure 77).

Denmark introduced the Energy 21 program—its fourth energy policy plan—in 1996 [86]. Energy 21 sets a national objective to reduce the country's carbon emissions by 20 percent below their 1988 level by 2005. The Danish government since 1992 has imposed a carbon tax of about $14.20 (DKr 100) per metric ton of carbon dioxide emitted, which was fully refundable to industrial consumers at first but was limited to 50 percent in 1993. Energy 21 includes a target for installation of 1,500 megawatts of wind capacity by 2005. By 1998, Denmark had already installed an estimated 1,467 megawatts of wind power, representing 12 percent of the country's total electricity consumption [87]. Energy 21 sets a target of 5,500 megawatts of installed wind capacity by 2030, with 4,000 megawatts slated for offshore installation.

A Green Tax Package for industrial consumers was also introduced by the Danish government in 1996, part of which included the taxation of space heating for carbon emissions and sulfur dioxide emissions, in combination with a refund in the form of subsidies for installation of energy-saving measures. The tax package will be fully phased in by the end of 2000.

Germany has substantially increased its wind-generated electricity production in recent years. In 1998, consumption of electricity from wind power in Germany exceeded that in the United States for the first time, as Germany installed 800 megawatts of new wind capacity to bring its total to 2,800 megawatts [88]. Germany was the world leader in wind capacity additions in 1998. Although no specific targets have been set for increasing wind capacity, the government has set a target to reduce carbon dioxide emissions by 25 percent relative to 1990 levels by 2005 and believes that wind will contribute to meeting that goal [89]. Two German states, Lower Saxony and Schleswig-Holstein, have plans to increase wind capacity to 1,000 megawatts by 2000 and 1,200 megawatts by 2010, respectively. A number of government programs support the development of renewables in Germany, and the Electricity Feed Law

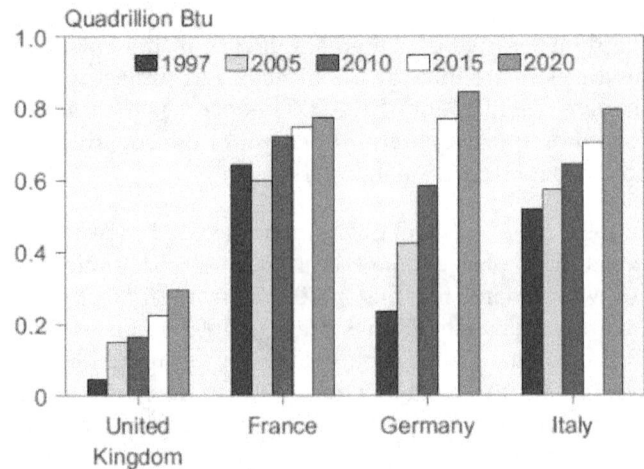

Figure 77. Consumption of Hydroelectricity and Other Renewable Resources in Western Europe, 1997-2020

Sources: **1997:** Energy Information Administration (EIA), *International Energy Annual 1997*, DOE/EIA-0219(97) (Washington, DC, April 1999). **Projections:** EIA, World Energy Projection System (2000).

fixes "buy-back" prices for approved renewables (including wind) at 90 percent of the average private consumer tariff.

At the end of 1998, Spain had installed more than 820 megawatts of wind capacity, adding an additional 650 megawatts in 1999 [90]. The Spanish National Energy Plan targets a 25-percent increase in renewable energy use over 1990 levels by 2000 [91]. The December 1998 Royal Law 2828/1998 sets the target that renewables should account for at least 12 percent of the country's energy demand by 2010.

In order to compensate for electricity that will be lost when Sweden closes its Barsebäck nuclear reactors by mid-2001, the country expects to increase the use of renewables, along with promoting energy conservation [92]. To that end, some $1.3 billion (1996 U.S. dollars) will be invested in long-term development of biofuels, ethanol, wind, solar, and other renewable sources.

By mid-1999, the United Kingdom had installed some 340 megawatts of wind capacity. The country has an official goal of generating 10 percent of its electricity demand from renewable energy sources by 2010; however, a 1999 report from the UK Parliament's House of Lords stated that while it would be "technically feasible" to achieve this goal, "present policies will not deliver them" [93]. The report states that achieving the 2010 target would require a sevenfold increase over the next 10 years in the rate of expanding renewable energy generation.

To help stimulate the development of renewable electricity generation, the UK government requires electricity suppliers to provide a portion of their supply from renewable sources. The requirement is outlined in the Government's Renewable Energy Obligations, with separate obligations for England and Wales (the Non-Fossil Fuel Obligation—NFFO), Scotland (the Scottish Renewables Obligation—SRO), and Northern Ireland (NI-NFFO). Five bids have been held under the NFFO to expand the amount of renewables installed in the country, two under the SRO (with a third expected in 1999), and two under the NI-NFFO [94].

The NFFO was enacted as part of the UK's Electricity Act of 1989 [95]. It was conceived as a way to support mostly the nuclear power industry when the electricity supply industry was privatized, but nuclear power remained in the public sector because it was thought that the nuclear facilities would not be an attractive investment for shareholders. A fossil fuel levy of 10 percent was originally applied to the price of electricity, and most of the money was used to subsidize the nuclear industry, with a small amount going to support renewables. In 1998, the nuclear subsidy was deemed no longer necessary, and the levy was reduced to 2.2 percent with all of the proceeds going to support renewables.

Even with support of the fossil fuel levy, installation of wind projects has been somewhat disappointing in the United Kingdom, particularly in comparison with some of the other Western European countries. In 1998, 13 megawatts of wind capacity was installed in the United Kingdom, compared with 395 megawatts in Spain, 235 megawatts in Denmark, and 800 megawatt in Germany [96, 97]. One of the major problems for increasing wind power in the United Kingdom is that it is often difficult to obtain planning approval. Almost all the wind projects submitted for planning approval in 1998 have failed to secure approval, and the British Wind Energy Association has estimated that the process usually takes more than 2 years and costs developers more than $160,000. Of the 20 appeals heard since the beginning of 1994, only 4 have been successful.

Eastern Europe and the Former Soviet Union

Little expansion in renewable use is projected for the countries of Eastern Europe and the former Soviet Union (EE/FSU) over the forecast period. Most of the development of hydroelectricity is expected to occur in the way of expanding or refurbishing existing hydroelectric plants. This is particularly the case in the FSU, where economic problems have persisted since the collapse of the Soviet Union. While the economies of the FSU are expected to begin recovery over the course of the projection period, they are expected to rely more on natural gas to displace the use of coal and nuclear for electricity generation than on renewable energy sources, given the large gas reserves available from Russia and several of

the other FSU republics. In the *IEO2000* reference case, consumption of hydroelectricity and other renewable resources in the former Soviet republics is projected to grow by less than 1 percent annually over the 1997 to 2020 time period, rising from 2.2 to 2.6 quadrillion Btu (Figure 78).

The economies of Eastern Europe have recovered much more quickly in the transition from planned to market-based economies than those of the FSU, and their prospects for development of renewable energy sources are therefore much brighter. In the reference case, the use of renewable energy in Eastern European triples over the 23-year projection period. As in the FSU, much of the increase in the region's energy consumption is expected to be in the form of natural gas use to displace coal and nuclear generation, but systematic growth of hydroelectricity is also expected in countries like Slovenia and the former Yugoslavian Republics of Croatia, Serbia-Montenegro, and Macedonia, where undeveloped hydroelectric potential still exists [98].

In 1999, Montenegro announced its intention to reduce the country's energy dependence on Serbia—from which it currently imports most of its electric power—by developing its hydroelectric resources [99]. The Montenegrin Economy ministry and state power utility EPCG announced in July that an international tender would be issued for construction and finance contracts for a series of hydroelectric plants on the River Moraca, including the 37-megawatt Zlatica, the 127.4-megawatt Andrijevo, the 37-megawatt Raslovici, and the 37-megawatt Milutinovici plants, for which tenders have

Figure 78. Consumption of Hydroelectricity and Other Renewable Resources in Eastern Europe and the Former Soviet Union, 1997-2020

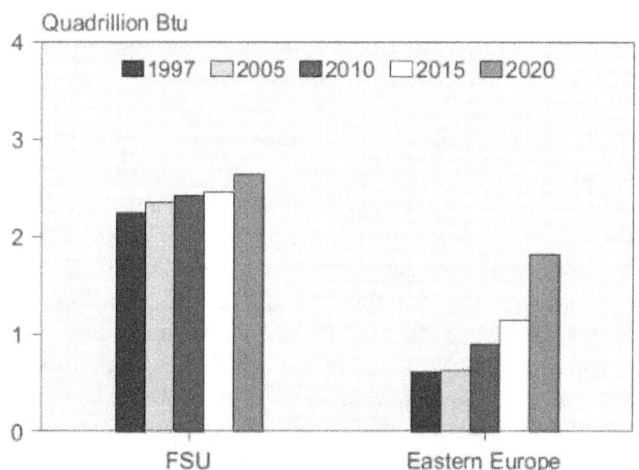

Sources: **1997:** Energy Information Administration (EIA), *International Energy Annual 1997,* DOE/EIA-0219(97) (Washington, DC, April 1999). **Projections:** EIA, World Energy Projection System (2000).

already been prepared. There are also plans to develop a fifth plant on the Moraca River, the 600-megawatt Kostanica, and for the Buk Bijela station on the Drina River, which will be a joint venture with the Serbian Srpska Republic in Bosnia. The country must develop all six hydroelectric plants if it is to meet its present domestic electricity requirements. The Kostanica plant would produce an estimated 1.3 terawatthours—twice as much as the other four plants on the Moraca combined. The entire package of projects is estimated to be worth about $500 million and could be completed within 6 years.

The government of Slovenia plans to augment that country's renewable energy sector as a long-term strategic goal [100], focusing on increased use of hydroelectricity, biomass, geothermal, solar, and waste-to-energy sources for electricity generation. The government plans to use tax incentives to promote renewable energy resources. At present, hydroelectricity accounts for a significant portion of Slovenia's generating capacity, with 40 small units and several large ones along the Drava, Sava, and Soca rivers. The combined capacity of the units is 743 megawatts, but many of the smaller plants date from before World War II and need to be refurbished to remain operational.

Georgia has continued to move forward with efforts to modernize its power sector, and it too has announced tenders for the sale of a majority of the country's generation assets and distribution networks [101]. Assets to be sold off in the first of two separate lots include a 25-year lease on 100 percent of the Khrami 1 and 2 hydroelectric plants, with installed capacities of 1,128 megawatts and 220 megawatts, respectively. In the second lot, 25-year leases will be offered on five hydroelectric facilities in the western part of the country: the 111.6-megawatt Ladjanuri, 38.4-megawatt Shaori, 80-megawatt Tkibuli, 52-megawatt Rioni, and 67-megawatt Gumati I and II. Major modernization contracts for the 1,300-megawatt Enguri hydroelectric plant in the autonomous region of Abkhazia and others will be tendered sometime in 2000.

Hydroelectric power accounts for 43,000 megawatts of Russia's generating capacity, about one-fifth of the country's total capacity [102]. More than 70 percent is at 11 stations of more than 1,000 megawatts capacity, led by three of the four largest power stations in Russia: the 6,400-megawatt Sayano Shushenskoye station in Khakassia, the 6,000-megawatt Krasnoyarsk station in Krasnoyarsk province, and the 4,500-megawatt Bratsk station in Irkutsk province. Although these stations account for a large portion of Russia's power generation, they utilize less than 20 percent of the country's estimated hydroelectric potential.

Although the Russian government has made some attempts to introduce competition into power markets, hydropower remains under state control. This has hurt the industry. In addition, hydropower has been hurt by electricity pricing, which is determined not by market forces but by regional energy commissions that give lower cost hydropower production no advantage over higher cost production.

There has been only limited success in establishing geothermal energy use in Russia. The country has only one 11-megawatt geothermal plant—which was constructed in 1966—operating at Pauzhetskaya in the Kamchatka region. There are plans to expand the Pauzhetskaya facility by another 7 megawatts before 2010. Another 80-megawatt geothermal power plant is under construction at Mutnovsk in the Kamchatka region, financed in part by the European Bank of Reconstruction and Development, which has signed a $100 million agreement for the construction of the first stage of the station. Total costs are expected to reach $500 million for the power plant and $120 million for the steam pipeline. Finally, a 30-megawatt geothermal plant is planned for Iturup Island in the Kuril Archipelago.

Electric power in Tajikistan comes primarily from hydroelectric dams, which Tajikistan's mountainous geography makes possible [103]. Tajikistan's total electricity generating capacity in 1997 was 4,400 megawatts. A new hydroelectric power dam, Sangtuda, is under construction with Russian and Iranian financing. The project, which was started during the Soviet period, remained unfinished for several years because of a lack of funds. Once completed, Sangtuda should satisfy northern Tajikistan's capacity needs.

Africa and the Middle East

In Africa and the Middle East, renewable energy resources remain largely undeveloped. Hydroelectricity constitutes a major portion of the electricity capacity in some African countries because of the general lack of electricity infrastructure. For instance, Kenya is almost completely dependent on hydroelectric power, but it has only 1,000 megawatts total generating capacity installed (as compared with 70,000 megawatts installed capacity in Turkey). In the Middle East, aside from Turkey and Iran, hydroelectricity constitutes only a modest amount of the installed electric capacity; and in all countries, geothermal and other renewable energy sources constitute almost none. There are opportunities for renewables in these regions in niche areas—particularly, in remote areas that cannot be accessed through national grids. *IEO2000* projects that on-grid renewables will increase by 4.3 percent per year in the Middle East between 1997 and 2020—growth that is expected to be fueled mainly by large-scale hydroelectric projects planned for Iran and Turkey (Figure 79). In Africa, renewables are projected to increase by 2.0 percent per year.

Figure 79. Consumption of Hydroelectricity and Other Renewable Resources in the Middle East and Africa, 1997-2020

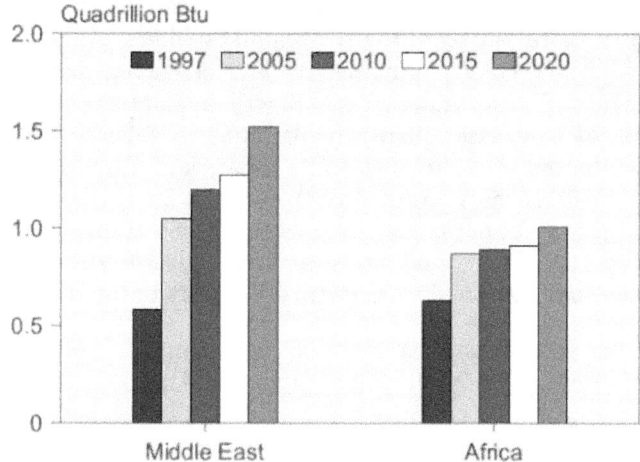

Sources: **1997:** Energy Information Administration (EIA), *International Energy Annual 1997*, DOE/EIA-0219(97) (Washington, DC, April 1999). **Projections:** EIA, World Energy Projection System (2000).

In Tanzania, the government has begun offering incentives for renewable energy projects, in particular for investment in solar, wind, and hydroelectric projects [104]. A 100-percent depreciation allowance will be extended to investors in the first year in which the renewable system operates, with exemptions granted from excise and customs duties and sales taxes on the import of materials and components used in renewable energy projects. Tanzania is also considering a proposal to force independent power producers to generate at least 5 percent of their electricity from renewables in remote areas of the country, making it possible for renewable energy sources to compete with other forms of energy.

Egypt's first solar power station is scheduled to come on line by 2001. Completion of the station will require more than $400,000 in assistance from Japan, which is supplying technical and economic feasibility studies and will provide the funds needed to construct the power plant in southern Egypt using its photovoltaic technology [105]. Once completed, the station will pump water and provide power.

Hydroelectricity provides the bulk of Egypt's electricity at present. In 1997, hydropower accounted for about 51 percent of the country's total electricity capacity [106]. Three large hydro projects currently operate at Aswan: the High Dam (2,100 megawatts), Aswan 1 (345 megawatts) and Aswan 2 (270 megawatts). Installed hydroelectric capacity is expected to grow only sightly in the next few years in Egypt. By 2000, the Esna Dam project is scheduled for completion, but after that, most additional electricity capacity is expected to be provided by natural gas.

There has been some movement in bringing wind power to Egypt. In 1999, the Japanese government announced that it would extend soft credits to fund a $120 million, 120-megawatt wind facility in the Zaafarana region on the Gulf of Suez [107]. The terms of the agreement include that it may be repaid over a 40-year period, with a 10-year grace period, at an annual interest rate of 0.75 percent. In March 1999, Spain committed to financing a 60-megawatt wind power plant in Zaafarana.

Hydroelectricity and other renewables provide less than 1 percent of all electricity generation in South Africa, but the country is making efforts to increase the amount of alternative renewables consumed [108]. Although the country has installed 500,000 square meters of solar water heater panels, this represents less than 1 percent of the potential market. Also, there are 5 megawatts of photovoltaic systems, 280,000 water-pumping windmills, and 60 megawatts of small-scale hydropower installed throughout the country. Eskom, the country's main supplier of electricity, hopes to use alternative renewable energy to supply remote areas of the countries that do not have easy access to national electricity grids. In 1998, with the help of a grant received from the Netherlands, Eskom electrified 90 non-grid schools [109].

Morocco's state-owned power company, Office National de l'Electricité (ONE), launched a $205 million tender in March 1999 for the installation of a pumping station at Afourer (in central Morocco) to boost electricity supplies during periods of peak demand [110]. The installation, when completed in 2003, will use cheap off-peak electricity to pump water uphill into the reservoir, then let it flow down again to drive turbines at peak hours.

ONE opened technical bids for the final lot of the Dchar El Oued Ait Messoud hydroelectric complex on March 17, 1999, and is now evaluating offers for construction of a turbine and turbo-reactor [111]. The 92-megawatt plant is due to be on line in 2002, producing some 200 million kilowatthours of electricity per year. The $153 million project will draw power from dams at Dchar El Oued and Ait Messoud. Part of the project cost will come from the Kuwait Fund for Arab Economic Development, which has signed a loan worth $22 million.

In 1999, Morocco advanced a number of renewable energy initiatives. In October 1999, ONE issued a tender for $200 million for the construction of two wind farms: a 140-megawatt project in Tangiers near the Strait of Gibralter and a 60-megawatt project in Tarfaya near the Western Sahara [112]. The projects are part of a national renewable energy program aimed at increasing the country's energy diversity—Morocco currently generates 80 percent of its electricity with oil. ONE is conducting a pre-feasibility study for a combined gas-fired and solar plant to be constructed as an independent power project [113]. Having abandoned sites at Ouazarzate and Taroudant in 1996, a consortium of the local Centre des Energies Renouvelables, Germany's Flagsol and Spain's Endesa has examined sites at Jerrada and Ain Beni Mathar, near Guercif. The site would require sufficient sunshine, proximity to a gas pipeline, and availability of water for cooling. A study by Pilksolar in May 1998 recommended construction at Ain Beni Mathar of a 178-megawatt integrated thermo-solar combined-cycle plant system. At Jerrada, a solar unit would be integrated with an existing thermal plant.

The cost of the electricity will be the determining factor, with Global Environmental Fund (GEF) subsidies covering the difference in cost over regular power. GEF would be expected to provide $40 to $60 million of the total $120 to $160 million cost. Consortium members would fund 20 percent of the remainder; 45 percent would be met by the Abu Dhabi Fund for Economic Development, 28 percent by the World Bank's International Finance Corporation, 9 percent by domestic loans, and 18 percent by Germany's Kreditanstalt für Wiederaufbau.

In Kenya, tenders have been issued for construction of the Olkaria II geothermal package [114]. The plan for the geothermal project is that high-voltage transmission lines will transmit power from the Olkaria field—located about 55 miles west of Nairobi in Kenya's Central Rift Valley—to the national grid. The Olkaria natural steam field is being developed by Kenya Electricity Generating Company (KenGen). Steam power from the northeast area of the field in the second phase of the project is to be used to drive two 32-megawatt turbines connected to 11-kilovolt generators that will feed into the national grid and sustain 64 megawatts of power for at least 30 years. The project is being funded by the World Bank's International Development Association, the European Investment Bank, KenGen, and Germany's Kreditanstalt für Wiederaufbau, which recently signed an agreement to provide $13.8 million. The first unit of the project is scheduled for completion in July 2001; the second in September 2001.

Kenya has an estimated 2,000 megawatts of potential geothermal capacity, with Olkaria alone estimated to have potential reserves in excess of 220 megawatts. The country's geothermal power production will double in 2003 when the Olkaria II and III projects are completed. Olkaria III, based in the southwest part of the steam field, is expected to come on line 2 years after Olkaria II. Kenya now produces 45 megawatts of electric power from the Olkaria I geothermal station, which constitutes 5 percent of national output. The total will rise to around 173 megawatts with the two additional projects.

Much of the developed hydroelectric capacity in the Middle East is in Turkey and Iran—45 percent and 38 percent of the region's total, respectively [115]—and both countries have plans to add large-scale hydroelectric projects in the near term. Turkey has significant hydroelectric power resources and is developing a great deal more as part of the $32 billion Southeast Anatolia (so-called GAP) hydroelectric and irrigation project [116]. When completed, GAP will include 21 dams, 19 hydroelectric plants (generating 27 billion kilowatthours of electricity), and a network of tunnels and irrigation canals. In April 1999, Turkey's Energy Undersecretary announced that several U.S. firms were negotiating to build an additional 9 or 10 small to medium-sized hydroelectric plants and dams in Turkey.

Iran expects several hydroelectric plants to become operational by 2004, including Shahid Abaspour (1,000 megawatts of capacity), Karun 3 (2,000 megawatts), Masjed-Soleyman (2,000 megawatts), and Karkheh (400 megawatts) [117]. Standard & Poor's Platt's has estimated that some 6,000 megawatts of new hydroelectric capacity could be installed in Iran by 2020.

Israel too has plans to increase the use of alternative renewable energy resources in the future, but it seems clear that renewables will not measurably displace thermal electricity generation. The country has plans to add nearly 250 megawatts of renewable energy projects (primarily solar and wind projects), along with 30 megawatts of waste-to-energy projects [118].

References

1. MSNBC, "Dam Coming Down To Help Fish Return: First Time Ecological Damage Used To Force Removal" (July 1, 1999), web site www.msnbc.com.

2. American Wind Energy Association, "World Wide Wind Capacity Tops 10,000 Megawatts," News Release (April 22, 1999), web site www.awea.org/news/news990422tenk.html.

3. American Wind Energy Association, "Wind Energy Tax Credit Extended," News Release (November 24, 1999), web site www.awea.org.

4. National Renewable Energy Laboratory and the International Energy Agency, *IEA Wind Energy Annual Report 1998* (Golden, CO, April 1999), p. 156.

5. American Wind Energy Association, "U.S. Wind Industry Finishes Best Year Ever, By Far," News Release (July 3, 1999), web site www.awea.org/news/wpa14.html.

6. American Wind Energy Association, "1999 Best Year Ever for Global Wind Energy Industry," News Release (December 23, 1999), web site www.awea.org/news/news991223glo.html.

7. "Drought Hits Hydro Power," *Financial Times: Power in Latin America*, No. 48 (June 1999), p. 18.

8. "Chile's Drought Challenges Regulators," *Financial Times: Power in Latin America*, No. 48 (June 1999), p. 1.

9. Energy Information Administration, *International Energy Annual 1997*, DOE/EIA-0219 (Washington, DC, April 1999), p. 94.

10. Energy Information Administration, *Annual Energy Outlook 2000*, DOE/EIA-0383(2000) (Washington, DC, November 1999), pp. 69, 139.

11. "Workers Begin Removal of Edwards Dam," *Hydro•Wire*, Vol. 20, No. 14 (July 19, 1999), p. 6.

12. S. H. Verhovek, "Returning River to Salmon, and Man to the Drawing Board," *The New York Times*, Vol. 149, No. 51,657 (September 26, 1999), pp. A1, A28.

13. "Lawmakers Endorse $12 Million for Elwha Dams Removal Program," *Hydro•Wire*, Vol. 20, No. 21 (October 25, 1999).

14. S. H. Verhovek, "Returning River to Salmon, and Man to the Drawing Board," *The New York Times*, Vol. 149, No. 51,657 (September 26, 1999), p. A28.

15. "Snake River Dam Resolution Called Unfair, *Environmental News Network* (July 26, 1999), web site www.enn.com.

16. Bonneville Power Administration, "The December 1999 Multi-Species Biological Assessment of the Federal Columbia River Power System (FCRPS) Is Now Available," Federal Caucus web site, www.bpa.gov (December 23, 1999).

17. Energy Information Administration, *Annual Energy Outlook 2000*, DOE/EIA-0383(2000) (Washington, DC, November 1999), pp. 11, 20.

18. Energy Information Administration, *Annual Energy Outlook 2000*, DOE/EIA-0383(2000) (Washington, DC, November 1999), pp. 21.

19. American Wind Energy Association, "Wind Energy Tax Credit Extended," News Release (November 24, 1999), web site www.awea.org.

20. "US Wind Industry Chalks Up Best Year, Tallies $1B in New Equipment," *The Solar Letter*, Vol. 9, No. 15 (July 16, 1999), p. 189.

21. "World's Largest Wind Power Facility Dedicated," *DOE This Month* (October, 1999), p. 4.

22. "US Wind Industry Chalks Up Best Year, Tallies $1B in New Equipment," *The Solar Letter*, Vol. 9, No. 15 (July 16, 1999), p. 189.

23. Energy Information Administration, *International Energy Annual 1997*, DOE/EIA-0219(97) (Washington, DC, April 1999), p. 94.

24. "News Briefs," *Hydro•Wire*, Vol. 20, No. 13 (June 28, 1999), pp. 8-10.

25. "Powerhouse, Quebec Diversion Cut from Churchill River Project," *Hydro•Wire*, Vol. 20, No. 13 (June 28, 1999), pp. 3-4.

26. "Churchill Project Delayed to Permit Innu Participation," *Hydro•Wire*, Vol. 20, No. 17 (August 30, 1999), pp. 6-7.

27. "News Briefs," *Hydro•Wire*, Vol. 20, No. 19 (September 27, 1999), p. 6.

28. "Utility, Indian Band Forge Partnership in New Hydro Project," *Hydro•Wire*, Vol. 20, No. 15 (August 2, 1999), p. 1.

29. "News Briefs," *Hydro•Wire*, Vol. 20, No. 17 (August 30, 1999), p. 7.

30. "News Briefs," *Hydro•Wire*, Vol. 20, No. 13 (August 16, 1999), pp. 5.

31. Canadian Association for Renewable Energies, "Harnessing the Wind," *Trends in Renewable Energies*, No. 91 (August 2, 1999), web site www.renewables.ca.

32. "Canadian Hydro Seals Deal for Cowley Ridge Wind Power, 18.9-MW Facility," *The Solar Letter*, Vol. 9, No. 15 (July 16, 1999), p. 188.

33. "Toronto Hydro To Build Two 20-Story Turbines for Green-Pricing Program," *The Solar Letter*, Vol. 9, No. 14 (July 2, 1999), p. 172.

34. Energy Information Administration, *International Energy Annual 1997*, DOE/EIA-0219(97) (Washington, DC, April 1999), p. 94.

35. "Drought Hits Hydro Power," *Financial Times: Power in Latin America*, No. 48 (June 1999), p. 18.

36. National Renewable Energy Laboratory and International Energy Agency, *IEA Wind Energy Annual 1998* (Golden CO, April 1999), pp. 105-107.

37. "Business and Technological Briefs," *Solar & Renewable Energy Outlook*, Vol. 25, No. 11 (October 8, 1999), p. 262.

38. Canadian Association for Renewable Energies, *Trends in Renewable Energies*, No. 90 (July 26, 1999), web site www.renewables.ca.

39. "Brazil's Experiment in Rural Power," *Financial Times: Power in Latin America*, No. 51 (September 1999), pp. 7-8.

40. "Brazil Renews Effort To Produce Cars Powered by Sugar Cane Fuel," *The Solar Letter*, Vol. 9, No. 13 (June 18, 1999), p. 166.

41. "Brazil Renews Effort To Produce Cars Powered by Sugar Cane Fuel," *The Solar Letter*, Vol. 9, No. 13 (June 18, 1999), p. 166.

42. "Brazilian Automakers Renew Interest in Alcohol," *Alternative Fuels Today* (September 17, 1999).

43. "Chile's Drought Challenges Regulators," *Financial Times: Power in Latin America*, No. 48 (June 1999), p. 1.

44. "Ralco Project Under Review," *Financial Times: Power in Latin America*, No. 50 (August 1999), p. 16.

45. "New Hydro Project Faces Delay," *Financial Times: Power in Latin America*, No. 48 (June 1999), p. 13.

46. NewsPage, "Reform of Machu Picchu Hydroelectric To Cost US$55mil" (August 2, 1999), web site www.newspage.com.

47. "Machu Picchu Rehab Underway," *Financial Times: Power in Latin America*, No. 50 (August 1999), p. 15.

48. "Energia Global International Christens 20-Megawatt Windfarm in Costa Rica," *The Solar Letter*, Vol. 9, No. 17 (August 13, 1999), p. 213.

49. Canadian Association for Renewable Energies, "Major Windfarm Starts in Central America," *Trends in Renewable Energies*, No. 92 (August 9, 1999), web site www.renewables.ca.

50. "Guatemala Gets Second Geothermal," *Financial Times: Power in Latin America*, No. 50 (August 1999), p. 23.

51. "Wind Plan Seeks Government Ok," *Financial Times: Power in Asia*, No. 291 (November 29, 1999), p. 27.

52. "Largest Wind Plant On Line," *Financial Times: Power in Asia*, No. 290 (November 15, 1999), p. 28.

53. BBC World Monitoring, China, "Workers Pouring Concrete at Three Gorges Dam" (August 31, 1999), web site www.nextcity.com.

54. Muzi Lateline News, "China Flooding Losses Significantly Down From Last Year" (August 19, 1999), web site dailynews.muzi.net.

55. BBC News, World: Asia-Pacific, "Chinese Dam Project Needs More Money" (March 16, 1999), web site news2.thls.bbc.co.uk.

56. "The Three Gorges Dam-II: The Super Dam Along the Yangtze River," web site welcome-to-china.com (September 16, 1996).

57. "$54b Plan for Gorges Power Grid," *Hong Kong Standard China* (October 6, 1997), web site www.htkstandard.com.

58. Energy Information Administration, *International Energy Annual 1997*, DOE/EIA-0219(97) (Washington, DC, April 1999), p. 90.

59. T. Poole, "China's Greatest Ambition May Prove Her Greatest Folly," web site www.virgin.net (not dated).

60. BBC World Monitoring, China, "Workers Pouring Concrete at Three Gorges Dam" (August 31, 1999), web site www.nextcity.com.

61. BBC News, World: Asia-Pacific, "China Dam Faces Cash Flow Crisis" (March 16, 1999), web site news2.thls.bbc.co.uk.

62. NewsEdge, "Director of China's Three Gorges Dam Calls for Foreign Inspectors" (April 20, 1999).

63. "Chinese Commission To Build Two Hydropower Plants in North Yangtze," *The Solar Letter*, Vol. 9, No. 14 (July 2, 1999), p. 177.

64. CEInet Economic Forecast, "Yangtze To Get 2 Dams, Power Plants" (August 5, 1999).

65. "China Oks Yellow River Power," *Financial Times Power in Asia*, No. 290 (November 15, 1999), p. 27.

66. "Record WB Renewable Power Loan for China," *Financial Times: Power in Asia*, No. 280 (June 28, 1999), pp. 1-3.

67. NewsPage, "Solar Energy Will Be Widely Used in Northwest China" (October 12, 1999), web site www.newspage.com.

68. "Peaking Shortage To Soar Over Next Decade," *Financial Times: Power in Asia*, No. 288 (October 18, 1999), p. 12.

69. D. D'Monte, "India Government Progresses With 12 Hydropower Projects," *Solar Letter*, Vol. 8, No. 20 (September 25, 1998), p. 351.

70. "Consortium Chosen for Chamera," *Financial Times: Power in Asia*, No. 283-284 (August 9, 1999), p. 12.

71. "Asian Opportunities: India," *Financial Times: Power in Asia*, No. 288 (October 18, 1999), p. 10.

72. "Bengal Pumped Storage Trouble," *Financial Times: Power in Asia*, No. 287 (October 4, 1999), pp. 13-14.

73. "Boost for Wind Power," *Financial Times: Power in Asia*, No. 282 (July 26, 1999), p. 14.

74. "Queue Lengthens for Viet Schemes," *Financial Times Power in Asia*, No. 283-284 (August 9, 1999), pp. 4-5.

75. "Vietnam Pushes Ahead With Giant Hydro Scheme," *Financial Times: Power in Asia*, No. 280 (June 28, 1999), pp. 22-3.

76. International Rivers Network, "Bakun Dam 'Delayed Indefinitely,'" Press Release (September 5, 1997), web site www.irn.org.

77. "Bakun Opposition," *Financial Times: Power in Asia*, No. 280 (June 28, 1999), pp. 15-16.

78. "Thailand/Demand: Consumption Inches Up," *Financial Times: Power in Asia*, No. 285 (September 6, 1999), pp. 18-19.

79. "New Accord To Buy Hydropower," *Financial Times: Power in Asia*, No. 285 (September 6, 1999), p. 17.

80. A. Airy, "Four Hydro Power Jvs With Bhutan on Cards," *The Financial Express* (October 21, 1998), web site www.financialexpress.com.

81. "Hydropower Schemes Tendered," *Financial Times: Power in Asia*, No. 290 (November 15, 1999), p. 14.

82. International Rivers Network, "San Roque Dam, Philippines: San Roque Currents" (September 24, 1999).

83. Milbank, Tweed, Hadley, & McCloy, "San Roque $937 Million Power Plant Financing Marks the First Independent Power Plant Financing Since the Asian Financial Crisis," News Release (December 1998), web site www.milbank.com.

84. International Rivers Network, "San Roque Dam, Philippines: San Roque Currents" (September 24, 1999).

85. Communication from Christine Kucera, Political/Economic Section, American Embassy, Kathmandu, Nepal.

86. Standard & Poor's Platt's, *European Outlook, Volume I* (Lexington, MA, 1999), p. 3.

87. National Renewable Energy Laboratory and International Energy Agency, *IEA Wind Energy Annual Report 1998* (Golden, CO, April 1999), p. 45.

88. American Wind Energy Association, "Wind Energy Is Fastest Growing Energy Source in World, Again," News Release (January 7, 1999), web site www.awea.org.

89. National Renewable Energy Laboratory and International Energy Agency, *IEA Wind Energy Annual Report 1998* (Golden, CO, April 1999), p. 65-6.

90. American Wind Energy Association News Release, "1999 Is Best Year Ever for Global Wind Energy Industry" (December 23, 1999), web site www.awea.org.

91. National Renewable Energy Laboratory and International Energy Agency, *IEA Wind Energy Annual Report 1998* (Golden, CO, April 1999), p. 127.

92. Standard & Poor's Platt's, *European Outlook, Volume I* (Lexington, MA, 1999), p. 91.

93. British Wind Energy Association News Release, "Lords Report on Electricity From Renewables" (July 7, 1999), web site www.bwea.com.

94. National Renewable Energy Laboratory and International Energy Agency, *IEA Wind Energy Annual Report 1998* (Golden, CO, April 1999), pp. 147-148.

95. British Wind Energy Association, "The Government's Policy for Renewables," Fact Sheet No. 6 (not dated), web site www.bwea.com.

96. National Renewable Energy Laboratory and International Energy Agency, *IEA Wind Energy Annual Report 1998* (Golden, CO, April 1999), p. 147-8.

97. American Wind Energy Association News Release, "Wind Energy Is Fastest Growing Energy Source in World, Again" (January 7, 1999), web site www.awea.org.

98. PlanEcon, Inc., *PlanEcon Energy Outlook for Eastern Europe and the Former Soviet Republics* (Washington, DC, October 1998), pp. 41-49.

99. "Montenegro Looks to Hydro for Power Independence," *Financial Times: East European Energy Report*, No. 94 (July 1999), pp. 5-6.

100. U.S. Department of Energy, Fossil Energy International, "An Energy Overview of Slovenia" (July 15, 1999), web site www.fe.doe.gov/international/slvnover.html.

101. "Georgia Pushes Ahead With Wholesale Power Privatisation," *Financial Times: East European Energy Report*, No. 94 (July 1999), pp. 4-5.

102. Energy Information Administration, *Country Analysis Briefs: Russia* (October 1998), web site www.eia.doe.gov.

103. Energy Information Administration, *Country Analysis Briefs: Tajikistan* (September 1999), web site www.eia.doe.gov.

104. "Tanzania Government Offers Financial Incentives for Renewables Development," *The Solar Letter*, Vol. 9, No. 15 (July 16, 1999), p. 187.

105. Canadian Association for Renewable Energies, "Egypt Prepares Its First Solar Power Project," *Trends in Renewable Energies*, No. 90 (July 26, 1999), web site www.renewables.ca.

106. Standard & Poor's Platt's, *World Energy Service: Africa/Middle East* (Lexington, MA, 1999), p. 49.

107. "Egypt: Japan, Spain Commit Soft Credits for Wind/Solar Schemes," *Financial Times: African Energy*, No. 14 (May 1999), p. 21.

108. "South Africa Backs Renewables in White Paper," *Trends in Renewable Energies*, No. 90 (July 26, 1999), web site www.renewables.ca.

109. Eskom, "Alternative Energy Sources," News Release (not dated), web site www.eskom.co.za.

110. "Morocco: Pumping Station Bids," *Financial Times: African Energy*, No. 14 (May 1999), pp. 7-8.

111. "ONE Opens Dchar El Oued Bids," *Financial Times: African Energy*, No. 14 (May 1999), pp. 7-8.

112. "Morocco Issues Call for Bids To Build 200 MW of Wind Power," *Wind Energy Weekly*, Vol. 18, No. 870 (October 29, 1999).

113. "Moroccan Gas/Plant Solar Update,"*Financial Times: African Energy*, No. 14 (May 1999), p. 8.

114. "Kenya: Olkaria II Geothermal Tenders Out Soon,"*Financial Times: African Energy*, No. 14 (May 1999), pp. 8-9.

115. Energy Information Administration, *International Energy Annual 1997*, DOE/EIA-0219(97) (Washington, DC, April 1999), pp. 94-95.

116. Energy Information Administration, *Country Analysis Briefs: Turkey* (August 1999), web site www.eia.doe.gov.

117. Standard & Poor's Platt's, *World Energy Service: Africa/Middle East* (Lexington, MA, 1999), p. 153.

118. Standard & Poor's Platt's, *World Energy Service: Africa/Middle East* (Lexington, MA, 1999), p. 177.

Electricity

Electricity consumption nearly doubles in the IEO2000 projections. Developing nations in Asia and in Central and South America are expected to lead the increase in world electricity use.

Worldwide electricity consumption in 2020 is projected to be 76 percent higher than its 1997 level. Long-term growth in electricity consumption is expected to be strongest in the developing economies of Asia, followed by Central and South America. The projected growth rates for electricity consumption in the developing Asian nations are close to 5 percent per year over the *International Energy Outlook 2000 (IEO2000)* forecast period (Table 20), and the growth rate for Central and South America averages about 4.2 percent per year. As a result, the developing nations in the two regions are expected to account for 35 percent of total electricity consumption in 2020, compared with only 22 percent in 1997. Rapid population growth and economic growth, along with greater industrialization and more widespread household electrification are responsible for the increase. Because much of the world's population still has limited access to electricity, future growth in electricity consumption will depend in large part on progress in connecting more of the world's population to the electricity grid.

For the entire range of the projection period, world electricity consumption is expected to rise from 12 trillion kilowatthours in 1997 to 22 trillion kilowatthours in 2020. As was seen for other forms of energy supply, the recent short-term trend in demand for electricity has been greatly affected by world economic and financial developments; however, an economic turnaround in Asia from the 1997-98 economic crisis has occurred much more rapidly than was expected one year ago.

Annual growth in electricity consumption for the industrialized economies is expected to average around 1.5 percent for the 1997 to 2020 forecast period, primarily as a result of more widespread applications of electrical devices. The growth will be counterbalanced in part, however, by slowing population growth and increases in energy efficiency.

For the developing nations of Africa and the Middle East, both economic growth and electricity consumption growth are expected to fall midway between those projected for the industrialized economies and the developing economies of Asia and Central and South America. In the Middle East, increases in electricity demand will result more from rapid population growth than from per capita increases in electricity usage. The story for Africa is expected to be similar, although growth in Africa will be slightly higher as more households on the continent gain access to the electricity grid. Both the Middle East and Africa are highly dependent on extractive industries for economic growth. In both regions, economic growth and electricity consumption growth rates will follow developments in the supply and demand for raw materials and particularly, in the Middle East, for petroleum.

Highlights of recent electricity developments around the world are as follows:

- Electricity demand and investment in electric power infrastructure have been positively affected by the recent net improvement in global economic conditions. Economic difficulties that started in Asia in 1997 and then moved to South America and Russia have begun to recede in developing Asia but continue in many of the economies of Central and South America. The U.S. economy has shown surprising strength long into its current economic expansion, and economic growth continues in Western Europe although at a slower rate. The former Soviet Union (FSU) region is also expected to see positive growth over the next several years—something that has not occurred since the dissolution of the Soviet Union.

- As the new millennium begins, the scope of many electricity companies has become increasingly global. Through mergers, acquisitions, joint ventures, and strategic alliances, many of the world's electricity companies have also become more integrated—and much larger. Regional electricity companies have become multinational electricity companies. Electricity companies have become natural gas companies, and vice versa. Several companies have also chosen to specialize, and some electricity companies have shed their generation assets to become "wires only" (that is, transmission and distribution only) concerns. Others have chosen to focus solely on generation. Some have even decided to focus on nuclear power. Electricity companies have also made acquisitions of companies wholly outside the energy arena, in areas such as telecommunications, water, cable television, sewage, and other industries. What have been the driving forces behind these transformations? It is in part due to a number of policy and market-related developments, such as

Table 20. World Net Electricity Consumption by Region, 1990-2020
(Billion Kilowatthours)

Region	History		Projections				Average Annual Percent Change, 1997-2020
	1990	1997	2005	2010	2015	2020	
Industrialized Countries	6,353	7,287	8,252	8,960	9,628	10,255	1.5
United States.	2,817	3,279	3,647	3,909	4,155	4,350	1.2
EE/FSU	1,906	1,484	1,550	1,720	1,873	2,115	1.6
Developing Countries	2,265	3,489	4,911	6,145	7,328	9,203	4.3
Developing Asia	1,260	2,103	3,071	3,899	4,707	5,957	4.6
China	551	956	1,521	2,045	2,588	3,450	5.7
India	257	397	626	788	937	1,154	4.7
South Korea	95	197	234	269	299	337	2.4
Other Developing Asia	357	552	690	796	883	1,016	2.7
Central and South America	448	624	875	1,092	1,272	1,619	4.2
Total World	10,524	12,260	14,713	16,826	18,828	21,574	2.5

Note: EE/FSU = Eastern Europe and the former Soviet Union.

Sources: **History:** Energy Information Administration (EIA), *International Energy Annual 1997*, DOE/EIA-0219(97) (Washington, DC, April 1999). **Projections:** EIA, World Energy Projection System (2000).

privatization, deregulation, and technological advances in natural gas exploration and electrical turbine efficiency.

- Privatization has and will continue to play a role in various nations' electricity sectors (see box on pages 120-122). In the developing world, $142 billion in private capital has flowed into electricity projects since 1990 (Figure 80). Among the developing regions, Latin America has been most aggressive in privatizing electricity assets and has also been among the largest targets of foreign direct investment in electricity. Raising capital to meet the need for increased generation capacity in the face of rapidly growing demand has motivated many developing countries to encourage various forms of private investment. Budgetary constraints have often prohibited the internal raising of such funds. Methods of privatizing have differed. For instance, most developing Asian nations have pursued a more cautious approach to foreign investment, in general preferring limited foreign investment in greenfield generation projects over direct ownership of electricity companies.[19] Some nations have preferred negotiated deals and others competitive bidding.

- The industrialized world has also seen a surge in cross-border electricity investments. Since the mid-1990s the United Kingdom and Australia, in particular, have been the most frequent targets of foreign investment in electricity. During the late 1990s, the United States also saw some sizable acquisitions of electric utilities by overseas companies.

Figure 80. Investment in Electricity Projects in Developing Countries with Private Participation, 1990-1998

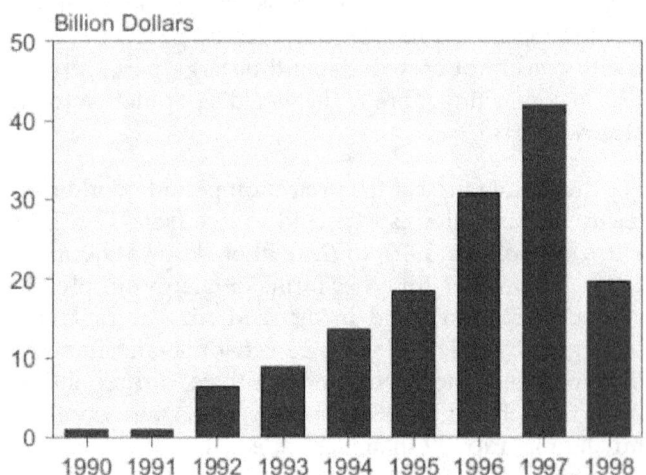

Source: World Bank, Private Participation Infrastructure Data Base.

- Deregulation leading to lower prices continues to be one of the major worldwide developments in the electricity industry. Germany, for instance, saw a dramatic drop in prices in 1999, especially for industrial customers, after the government's decision to expedite electricity reforms in compliance with a European Community directive. In the United States, State-initiated deregulation continues, and a Federal regulatory reform package, the Comprehensive Electricity Competition Act, was proposed by the Clinton Administration in 1999.

[19]A "greenfield" project is an industrial development in a rural area with no established infrastructure.

• Technological advances have greatly improved the position of natural gas as a fuel for electricity generation, and continued improvement is expected over the forecast period. The advances to date involve both improved technologies for the extraction of natural gas and improvements in natural gas turbines.

Primary Fuel Use for Electricity Generation

Natural Gas

Natural gas is increasingly becoming the fuel of choice for new electricity projects around the globe (Table 21). Over the next two decades world natural gas use in electricity generation is expected to more than double [1], with particularly strong growth in North America, Western Europe, and Central and South America (Figure 81). Over the 1997-2020 projection period, gas in North America is expected to gain usage relative to coal and nuclear generation. South America is expected to

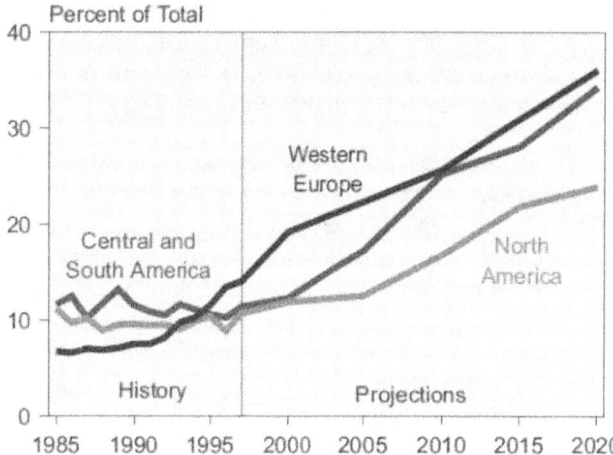

Figure 81. Natural Gas Use for Electricity Generation by Region, 1985-2020

Sources: **History:** Derived from Energy Information Administration (EIA), *International Energy Annual 1997*, DOE/EIA-0219(97) (Washington, DC, April 1999). **Projections:** EIA, World Energy Projection System (2000).

Table 21. World Energy Consumption for Electricity Generation by Region and Fuel, 1990-2020
(Quadrillion Btu)

Region and Fuel	History		Projections		
	1990	1997	2010	2015	2020
Industrialized	70.6	78.6	94.8	99.2	101.9
Oil	5.9	5.2	5.4	5.6	5.8
Natural Gas	6.2	8.6	15.6	19.8	22.5
Coal	27.3	28.0	32.7	33.4	33.9
Nuclear.	16.3	19.6	20.0	18.0	16.2
Renewables	15.0	17.2	21.1	22.3	23.5
EE/FSU	31.2	24.9	28.1	29.3	32.2
Oil	3.8	2.7	3.3	3.9	4.6
Natural Gas	10.8	10.0	13.3	14.6	17.2
Coal	10.9	6.5	5.1	4.0	3.3
Nuclear.	2.9	2.8	3.0	3.1	2.7
Renewables	2.8	2.8	3.3	3.6	4.5
Developing	27.6	40.4	67.8	76.4	90.6
Oil	4.4	5.5	8.0	9.4	10.7
Natural Gas	3.1	5.5	11.4	13.2	17.5
Coal	11.0	17.4	29.5	33.5	39.8
Nuclear.	1.1	1.6	3.0	3.5	3.6
Renewables	8.1	10.5	15.9	16.9	19.0
Total World	129.5	144.0	190.7	204.8	224.3
Oil	14.1	13.4	16.7	18.9	21.1
Natural Gas	20.0	24.2	40.3	47.6	57.2
Coal	49.1	51.9	67.3	70.9	76.9
Nuclear.	20.4	24.0	26.0	24.6	22.5
Renewables	25.9	30.6	40.4	42.8	47.0

Note: EE/FSU = Eastern Europe and the former Soviet Union.

Sources: **History:** Derived from Energy Information Administration (EIA), *International Energy Annual 1997*, DOE/EIA-0219(97) (Washington, DC, April 1999). **Projections:** EIA, World Energy Projection System (2000).

increase natural gas generation to supplement its large base of hydroelectricity generation. Western Europe is moving from nuclear to greater reliance on gas. Eastern Europe is expected to move from coal to gas. And a major share of capacity expansion in Asia and the Middle East is also expected to rely on natural gas.

In several countries, policy developments in natural gas and electricity have encouraged greater integration of the two industries. Natural gas deregulation has often accompanied or preceded the deregulation of electric power. In both the United States and the United Kingdom, natural gas deregulation preceded electricity deregulation by several years. Natural gas regulatory reforms in both countries provided a template for subsequent electricity reform.

The efficiency of natural-gas-fired generation, particularly for cogeneration or combined-cycle units, has increased greatly. The advances have also improved the competitiveness of gas relative to other fuels, including coal. Another factor that has worked in favor of natural-gas-fired generation is falling capital costs. For instance, the capital cost per kilowatt of capacity for the current generation of combined-cycle power plants is $449 (1998 dollars), compared with $1,102 for a similar coal unit [2].

Natural gas is expected to account for 25 percent of the global electricity fuels market in 2020, up from 17 percent in 1997. The growing popularity of gas has resulted, in part, from increased confidence in the future availability of natural gas supplies. Significant improvements in natural gas turbine technology and the environmental advantage of natural gas over other fossil fuels for electricity generation have also contributed.

Increases in the availability of liquefied natural gas (LNG) will also lead to more use of natural gas in power generation. Although currently accounting for 5 percent of world gas consumption, LNG exports have grown by 40 percent since 1992 [3]. Currently, Algeria, Indonesia, and Malaysia are the largest exporters of LNG, and Japan, South Korea, and France are the largest importers [4].

Pipeline trade in natural gas, which is nearly three times as large as LNG trade, is also a rapidly growing industry. In recent years, exports of natural gas from Canada into the United States, from Norway and Russia into Western Europe, and from Algeria into Italy and Spain have contributed to the increase in world trade. In addition, the recent completion of a number of pipeline projects under construction in South America means that Argentina and Bolivia will become major exporters of natural gas, with Chile and Brazil becoming major importers.

Nuclear Power

Nuclear power's share of the electricity market is expected to drop sharply over the forecast period, to 10 percent of the global power market in 2020 from 17 percent in 1997. Factors contributing to the decline include past cost overruns in the construction of nuclear facilities, the high costs of plant decommissioning and spent fuel retirement, and safety and environmental concerns.

Both Sweden and Germany are committed to the gradual phaseout of their nuclear power programs. The German government is currently committed to begin its nuclear phaseout in 2002. Only France and Japan are expected to continue to rely on nuclear power to the extent that they have in the past, although a 1999 accident at a nuclear fuel facility in Japan casts some doubt on the future of that country's nuclear industry. Most of the other industrialized nations are expected to reduce their reliance on nuclear power. In Canada and the United States, for instance, nuclear power provided 15 percent and 19 percent, respectively, of total electricity production in 1997; those shares are expected to fall to 8 percent and 10 percent, respectively, by 2020. The United Kingdom is also expected to reduce its reliance on nuclear power, from 34 percent of its electricity supply in 1997 to 18 percent in 2020.

Coal

Coal is expected to continue dominating electricity fuel markets in the future, as it has for the past two decades, despite a slight loss of market share over the forecast period. In 2020, coal is projected to account for nearly 34 percent of the world's electricity consumption, compared with 36 percent in 1997.

In the *IEO2000* projections, China has the world's highest growth rate for coal demand, and its share of world coal consumption for electricity generation increases from 15 percent in 1997 to nearly one-third in 2020. China has been the world's leading consumer of coal since 1982, followed by the United States. Rapid growth in coal consumption is also projected for India, despite the government's desire to increase natural gas use for electricity production; and U.S. coal consumption is expected to increase by roughly one-third over the forecast period.

Coal consumption is expected to decline in the nations of Western Europe, which are turning to increasingly available natural gas supplies for future growth in electricity production. The elimination of subsidies in the United Kingdom was largely responsible for a 50-percent loss in its domestic coal production between 1992 and 1998 and a greatly reduced role for coal in electricity generation. Both coal production and consumption have declined sharply in recent years in the nations of Eastern Europe and the FSU (EE/FSU), as their economies have moved away from central planning. Over the forecast period, EE/FSU coal use for electricity generation is expected to decline further. In large measure, coal's lost share of the EE/FSU electricity market is expected to be taken over by natural gas.

Hydroelectricity and Other Renewables

The use of renewables (largely, hydropower) to fuel electricity generation is expected to remain stable over the forecast period, accounting for 21 percent of the world's electricity supply in 2020, as it did in 1997. For the world to maintain its present degree of reliance on hydroelectric power will require substantial capacity expansion, most of which is expected to occur in Asia, and especially in China.

Currently, no other region is as dependent on hydroelectric power as Central and South America, which accounts for only 5 percent of the world's electricity generation but 18 percent of its hydroelectric power generation [5]. Central and South America is expected to increase its output of renewable-based electricity from 5.4 quadrillion Btu in 1997 to 7.4 quadrillion Btu in 2020. Still, despite sizable investments in new hydroelectric facilities, the region will be significantly less reliant on renewables for electricity generation in 2020 than today. In recent years, some very large hydroelectric power projects in South America have seen significant construction delays and major cost overruns as a result of increasing environmental concerns and the unwillingness of indigenous people to be uprooted for new hydroelectric projects.

The relative importance of the consumption of hydroelectricity and other renewables for electricity generation is expected to remain relatively stable in the United States and Canada and to increase in Western Europe. The 1997 renewable shares of U.S. and Canadian electricity markets—20 percent and 62 percent, respectively—are not projected to change significantly over the forecast period. In Western Europe, however, renewable energy sources are projected to provide 24 percent of the region's total electricity supply in 2020, up from 20 percent in 1997.

China and India account for almost two-thirds of renewable electricity generation among the developing countries of Asia. In China, renewable energy use will be bolstered by the completion of the 18,200-megawatt Three Gorges Dam and several other large hydropower projects. The Three Gorges project is scheduled to be fully operational by 2009. Although India currently produces far less hydroelectricity than China, hydropower still accounted for 22 percent of its installed electricity capacity in 1997 [6].

Regional Highlights

Developing Asia

Of all world regions, Asia is expected to show the most robust rate of growth in electricity consumption over the forecast period. Electricity demand in developing Asian nations is expected to grow at nearly a 5-percent annual rate between 1997 and 2020. Developing Asia accounted for 17 percent of worldwide electricity consumption in 1997, and by 2020 it is expected to account for 28 percent.

Coal, which supplied 60 percent of the fuel used to generate electricity in Asia in 1997, is expected to maintain that level through 2020, and in the rapidly growing Asian energy market, coal consumption in absolute terms is expected to more than double over the same period. Nuclear, renewables, and oil are expected to lose market share. Natural gas is the only fuel that is expected to increase its share of the Asian electricity market, from 8 percent in 1997 to 12 percent in 2020.

The financial and economic crisis that started in Thailand and quickly spread to other economies of Southeast Asia in mid-1997 has eased considerably. By 1999, most Asian nations began to show positive rates of economic growth, and developing Asia as a whole is expected to reestablish its trend of economic growth soon.

China

Overall, China is expected to add more to its electricity generation capacity between 1997 and 2020 than any other nation in the world—for example, more than twice the capacity additions projected for the United States. China is far and away developing Asia's largest economy, accounting for roughly one-third of the region's economic activity. China has also had the region's fastest rate of economic growth in recent years. Although its rate of economic growth has slowed over the past year or two, the Chinese economy was not dramatically affected by Asia's economic crisis.

China's current 237,000 megawatts of installed electricity capacity is second only to that of the United States [7]. Electricity consumption is expected to grow at a 5.7-percent annual rate over the 1997-2020 period. China's fast pace of future electricity consumption growth is due in part to its current underdeveloped electricity sector. Per capita consumption of electricity is currently one-twentieth of that in the United States.

Coal currently accounts for 75 percent of China's electricity fuels market, and its share is expected to remain stable through 2020. Clearly, however, if the Kyoto Climate Change Protocol is ratified, China could become an ideal candidate for joint implementation agreements to mitigate growth in carbon emissions.

China has the world's second largest coal reserves and is both the world's largest producer and consumer of coal. However, its coal reserves generally lie in the interior region of the country, far away from coastal economic activity. China is currently promoting the building of minemouth electricity plants rather than constructing additional rail lines to transport coal to eastern regions [8].

After coal, renewables account for the second largest share of China's electricity market, with an 18 percent overall share in 1997. Hydroelectric capacity in the country is expected to double between 1997 and 2010 and to increase its share of China's total electricity market. By the time it becomes fully operational in 2009, the $30 billion Three Gorges Dam will have an installed capacity of 18,200 megawatts of power. After 2010, growth in renewable energy is expected to moderate, and its share of the electricity market is expected to start to fall.

Although nuclear power currently accounts for a very small share of China's electricity market (approximately 1 percent in 1997), the Chinese government has an ambitious plan for additional nuclear power over the next two decades. By the end of the forecast period, nuclear power plants are expected to supply nearly 4 percent of the electricity used in China.

During the late 1980s, China implemented electricity reforms aimed at reducing government's managerial role in electricity supply [9]. The government allowed for a "fuel cost rider" in 1987, permitting generation companies to pass on higher fuel input costs to consumers [10]. More recently, price reforms have been undertaken to increase the attractiveness of investments in China's electricity sector, which had periodically suffered from capacity shortages. One such reform was implemented in 1996 during the financing negotiations surrounding the Laiban B project (a 700-megawatt coal plant). In awarding the contract for the financing of Laiban B, rather than negotiating an allowable rate of return, China's government chose to auction off the project to bidders offering the lowest tariff per kilowatt. Before the Laiban B deal, foreign investors had often criticized China's allowable rates of return on electricity investment for being too low.

Price reform is another means by which the Chinese government has attempted to attract private capital investment in electricity. In 1998, China deregulated electricity prices for rural areas [11]. In 1999, China's government announced plans to allow generators to bid competitively for access to power networks [12].

India

Second only to China among developing countries in terms of population and economic activity, India is expected to increase its consumption of electricity at a 4.7-percent annual rate over the forecast period. Heavy reliance on coal as an electricity fuel is expected to lessen somewhat, with coal's share of the market declining from 78 percent in 1997 to 63 percent in 2020. Natural gas and renewables will largely make up for coal's lost share. In 2020, natural gas is expected to account for 14 percent of India's electricity fuels market, up from 6 percent in 1997. The renewable fuel share of the electricity market is also expected to rise, from 12 percent to 17 percent, and the nuclear share is expected to increase from 2 percent in 1997 to 4 percent in 2020.

As in China, foreign investment will play a key role in the financing of India's power sector expansion. The Indian government opened up the power sector to private investment in 1991 with the passage of an amendment to the 1948 Electricity Supply Act that allowed for the construction of independent power projects.

In December 1996, the Indian central government announced its policy for electricity development [13]. Called the "Common Minimum National Plan for Power," the policy intends to restructure and corporatize the state electricity boards, to allow them greater autonomy, and to allow them to operate along commercial lines. The plan also attempts to ease the approval process for private power projects selected for competitive bidding by the central government. In June 1998, the central government went several steps further and eased its rules for foreign investment in the power sector. Automatic approval is to be given to projects costing in excess of 15 billion rupees (about $355 million) that involve 100 percent foreign equity.

The removal of subsidies flowing from urban electricity consumers to rural users has been a serious issue as India has undertaken electricity reform. The subsidies have been substantial, and their removal would in some Indian regions lead to sizable increases in rural electricity rates. The Indian government's Electricity Regulatory Commission issued an ordinance in 1998 directed at rationalizing electricity tariffs and subsidy policies. Under the new ordinance, the state regulatory entities would have the authority to remove rural subsidies [14].

India is also in dire need of an upgrade of its transmission system. Currently, as much as 20 percent of India's electricity is lost [15], much of it through "nontechnical" losses from theft or leakages and from errors in meter reading, accounting, and billing procedures [16].

Other Developing Asia

Developing Asian nations other than China and India also are expected to see rapid growth in electricity consumption over the coming years. Although in 1997 and 1998 many Asian economies slipped into recession— some for the first time in recent memory—the previous economic growth trend is expected to be reestablished over the next few years. Most of developing Asia had positive growth in 1999. Electricity consumption for the collective region is expected to grow at a 3-percent annual rate between 1997 and 2020.

In 1997, the region as a whole depended most heavily on coal (which supplied 31 percent of electricity) and oil (20 percent). By 2020, oil's share is expected to fall to 16 percent. No other world region outside the Middle East currently depends so heavily on oil as a source of electricity generation. Renewables are also projected to decline in importance, falling to 13 percent of the electricity fuels market by 2020 from 21 percent in 1997. Little additional nuclear capacity is expected to be built in other developing Asia, except in South Korea, and the nuclear share is expected to fall from 18 percent in 1997 to 13 percent in 2020.

Natural gas is expected to supplant oil and renewables in large measure. From 19 percent of the region's electricity fuels market in 1997, the natural gas share is expected to increase to 31 percent by 2020. In the near term, growth in natural-gas-fired generation is hampered by a lack of transportation infrastructure. For instance, virtually all of Taiwan's natural gas demand is met by imported LNG. In the long term, natural gas supplies might arrive via pipelines connecting the Caspian sea region with China and perhaps Japan, and natural gas pipelines may some day connect gas reserves in Indonesia to electric power plants in other Southeast Asian nations.

Japan

In Japan, electricity consumption is projected to grow by 1.3 percent per year over the *IEO2000* forecast period, reflecting of the nation's advanced level of economic development and slow population growth. Among the industrialized nations, only France relies more heavily than Japan on nuclear, which currently provides more than one-third of Japan's electricity. Japan is expected to continue construction of nuclear power plants and to increase its reliance on nuclear power from 35 percent of its total electricity needs in 1997 to 36 percent in 2020.

Growing public opposition to nuclear power in Japan could intensify in the future and perhaps reverse the national commitment to expansion of nuclear capacity. On September 30, 1999, Japan's worst nuclear accident occurred when workers at a nuclear fuel facility in Tokaimura set off an uncontrolled nuclear reaction that resulted in the death of one worker from radiation exposure.

Natural gas consumption in Japan is mostly in the form of LNG use, which is expected to grow slightly over the forecast period. Japan is by far the world's largest importer of LNG, most of which comes from Indonesia and Malaysia. The possibility of constructing a pipeline to import Russian natural gas from Sakhalin Island has also been considered. In contrast, Japan's dependence on oil as a generating fuel is expected to drop sharply by 2020. Coal, which accounted for 15 percent of Japan's electricity fuels market in 1997, is expected to claim a 18-percent share by 2020.

In the near term, Japan's inability to extricate itself from its current economic difficulties has raised some doubts about when the Japanese economy will return to its earlier growth rate. The current economic problems stem in part from a financial system that had grown dangerously over-leveraged and averse to reform. Although reform measures in the past have often been insufficient to produce the intended results, current efforts involving major fiscal stimulus packages are given better odds because of their breadth and scope. Government expenditures in the first quarter of 1999 produced a temporary surge in economic activity, but the pace of growth quickly abated in the second and third quarters of the year. At the beginning of 2000, it remains uncertain when and whether Japan will return to its long-run economic growth trend.

Electricity prices in Japan are among the highest in the world. As a result, in April 1995, Japan amended its Electricity Business Act, seeking to force open access in generation and to allow nontraditional suppliers to engage in direct sales. Before the amendment, sales by nontraditional suppliers required approval from one of 10 traditional generation companies. The amendment also allowed for tariff reform, giving electricity suppliers more discretion in setting prices. Wholesale wheeling was also introduced. Reforms were furthered in 1998, when the Ministry of Trade and Industry allowed industrial companies (nonutilities) to sell electricity directly to large consumers. In March 2000, industrial consumers of electricity with consumption exceeding 2 megawatts of power will be allowed greater choice in selecting suppliers. In May 1999, the Japanese Diet approved measures to implement limited retail competition.

Central and South America

In the *IEO2000* projections, the growth in electricity consumption in Central and South America is second only to that in developing Asia. Electricity consumption growth is expected to average 4.2 percent per year over the 1997-2020 forecast period. Brazil, which accounts for about half the region's economic activity and population, is expected to see electricity consumption growth of 4.9 percent per year. In the very near term, however, the developing countries of Central and South America remain the greatest casualties of the Asian financial crisis. The region as a whole experienced negative economic growth in 1999, although the economic recession in Brazil was not as pronounced as many analysts feared at the beginning of 1999.

Foreign Investment in the Electricity Sectors of Asia and South America

Two regions of the world, developing Asia and South America, have been particularly active in attracting foreign investment in their electricity sectors, which will have profound impacts on the landscape of their electric power industries. In developing Asia, foreign participation in the electric power industry is largely restricted to greenfield generation projects and joint ventures. Although the emerging economies of developing Asia currently are attracting the greatest amount of foreign investment, privatization of the electric power industry in many South American countries makes the region, in some ways, a more attractive market for international companies.

In order to raise the investment capital needed to support Asia's rapid growth in electricity demand, many Asian nations have undertaken the privatization of their electricity assets. Developing Asian nations have been typically less open to outright acquisition of electricity assets. Most private investment has been limited to greenfield electricity projects (see figure), typically under build-operate-transfer (BOT)[a] or build-operate-own (BOO) arrangements. Malaysia, Indonesia, Pakistan, and Thailand have favored the BOO structure, whereas China and the Philippines have favored BOT arrangements.[b]

For the 1990-1997 period, the world's top 10 national targets of foreign investment in electricity included China (ranked second), the Philippines (ranked fourth), Indonesia (fifth), India (sixth), Pakistan (seventh), Malaysia (eighth), and Thailand (tenth).[c] The focus of foreign investment in developing Asia's electricity projects has been on the construction of new generation facilities. Between 1990 and 1997, foreign participation was involved in 57 percent of the total investment in the region's greenfield generation projects.[d]

Foreign investment has played a critical role in financing the expansion of China's electric power infrastructure and is expected to play an even more important role in the future. In China, most foreign investment in electricity has been restricted to joint ventures. A World Bank study found that private participation in relatively large electricity projects has generally

involved less than 50 percent ownership.[d] For the 1997-2000 period, China expects foreign investment to supply 20 percent of its electric power investment capital.[e] To increase their access to overseas capital, several of China's electricity companies have recently acquired listings on the New York Stock Exchange, as well as stock exchanges in London and Hong Kong.

Some of the largest foreign investments in developing Asia's electricity projects have been marked by controversy. For example, a much publicized dispute surrounding Enron's building of the $2.8 billion Dabhol power plant in the Indian state of Maharashtra underscores some of the potential conflicts that arise between foreign investors and host governments. In 1995, a newly elected nationalist state government decided to cancel the Dabhol project after Enron and its partners (Bechtel Enterprises and General Electric) had spent $300 million. The new government alleged that the previous government had secretly negotiated the contract with Enron under terms that disadvantaged consumers and unfairly advantaged Enron. At the time, the cancellation had the effect of jeopardizing the credibility of India's economic reform program. Enron later

(continued on page 121)

Investment in Greenfield Electricity Projects by Region, 1990-1998

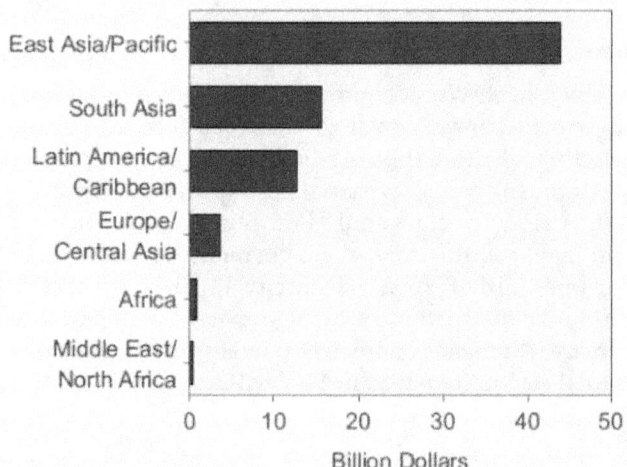

Source: World Bank, Private Participation Infrastructure Data Base.

---

[a]BOT agreements allow foreign investors to build a plant and operate it for a set period of time before transferring the investment to the host country.

[b]T. Soutar and J. Hanson, "Government Support Assists Asian Power Projects," *International Financial Law Review* (London, UK: Euromoney Publications, 1997).

[c]The countries were ranked in the following order: Brazil, China, Argentina, Philippines, Indonesia, India, Pakistan, Malaysia, Colombia, and Thailand.

[d]A.K. Izaguirre, "Private Participation in the Electricity Sector—Recent Trends," *Public Policy for the Private Sector* (Washington, DC: World Bank, September 1998).

[e]Speech delivered by Mr. Xie Songlin, Chief Economist, State Power Corporation, People's Republic of China, McGraw-Hill's 13th Annual Global Power Market Conference (New Orleans, LA, March 1998).

Foreign Investment in the Electricity Sectors of Asia and South America (Continued)

successfully renegotiated a deal with the state government in 1996, which called for a 22-percent reduction in electricity prices. Enron also sold the state government a 30-percent share in the project. The power plant, which is the company's largest overseas operation and India's largest foreign investment, is expected to have a capacity of 2,450 megawatts.[f]

Pakistan's Hub power plant proved even more controversial than the Dabhol project. Hubco owns more than 10 percent of Pakistan's generation capacity, and Hub was Pakistan's first major electricity project in recent times that involved foreign investment. The Hub deal was completed in 1995, and 1,292 megawatts of capacity came on line in 1996. As in Dabhol, a newly elected government leveled corruption allegations against the Hubco company and the former government that negotiated the Hubco project. The main customer of Hub, the Water and Power Development Authority of Pakistan, subsequently refused to make the full negotiated payment for its electricity purchases. National Power of the United Kingdom owns 25 percent of Hubco.

In Latin America, in contrast to Asia, foreign investment in electricity projects has proceeded more aggressively and with limited controversy. The need to attract foreign capital in order to expand and upgrade electricity infrastructure in developing countries has inspired a wave of privatization, particularly among Latin American nations. For many nations privatization has become the only effective method of raising capital on favorable terms. High levels of past public sector borrowing have saddled many nations with large levels of debt. As a consequence, many nations have had little recourse but to sell state assets to reduce debt, generate revenue and raise investment capital.[g] The figure opposite shows the dollar value of foreign investment in "divested" electricity assets by region. By this measure, Latin America exceeds the rest of the developing world combined. In large part, this is due to the relative openness of Latin American governments in the 1990s to foreign investment in electricity. Brazil has been the largest target of U.S. investment in South America, followed by Argentina.[h] On a per capita basis, Chile and Argentina have been the lead targets for such investments.

South America has also undergone a wave of energy deregulation and privatization and has turned into

a major competitive arena for some of the world's largest multi-scope multinational energy companies. Although Chile was the first nation in the region to embark upon energy privatization and deregulation, it was Argentina's move toward energy reform and privatization, which occurred 10 years after Chile's, that precipitated a continent-wide sea change in energy policies and a massive inflow of foreign investment. In Argentina, privatization also included steps to remove restrictions on foreign investment—another major development in Latin American energy policy.

The corporate response to privatization and deregulation of South American energy has been nothing short of historic. In the mid- to late 1990s, the continent saw a virtual swallowing up of newly privatized South American energy companies, many by newly privatized and/or deregulated energy companies from abroad—in particular, from the United Kingdom and United States. One major effect of electricity deregulation (in the United States) and electricity privatization (in the United Kingdom) was the removal of restrictions on foreign investments by electricity companies. Perhaps Latin America illustrates why the most compelling reason for the extraordinary flows of capital into overseas electricity projects is the most elementary one. Public policies were adopted on both sides of the Atlantic that made it possible. As barriers to foreign

(continued on page 122)

Investment in Privatized Electricity Projects by Region, 1990-1998

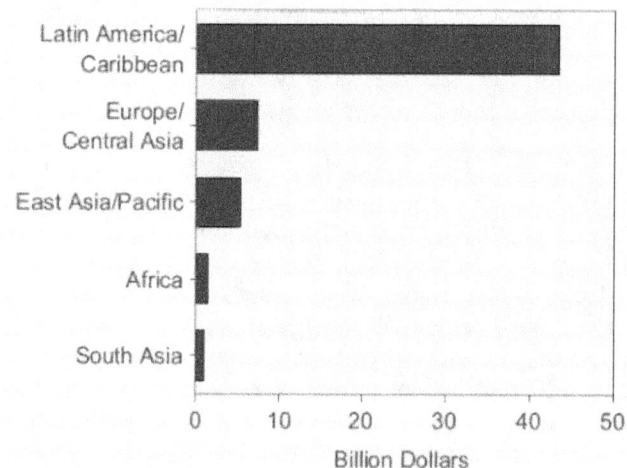

Source: World Bank, Private Participation Infrastructure Data Base.

---

[f]J. Karp and K. Kranhold, "Power Politics: Enron's Plant in India Was Dead; This Month It Will Go on Stream," *The Wall Street Journal* (February 5, 1999).

[g]World Bank, *The World Bank Report of 1994* (New York, NY: Oxford University Press, 1995).

[h]A.K. Izaguirre, "Private Participation in the Electricity Sector—Recent Trends," *Public Policy for the Private Sector* (Washington, DC: World Bank, September 1998), p. 4.

Energy Information Administration / International Energy Outlook 2000   121

investment in energy have fallen across the globe, billions of dollars in foreign investment have followed.

Many of the same U.S. natural gas and electricity corporations that were active in the UK market after privatization there have since moved aggressively into South America, as have many of the recently privatized UK energy companies. Privatization and an opening to foreign investment has opened doors to cross-border investments by other European companies and by indigenous South American companies. For example, Endesa, the recently privatized Spanish utility that is Spain's largest generation company, and Chilgener, the privatized Chilean utility, have also made major cross-border energy investments in South America.

Between 1990 and 1997, foreign investors channeled more than $45 billion into Latin American electricity investments,[i] and over the next few years Latin America could surpass Asia as the largest target of foreign investment in electricity. The privatization of Brazil's electricity industry alone, which began only in 1997, will, once completed, attract an estimated $60 billion.[j]

Privatization of electricity in Brazil has been taking place in the context of a concerted effort at overall reform of the Brazilian economy and the privatization of several other state-owned industries. The privatization of electricity in Brazil will be one of the world's largest privatization efforts on record, yielding billions of dollars to Brazil's federal and state treasuries. Privatization-related Brazilian electricity asset sales were the largest energy-related financial transaction (and privatization-related transactions) in 1997, 1998, and 1999. By late 1997, roughly $60 billion in Brazilian electricity assets has been slated for privatization.[j] Privatization of electricity in Brazil has attracted billions of dollars in foreign investment, the largest portion of which has come from U.S.-based companies.

In October 1999, a setback to foreign investment in Brazil's electricity industry occurred when the governor of the Brazilian state of Minas Gerais suspended the directors representing the Southern Company (a U.S. corporation) from the board of the electric utility company, Cemig, in which Southern had purchased shares in 1997.

[i]A.K. Izaguirre, "Private Participation in the Electricity Sector—Recent Trends," *Public Policy for the Private Sector* (Washington, DC: World Bank, September 1998).

[j]R.D. Feldman, "Brazil's Power Privatization: Prelude to New Infrastructure Development Approaches," *The Journal of Project Finance*, Vol. 3, No. 3 (Fall 1997), p. 13.

Much of Central and South America's electricity consumption growth will stem from expanded access to national electricity grids for a growing segment of the population. Currently, roughly 30 percent of Central and South America's population has no access to the grid, and per capita electricity consumption for the region is roughly 12 percent of that in the United States.

For several decades, hydroelectric power has dominated electricity supply in Central and South America. In the region as a whole, renewables accounted for 77 percent of electricity supplied in 1997. The region's reliance on hydropower is expected to lessen over the forecast period, with renewable energy (largely hydropower) accounting for only 54 percent of total electricity generated in 2020. A lack of suitable sites, long startup times, cost overruns, and concerns over displaced populations and environmental damage have all worked to diminish the attractiveness of hydropower investments. The reliability of hydroelectric power also became a growing concern during the drought years of the late 1990s.

Central and South America will rely increasingly on natural gas as a fuel for new electricity generation. The share of the electricity market supplied by generation from gas-fired power plants is expected to grow from 11

percent in 1997 to 34 percent in 2020. Oil, coal, and nuclear power currently account for 12.2 percent of the region's total electricity generation, and that share is expected to remain relatively stable over the forecast period.

The growing role of natural gas in electricity generation is contingent on the completion of several major pipeline operations linking producing countries, such as Argentina and Bolivia, with consuming countries, such as Chile and Brazil. In 1999, a 2,000-mile pipeline linking Bolivian natural gas fields to Brazil was completed, and a 600-mile pipeline running from the Noroeste Basin in Argentina to Mejillones, Chile, began commercial operation. The Argentina/Chile pipeline was a joint venture between CMS Energy (of the United States), Endesa (of Chile), and two Argentine gas producers, Pluspetro Energy and Astra.

An international electricity market is also evolving in South America. Currently, Argentina, Brazil, Venezuela, Chile, and Ecuador are completing a unified electricity transmission system. The system is being established in part to help diminish hydroelectricity shortages related to droughts, which have been fairly serious in recent years [17].

In Brazil, the region's largest energy-consuming country, electricity use is expected to grow by 4.9 percent per year from 1997 through 2020. Currently, hydroelectric power accounts for 95 percent of all electricity generated in Brazil. Over the forecast period, Brazil is expected to continue to invest heavily in hydroelectricity. By 2020, however, hydropower is expected to supply a slightly smaller share (87 percent) of the Brazilian market, largely as a result of the completion of nuclear, oil, and natural gas generation units. Historically, thermal electricity generation has been used in Brazil to offset seasonal swings in water power. Coal is expected to continue to play a small role in Brazil's future electricity fuels market, contributing about 1 percent of the total electricity supply for the country throughout the forecast.

Western Europe

Western Europe is expected to average roughly 1.7-percent annual growth in electricity consumption over the projection period, accompanied by a major shift in electricity sector fuel use. In 1997, nuclear power provided more than one-third of Western Europe's electricity supply—more than any other generating source. By 2020, however, the nuclear share is expected to drop to around one-quarter of the market. Natural gas will largely displace nuclear as the dominant electricity fuel in the European electricity market, accounting for 28 percent of the total in 2020. The growing availability of North Sea gas, as well as gas imports from Algeria and Russia, has provided a foundation for the expansion. Growing concerns over the damaging environmental effects of coal and oil combustion will also encourage the switch to gas.

The United Kingdom, with its ample North Sea gas supplies, is expected to see the most significant increase in reliance on natural gas as a generation fuel. By 2020, natural gas is projected to account for 49 percent of the electricity generation fuels market in the United Kingdom, up from 22 percent in 1997. Germany also is expected to reduce its reliance on nuclear power (from 33 percent in 1997 to 21 percent in 2020) and increase its reliance on gas (from 9 percent in 1997 to 17 percent in 2020). Ample natural gas supplies are available from Russia and the North Sea to accommodate the transition. France is the only European country expected to continue heavy reliance on nuclear power, although the nuclear share of its electricity fuel market falls from 80 percent in 1997 to 73 percent in 2020.

In part, Europe's growing alienation from nuclear power can be traced back to the 1986 accident at Chernobyl. In addition, falling prices for coal and natural gas have diminished the economic value of nuclear power. In Germany, the newly elected Social Democratic chancellor, Gerhard Shroeder, has proposed a phaseout of nuclear power generation starting in 2002. (A newly elected Social Democratic government made a similar pledge in Sweden in 1994; however, none of that country's nuclear power plant was closed until the Barsebaeck 1 unit was shut down on November 30, 1999.)

For more than two decades, Western Europe has been reducing its reliance on coal and oil as electricity generation fuels. In 1970, coal accounted for 40 percent of the generation market and oil 22 percent. In 1997, coal and oil accounted for 25 percent and 8 percent of the electricity market, respectively. By 2020, coal's share of the market is expected to slip to 16 percent, with oil's share remaining at 8 percent.

Western Europe is currently the scene of attempts to create a continent-wide wholesale market in electricity. A 1996 directive by the European Community required all signatories to open their domestic electricity markets to new suppliers starting in February 1999.[20] In the initial implementation period (during 1999), 23 percent of each participating nation's electricity market was to be opened to competition. In 2006, signatory countries will be required to open up one-third of their electricity markets to new suppliers. The directive allows for two methods of open competition: one method involves negotiations through third-party access; the other is a single-buyer model, which allows a single buyer (for example, Electricite de France) to be the purchasing agent for all electricity sold by new suppliers in the French electricity market. The more restrictive single-buyer approach is favored by France and Germany. The onset of competition has led to some transnational acquisitions by European electricity companies eager to engage in cross-border trade.

Recent changes in European governments have set new directions in energy policies. Shortly after being elected, the United Kingdom's new Labor prime minister, Tony Blair, announced a windfall profits tax on the country's newly privatized electricity companies. U.S. companies had purchased 9 of 12 recently privatized UK distribution companies. Largely due to imposition of the windfall profits tax, those companies took charges of over $3 billion. In 1998, Dominion Resources announced that it would sell its regional distribution company, East Midlands Electricity, to PowerGen, the second-largest electricity generation company in the United Kingdom. Similarly, Entergy, a Louisiana company, sold its UK distribution company, London Electricity, to Electricite de France.

[20] Greece and Ireland were allowed to delay compliance with the directive for 2 years.

Europe has also seen some moves to introduce consumer choice in the retail electricity market. The recently privatized electricity industry in the United Kingdom moved one step closer to full competition in September 1998, when London residents became eligible to nominate their preferred electricity suppliers. In June 1999, virtually all households were given the option of choosing electricity suppliers. The Nordic nations implemented fully competitive supply markets at the retail level starting in 1996, and today households in Finland, Norway, and Sweden are allowed to choose their electricity suppliers (see box on pages 125-126).

Perhaps the most surprising development in European electricity relates to the heightened level of competition among electricity suppliers in Germany. In 1995, German electricity prices were the highest in Europe, but industrial electricity prices there are estimated to have fallen by 37 percent between 1995 and 1999 (Figure 82). Part of the decline in German industrial prices is, however, attributable to the abolishment of the Kohlepfennig, a tariff on electricity used to support Germany's coal industry. The tariff, which amounted to 8.5 percent of the final price, was abolished in 1995. Electricity companies have established retail marketing affiliates, and prices to households have also fallen, although less dramatically. Electricity price wars have become commonplace in the German market. Excess German electricity capacity, as well as access to inexpensive electricity imports, may also have contributed to the price decline. For the most part, however, the sharp price decline has largely been attributable to recent German electricity deregulation initiatives. In April 1998, the German government decided to exceed the requirements laid out in the European Community electricity directive by allowing industrial consumers to buy electricity freely 1 year earlier than required.

One drawback to the decline in German electricity prices is the effect it might have on Germany's nascent renewables industry. Wind power, which currently receives a subsidy, might be most threatened by the decline in prices for electricity from conventional power sources. Wind power subsidies are currently based on electricity prices, and as prices fall so too does the subsidy. The head of Preussenelektra predicted that subsidy declines would discourage future investment in German wind power [18]. Currently Germany has a goal of using wind power to provide 10 percent of the nation's power requirements by 2010 [19].

Many analysts feel that intensified competition will lead to an eventual consolidation of Germany's highly fragmented electricity industry, which comprises some 900 different companies. Ninety percent of generation, however, is produced by 8 major "supraregional" utilities [20]. In 1999, the most notable European merger and

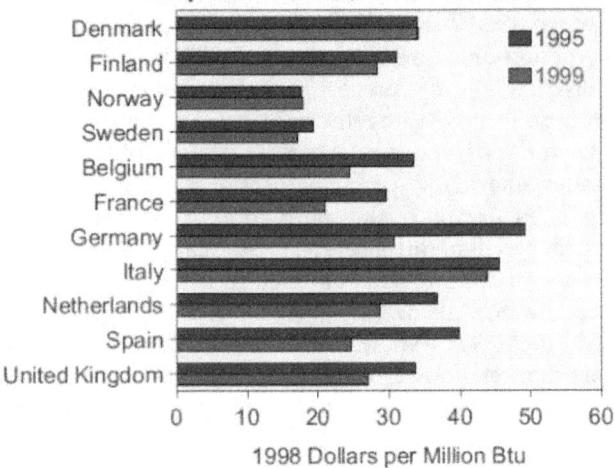

Figure 82. Industrial Electricity Prices in Western European Nations, 1995 and 1999

Note: 1999 data are projections.
Source: Standard & Poor's Platt's, *World Energy Service: European Outlook*, Vols. I, II, and III (Lexington, MA, 1999).

acquisition in electricity took place in Germany. In August 1999, two of Germany's electricity behemoths, Viag, which operates Bayenwerk, and Veba, which operates Preussenelektra, announced an intention to merge. Veba and Viag are ranked second and third in size among German electricity companies. Both companies are also among the eight German transmission companies [21].

Eastern Europe and the Former Soviet Union

In the FSU, natural gas is expected to displace coal as a generation fuel. Coal, which accounted for 16 percent of the fuels market in FSU countries in 1997, is expected to account for just 5 percent in 2020, with the natural gas share growing from 49 percent in 1997 to 61 percent in 2020. In 1997, no industrialized country in the world relied so heavily on natural gas as a generation fuel. The nuclear share of the FSU fuel market is expected to slip from 11 percent in 1997 to 8 percent in 2020, and the renewable energy share is expected to fall slightly, from 11 percent in 1997 to 10 percent in 2020. Oil's share is expected to increase slightly, from 12 percent in 1997 to 16 percent in 2020.

The Eastern European countries rely more heavily on coal as a generation fuel. Coal accounted for a 61-percent share of the Eastern European electricity market in 1997. Coal's share of the market is expected to fall to 29 percent by 2020, largely being replaced by hydropower and imported Russian natural gas. The role of nuclear power as an energy source in Eastern Europe is expected to fall steadily over the forecast period, from 13 percent in 1997 to 10 percent in 2020.

The FSU and much of Eastern Europe suffer from an antiquated electricity supply infrastructure. Although

Nordic Electricity Reform

Along with the United Kingdom, the nations of Nordic Europe (Sweden, Finland, Norway, and Denmark) have been at the forefront of electricity reform, which has had profound effects on the region's electricity market and industrial structure. On January 1, 1991, Norway's Energy Act became law. The goal of the Energy Act was nothing short of transforming the Norwegian electricity market from a highly regulated environment to one that would be, in large measure, deregulated. Swedish electricity reform followed 5 years later, and Denmark and Finland introduced reforms 6 months after Sweden's.

The driving forces behind Nordic electricity reform included resolving the problems of large regional price variations, noncompetitive business practices, suboptimal capacity usage, and excess productive capacity. The reforms involved the creation of pools for the trading of electricity; the separation of generation from transmission operations; unbundling of services and prices; and the requirement that transmission companies provide nondiscriminatory open access to their grids. They also provided retail customer choice—a first in global electricity reform.

With the exception of Denmark, the Nordic countries did not privatize their state-owned electricity companies as did the United Kingdom. The national companies in both Norway and Sweden are, for the most part, the dominant generation and transmission enterprises, whereas most of their distribution companies are municipally owned or cooperative ventures. For all the Nordic countries, the purpose of reform was to compel the industry to behave more as a commercial enterprise than as a government-sanctioned monopoly. As in the United Kingdom and the United States, however, electricity reform took place in the context of a national effort to introduce reform in industry in general and, more specifically, as an effort to introduce reform in state-owned and municipally owned industries, or those with monopolistic qualities. In Sweden, for instance, since the mid-1980s industries such as long-distance freight, postal services, air travel, and telecommunications have been deregulated.[a]

Unbundling of operations became a major goal of electricity reform in the Nordic countries. Generation and marketing were considered potentially competitive operations, whereas transmission and distribution were still considered natural monopolies and thus in need of continued regulation. In 1992, Norway's Statkraft, the government-owned electricity enterprise, was restructured. Statkraft was required to divest its transmission operation, which formed the foundation of a new company, Statnett SF. Statkraft was to become purely a generation company, retaining the former company's generation assets. Norwegian electricity prices were also unbundled for the sake of providing greater market transparency.

In 1992, Sweden's Vattenfall, formerly a public service utility, now a profit-maximizing state-owned corporation, was also separated into an electricity company with generation and distribution and a transmission enterprise. The former retained the name Vattenfall, the latter acquired the name Svenska Kraftnät. Vattenfall competes with other power generation companies; Svenska Kraftnät is responsible for managing the national grid. One aim of Sweden's reform was to provide open, nondiscriminatory access to transmission facilities by companies other than the large generators.[b]

The corporate responses to electricity reform were both anticipatory and reactionary. They included widespread mergers and acquisitions leading to greater industry concentration, increased vertical integration, inward and outward foreign investment, and the creation of a variety of new energy customer services.

Mergers and acquisitions did in fact start in the 1980s and increased around 1990, in part because industries and municipalities were selling off electricity assets, and partly in anticipation of the reform. In Sweden, the number of distribution companies fell from 525 in 1976 to 290 in 1990 and 224 in 1998.[c] In Norway, the number of distributors fell from around 230 before reform to roughly 200 by the mid-1990s.[d] Several of Sweden's larger generators, particularly Vattenfall and Sydkraft, have acquired regional distribution companies. Between 1993 and 1996, generating companies acquired 15 distribution companies.[d] Typically, mergers and acquisitions following deregulation in Norway, Sweden, and Finland have involved takeovers of regional distribution companies by large generation companies.

(continued on page 126)

---

[a]Swedish Competition Authority, *Deregulation of the Swedish Electricity Market*, Summary of Report of 1996:3 (Stockholm, Sweden, 1996).

[b]Svenska Kraftnät, "The Swedish Electricity Market Reform and Its Implications for Svenska Kraftnät" (not dated).

[c]"The Structural Changes in the Swedish Electricity Market," Sveriges El Leveratorer corporate web site, www.svel.se/english/market.htm (December 28, 1999).

[d]International Energy Agency, *Energy Policies of IEA Countries: Sweden 1996 Review* (Paris, France, 1996).

Foreign companies have also acquired electricity assets in the Nordic countries. Electricite de France obtained interests in Sweden's Graninge and Sydkraft, although it later sold off its Sydkraft holdings. Germany's Preussenelektra also acquired an interest in Graninge. In a further merging of German and Swedish corporate interests, Preussenelektra acquired an interest in Sydkraft at the same time that Sydkraft took a corresponding interest in VEBA, Preussenelektra's parent company. Preussenelektra has since increased its holdings in Sydkraft. A troika of national ownership emerged when Norway's Statkraft acquired substantial shares in Sydkraft. Currently, the voting majority of shares in Sydkraft lies in foreign hands. Until 1990, Sydkraft was a municipally dominated company. The U.S. corporation, Enron, has established itself as a Nordic electricity trading company, expanding its Nordic marketing business from 3 million megawatthours in 1996 to 8.4 million megawatthours in 1997.[e]

In the Nordic countries, deregulation has laid the foundation for several new electricity companies—much as it has in the United Kingdom and the United States. The new companies serve as brokers, traders, and marketers of energy services. For example, Sweden's largest electricity company, Vattenfall, sells electricity directly to households, industrial companies, and commercial businesses and also sells natural gas, heat, and other energy products and services.[f] Regional oil companies, such as Statoil and Neste, have also become Nordic electricity suppliers and have reclassified themselves as full-service energy providers.[g] Statoil expects to make substantial inroads into the household market with the recent abolishment of the meter requirement.[h] Norsk Hydro has created a subsidiary, Hydro Energy, which in addition to producing electrical power in Norway trades power in Norway and Sweden and electricity and natural gas in the United Kingdom and Belgium.[i] Stockholm Energi has discussed a cooperative arrangement with the telecommunications company, Glocalnet.[j]

[e]Enron Corporation, *Annual Report 1997* (Houston, TX, 1998), p. 21.

[f]"The Vattenfall Group," Vattenfall company web site www.vattenfall.se (not dated).

[g]J.P. Hirl, "Nordic Gas & Power; A Blueprint for the Rest of Europe," *Petroleum Economist*, Vol. 65, No. 3 (March 24, 1998), pp.12-13.

[h]The earlier requirement that consumers install hourly meters in order to change electricity suppliers was considered a major impediment to competition at the household level.

[i]"Countries in Which We Operate," Hydro Energy corporate web site www.hydro.com (not dated).

[j]"Information Regarding the Cooperation Between Stockholm Energi and Glocalnet AB," Press Release (August 11, 1998), web site www.glocalnet.com/press/pr-980811-en.stm.

electricity demand is expected to be 43 percent higher in 2020 than in 1997, the region is not expected to see much in the way of capacity expansion, although the fuel mix will involve a movement away from coal to natural gas. Rather, future investment will be directed in large part to upgrades, in efforts to bring the region's electricity industry up to the standards of those in the industrialized nations.

North America

United States

U.S. electricity consumption is projected to grow by an annual average of 1.2 percent over the 1997-2020 *IEO2000* forecast period. The projected increase is relatively small, because electricity intensity in the United States is already high relative to other industrial economies. Demand growth in the United States has slowed considerably since the 1960s, when electricity consumption was rising at a rate of 7 percent per year. Saturation of households with electronic appliances over time and efficiency improvements in appliances—partly as a result of new standards—are responsible for the slower growth in total electricity consumption [22].

The United States is expected to reduce its reliance on nuclear power as a source of electricity generation over the forecast period. Nuclear power, which accounted for 19 percent of the total energy consumed for electricity generation in the United States in 1997, is expected to have only a 10-percent share by 2020. Coal's share is expected to hold steady at roughly half the market. The renewable share of the fuels market is expected to fall from 20 percent in 1997 to 18 percent by 2020. Natural gas is expected to replace nuclear power in large part, with a 20-percent share in 2020 as compared with 10 percent in 1997. The last time a nuclear power plant came on line in the United States was 1996, and the last new order was placed in the late 1970s. No other units are expected to come on line in the future. Deregulation has forced decisionmakers to address the issue of stranded costs and could expedite the move away from nuclear power.

In 1999—a banner year for U.S. electricity company mergers and acquisitions as well as a number of alternative strategies being used to realign the industry (see box on page 128)—a record 26 electricity mergers and acquisitions were announced (Figure 83). The wave of activity resulted in part from major policy reforms whose origins go back to the late 1970s. The Public Utility

Figure 83. Mergers and Acquisitions in the U.S. Electricity Industry, 1992-1999

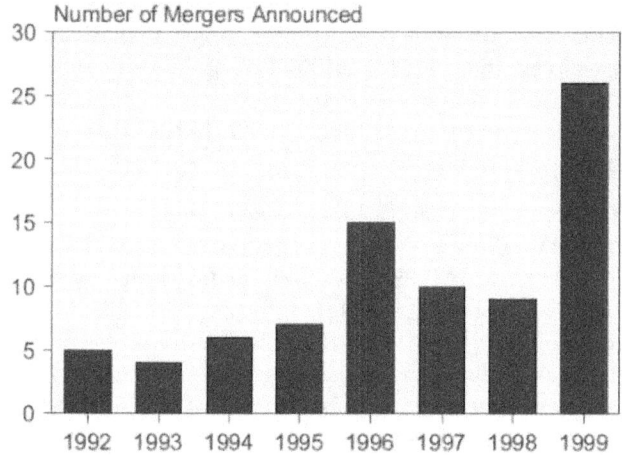

Number of Mergers Announced

Source: Data provided in a fax from Joan Esquivar of the Edison Electric Institute on February 22, 2000.

Regulatory Policy Act of 1978 (PURPA) mandated open access as a tool to achieve greater competition in electricity supply. More precisely, PURPA required transmission companies to "interconnect with and buy whatever capacity any facility meeting the criteria for a 'qualifying facility'[21] had to offer and had to pay that facility the utility's own incremental or avoided cost of production" [23].

The U.S. Government pushed open access a step further with the passage of the Energy Policy Act of 1992 (EPACT), which allows for wholesale power competition by creating a new class of wholesale generator and expanding the power of the Federal Energy Regulatory Commission (FERC) to order open transmission access [24]. EPACT also promoted eventual competition at the retail level. Based on the mandate derived from EPACT, the FERC issued Orders 888 and 889. Order 888 required electricity transmission lines to provide open access and unbundle power sales for transmission services. Order 889 required the establishment of an electronic trading system similar to the one that evolved in the natural gas market only a few years earlier.

The acceleration of activity in the late 1990s may also be in part the result of a change in the FERC's antitrust policy. In 1996, the FERC adopted the Department of Justice/Federal Trade Commission merger guidelines as a screening device to determine whether a proposed merger would cause an unacceptable increase in market power. In addition, the new policy uses a quantitative screen to derive the potential merger's impact on competition. The new policy also attempts to reduce the procedural steps involved in a review, along with a substantial reduction in the review time for most

mergers. Since its implementation of the new policy, more merger and acquisition approvals have been granted by the FERC, and the announcements of merger and acquisitions has accelerated [25].

The past few years have seen various types of mergers, both horizontal and vertical. There have been many geographic mergers between similar companies sharing contiguous territories, such as the one pending between American Electric Power (AEP) and Central and Southwest. AEP's territory covers Indiana, Kentucky, Michigan, Ohio, Tennessee, Virginia, and West Virginia; Central and Southwest operates in Arkansas, Louisiana, Oklahoma, and Texas. There have also been mergers between companies sharing overlapping territories. The Delmarva/Atlantic merger, the Kansas City Power and Light/Utilicorp merger, and the New England Electric/Eastern Utilities merger are examples.

Several recent electricity mergers and acquisitions have also resulted in more vertically integrated companies and companies that provide a host of energy and non-energy services. One example of a vertical acquisition is NGC Corp's takeover of Destec. NGC Corp markets energy in the United States and the United Kingdom and operates both the Natural Gas Clearinghouse and the Electric Clearinghouse. Destec is an independent power producer.

Some companies have decided to become strictly "wires" companies, while others have decided to become strictly generation companies. For instance, Amergen, a joint venture between Peco Energy and British Energy, the UK nuclear power company, has been buying up nuclear power generation assets. Similarly, Entergy has begun an attempt to acquire nuclear power assets. Economies of scale may produce some benefits in the operation of several nuclear power plants, as opposed to a single plant. For instance, the U.S. Nuclear Regulatory Commission sees greater safety in the operation of multiple nuclear units by a single company, in addition to economic benefits [26].

A number of companies over the past few years have announced plans to sell off all their generation units, or at least their non-nuclear units—nuclear units, apparently, being more difficult to sell. United Illuminating Company has announced that it will divest its fossil generation assets, and Unitil has said it will exit the generation business entirely. In 1997, Boston Edison and Montana Power also announced that they no longer intended to be in the power generation business.

Several companies have entered into entirely new lines of business that have the common element of connection to end users either by wire, pipe, or airwave. For

[21] A "qualifying facility" is defined as a cogeneration or small power production facility that meets certain ownership, operating, and efficiency criteria established by the FERC.

Alternatives to Mergers and Acquisitions in the U.S. Electric Power Industry

While 1999 was a particularly good year for mergers and acquisitions, there are also a number of alternative strategies that have been used in recent years to realign the U.S. electricity industry. One alternative has been to form joint ventures or strategic alliances. Joint ventures usually involve the creation of independent, standalone enterprises owned by the participating companies. Investor-owned utilities have entered into this form of business relation in order to engage in new areas of business, such as electricity trading. For example, Baltimore Gas and Electric has entered into a joint venture agreement with the Wall Street firm of Goldman Sachs, forming Orion Power Holdings to acquire merchant and other generating plants in the United States and Canada. Duke Power has formed a 50/50 partnership with Louis Dreyfus to provide energy services such as risk management, power marketing, and plant construction. Sonat and AlliedSignal have formed a strategic alliance to market and support AlliedSignal's TurboGenerator products. And the U.S. petroleum giant, Mobil, has formed a joint venture with PanEnergy to market natural gas at the retail level.

Another alternative that has been exercised is divestiture. Many investor-owned utilities have sold off large portions of their generation assets piecemeal. This may be based on several considerations. For one, as States deregulate, public utility commissions have become more concerned with market share and competitiveness in generation when reviewing merger and acquisition proposals. Secondly, several utilities have, as mentioned earlier, decided to specialize, some as power companies and others as wires companies. Thirdly, some companies have indicated a desire to exit the nuclear power business altogether, while others have decided to specialize in nuclear power.

Privatization has clearly been a driving force behind electricity mergers and acquisitions around the world. Had it not been for privatization, very few of the many cross-border utility acquisitions would have occurred. Until the early 1990s, most electricity companies operated almost solely within their national borders, largely because they had no other choice. Privatization has paved the way for billions of dollars of mergers and acquisitions in electricity and created opportunities for foreign investment in electricity that were largely absent before.

The figure below indicates just how sudden and substantial the flow of U.S. investment in overseas electricity has been. In 1998, overseas acquisitions by U.S. electricity companies totaled nearly $25 billion, up from $500 million a decade earlier. It should be noted that electricity merger and acquisition activity is part and parcel of a global trend toward increased international investment. Large-scale mergers and acquisitions have taken place in petroleum, banking, pharmaceuticals, telecommunications, and automobile manufacture in the late 1990s.

U.S. Direct Investment in Overseas Utilities, 1988-1998

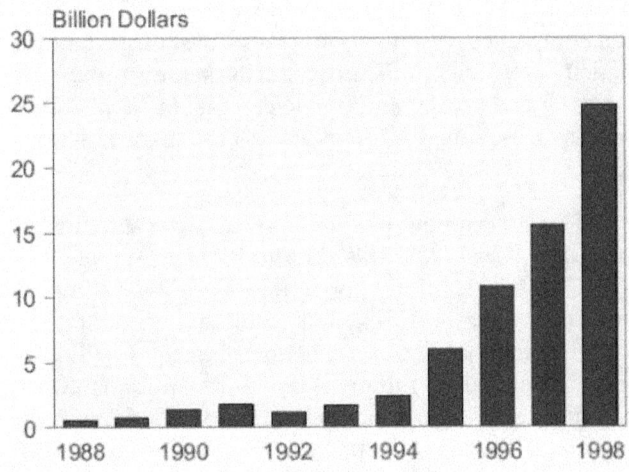

Note: These utilities include, in addition to electricity, natural gas distribution, and sanitary services. However, the sharp upward climb in these investments during the years 1994 through 1998 is almost entirely due to overseas electric utility investments by U.S. companies.

Source: U.S. Department of Commerce, Bureau of Economic Analysis, Survey of Current Business, (Washington, DC), various issues.

instance, Duke Energy provides its customers with water, natural gas, electricity, and telecommunications services. Electricity companies in the United Kingdom have made similar acquisitions in the decade following privatization and deregulation there. Nearly all the privatized distribution companies in the United Kingdom have formed or obtained telecommunications subsidies, and many have also merged with (or been acquired by) natural gas companies, water companies, and sewage companies.

Over the past few years, almost all the major U.S. natural gas transmission companies have merged with, acquired, or been acquired by electricity companies. Some examples include Duke and PanEnergy, Enron and Portland General, and the pending merger between Dominion Resources and Consolidated Natural Gas. At the distribution level there has also been some convergence of electricity and natural gas concerns. The merger between Houston Industries and NorAm is one example. Convergence mergers, whether they be between natural gas transmission companies and electricity companies or between natural gas distribution companies and electricity companies, generally have not received intensive regulatory review. In the cases of the Enron/Portland General and Duke/PanEnergy mergers, FERC approval came only 6 months after the merger applications were filed.

Another corporate response to policy changes has been the creation of a rapidly growing independent power industry, which has made for a more competitive environment in generation. In 1998, there were 109 independent power producers active in the United States, and they accounted for about 7 percent of existing capacity [27]. About half of all new capacity additions in the United States are currently being supplied by independent power producers.

Canada

Like the United States, Canada is also expected to see relatively slow growth in electricity consumption over the forecast period. Between 1997 and 2020, annual growth in Canadian electricity consumption is expected to average 1.2 percent. Canada also is expected to reduce its dependence on nuclear power over the coming years. Nuclear power, which accounted for 15 percent of total electricity generated in 1997, is expected to provide 8 percent of total supply in 2020. In fact, Canada may reduce its use of nuclear power even more dramatically. Currently, Ontario is reevaluating the safety of its nuclear power industry. In late 1997 and early 1998, Hydro Ontario shut down seven of its older power plants, or 17 percent (4,300 megawatts) of its operating capacity. At present, it remains uncertain whether the plants will be brought back on line sometime after 2000 as was intended. If the plants are prematurely retired, Canada's future dependence on nuclear power would be reduced even more. In addition, the loss of capacity could lead to a temporary reversal of electricity trade flows between the United States and Canada.

With its natural gas reserves being developed at a rapid pace, Canada will also depend less heavily on hydroelectricity in future years, although hydropower will still be its most important source of electricity in the 2020, maintaining roughly a two-thirds share of the nation's electricity supply. In recent years, opposition to the construction of new hydroelectric facilities has grown as a result of concerns about environmental damage and potential harm to native peoples. Natural gas is expected in large measure to make up for the projected reductions in nuclear power and hydroelectricity.

Mexico

Mexico, with a higher expected rate of future economic growth and a lower starting base in terms of per capita electricity consumption, is projected to lead North America in electricity consumption growth. Over the 1997-2020 forecast period, Mexico's electricity consumption is projected to grow by nearly 4 percent per year.

Plagued with a serious air pollution problem, Mexico has been moving aggressively away from oil-fired generation to natural gas. In order to finance future electricity infrastructure to meet the needs of a rapidly growing population and economy, the Mexican government is actively encouraging the development of private power projects. Several U.S. companies have undertaken investments in gas-fired generation facilities in Mexico. Although Mexico currently allows private investment in independent power production, all sales from such operations must be directly to the state-owned utility, Comision Federal de Electricidad (CFE).

Mexico has allowed foreign investment in power plants since 1992, but as of November 1999 only eight contracts had been awarded and only one had actually closed on financing [28]. In February 1999, the president of Mexico, Ernesto Zedillo, proposed the restructuring and privatization of CFE. The proposed restructuring would involve breaking up the state-owned company along functional lines—generation, transmission, and distribution—holding out the possibility that generation might one day be privatized [29].

Africa

South Africa accounts for almost two-thirds of the electricity generated on the African continent, and South Africa, Egypt, Algeria, Libya, and Morocco together account for roughly 90 percent of the continent's total electricity production. Africa as a whole is expected to see electricity consumption grow at a 3.7-percent annual rate over the 1997-2020 projection period. No other region has as little access to electric power as Africa. Coal provided roughly half of the region's electricity production in 1997, and in 2020 its share still is expected to be 38 percent.

Several African countries have recently opened their electricity sectors to private investment. In Morocco, CMS Energy and the Swedish/Swiss company, Asea Brown Boveri, began construction on the Jorf Lasfar power plant in 1997. The $1.5 billion plant will have a capacity of 1,360 megawatts upon completion in 2000 [30]. It is the largest plant of its kind to date in Africa, and eventually it will provide Morocco with about 30 percent of its electricity supply. The Egyptian cabinet in 1996 approved the startup of a BOT program involving 1,600 megawatts of power [31].

In 1990, the Ivory Coast began the first electricity sector reforms in Africa. The Ivory Coast is using a BOT arrangement to finance, build, and manage major infrastructure projects without increasing its debt level. Twelve projects have been proposed, and five have been awarded to private operators, including a new thermal electric generation facility near Abidjan [32]. Nigeria is also attempting to encourage foreign participation in electricity generation. In late 1998, Mobil, one of the largest producers of oil in Nigeria, announced that it had contracted to build a 350-megawatt natural-gas-fired independent power project in Nigeria [33].

In March 1999, Senegal announced the privatization of its electric power industry. In that same month, the Senegalese government sold 34 percent of the shares of the Société Nationale d'Électricité (SÉNÉLEC), to the French-Canadian consortium, Hydro-Quebec-International-ELYO (HQI-ELYO) for $69 million (U.S. dollars) [34]. As a result, the HQI-ELYO consortium became responsible for managing all electricity production, transmission, and distribution activities associated with SÉNÉLEC.

Middle East

Almost two-thirds of the Middle East region's economic output is accounted for by Iran and Saudi Arabia, along with half the region's electricity consumption. Iran is the most populous country in the Middle East, and Saudi Arabia has one of the highest per capita incomes. Other large users of electricity in the Middle East include Israel, Iraq, and Kuwait. Largely as a result of growth in the region's dominant economies, electricity consumption in the Middle East is expected to grow at a 3-percent annual rate over the projection period.

The Middle East depends heavily on petroleum to fuel its electricity generation. In 1997, oil-fired generation accounted for 36 percent of all electricity produced and natural gas 38 percent. That level of dependence is expected to continue over the forecast period. Over the next few years, Iran is expected to enter the league of nations owning nuclear power reactors, and by 2020 nuclear power is expected to account for 2 percent of the region's electricity production.

Among Middle Eastern nations, Israel took a step towards privatization recently. In 1996, Israel's parliament passed a new electricity law that will allow the Energy Minister to grant permits to independent power producers [35].

References

1. Energy Information Administration, *International Energy Outlook 1999*, DOE/EIA-0484(99) (Washington, DC, March 1999).

2. Energy Information Administration, *Assumptions to the Annual Energy Outlook 2000*, DOE/EIA-0554(2000) (Washington, DC, January 2000), p. 67.

3. BP Amoco, *BP Amoco Statistical Review of World Energy* (London, UK, June 1999), p. 28; and British Petroleum, *BP Statistical Review of World Energy* (London, UK, June 1993), p. 24.

4. British Petroleum, *BP Statistical Review of World Energy* (London, UK, June 1998), p. 28.

5. Energy Information Administration, *Annual Energy Outlook 1998*, DOE/EIA-0384(98) (Washington, DC, July, 1999), p. 295.

6. Energy Information Administration, *International Energy Annual 1997*, DOE/EIA-0219(97) (Washington, DC, April 1999), p. 95.

7. Energy Information Administration, *International Energy Annual 1997*, DOE/EIA-0219(97) (Washington, DC, April 1999), p. 95.

8. "AES Lands 2,100 MW Coal-Fired Deal in China," *Electricity Journal*, Vol. 10, No. 7 (August/September 1997), p. 15.

9. APEC's Energy Regulator's Forum, Electricity Regulatory Arrangements, "Summary Submission China," Department of Agiculture, Fisheries, and Forestry—Australia, web site http://dpie.gov.au/resources.

10. W. Chandler et al., *China's Electric Power Options: An Analysis of Economic and Environmental Costs*, PNWD-2433 (Washington, DC: Pacific Northwest Laboratory, Battelle Memorial Institute, June 1998), p. 10.

11. "China To Raise Electricity Prices," *Power Economics* (June 1998), p. 9.

12. "Beijing Bites Bullet on Power Sector Reforms," *Financial Times: Power in Asia*, No. 29 (November 1, 1999), p. 2.

13. I.M. Sahai, "A New Power Policy," *Independent Energy*, Vol. 27, No. 3 (April 1997), p. 33.

14. "The Private Sector: Cautiously Interested in Distribution in India," *The Electricity Journal*, Vol. 11, No. 5 (June 1998), pp. 23-24.

15. I.M.Sahai, "Privatizing India's Grid," *Independent Energy*, Vol. 28, No. 1 (January/February 1998), p. 26.

16. J.P. Banks, "The Private Sector: Cautiously Interested in Distribution in India," *The Electricity Journal*, Vol. 11, No. 5 (June 1998), p. 22.

17. "Business: Power to the People," *The Economist*, Vol. 346, No. 8061 (March 28, 1998), p. 61.

18. M. Hibbs, "German Backlash Forming Against Open Market," *Nucleonics Week*, Vol. 40, No. 39 (September 30, 1999), p. 38.

19. M. Hibbs, "German Backlash Forming Against Open Market," *Nucleonics Week*, Vol. 40, No. 39 (September 30, 1999), p. 38.

20. Standard and Poor's Platt's, *World Energy Service: European Outlook 1999*, Vol. 2 (Lexington, MA, 1999), p. 87.

21. A. Taylor, "Viag's Planned Merger With Veba Is Setting the Tone for Future Liberalization of German Power Supply," *Financial Surveys Edition* (September 23, 1999), p. 5.

22. Energy Information Administration, *Annual Energy Outlook 1998*, DOE/EIA-0383(98) (Washington DC, December 1997), p. 50.

23. Energy Information Administration, *The Changing Structure of the Electric Power Industry: An Update*, DOE/EIA-0562(96) (Washington DC, December 1996), pp. 27-29.

24. Energy Information Administration, *The Changing Structure of the Electric Power Industry: Selected Issues 1998*, DOE/EIA-0562(98) (Washington, DC, July 1998), p. 1.

25. Edison Electric Institute, *Financial Info, Mergers and Acquisitions* (Washington, DC, August 28, 1998).

26. R. Smith, "Fissionaries, Two Utility Executives See Potential Riches in Nuclear Stepchild," *Wall Street Journal* (October 28, 1999), p. A1.

27. McGraw Hill Companies, *Global Power Report* (July 24, 1998), p. 9.

28. "Financing Stalls in Mexico," *Financial Times: Power in Latin America*, No. 53 (November 1999).

29. "Financing Stalls in Mexico," *Financial Times: Power in Latin America*, No. 53 (November 1999).

30. Milbank, Tweed, Hadley & McCloy, LLP, "Largest Independent Power Plant Project Financing in Africa and the Middle East Completed," Press Release (Decenber 1997), web site www.milbank.com/press74.html.

31. DRI/McGraw-Hill, *World Energy Service: Africa/Middle East Outlook* (Lexington, MA, April 1998), p. 19

32. Energy Information Administration, *Country Analysis Briefs: Cote d'Ivoire* (March 1999), web site www.eia.doe.gov/emeu/cabs/cdivoire.html.

33. "Nigeria: Mobil To Build IPP," *Independent Energy*, Vol. 28, No. 8 (October 1998), pp. 6-10.

34. Hydro-Quebec, "Hydro-Québec International Becomes a Shareholder of the Société Nationale d'Électricité (SÉNÉLEC), Press Release (March 2, 1999), web site www.hydro.qc.ca.

35. DRI/McGraw-Hill, *World Energy Service: Africa/Middle East Outlook* (Lexington, MA, April 1998), p. 19.

Transportation Energy Use

Oil is expected to remain the primary fuel source for transportation throughout the world, and transportation fuels are projected to account for more than one-half of total world oil consumption from 2005 through 2020.

With little competition from alternative fuels, at least at the present time, oil is expected to remain the primary energy source for fueling transportation around the globe in the *International Energy Outlook 2000 (IEO2000)* projections. In the reference case, the share of total world oil consumption that goes to the transportation sector increases from 49 percent in 1997 to 55 percent in 2020 (Figure 84). The *IEO2000* projections group transportation energy use into three travel modes—road, air, and other (mostly rail but also including pipelines, inland waterways, and marine bunkers). Increases are expected for all travel modes, but road transport continues to constitute by far the largest share of transportation energy use (Figure 85).

Over the next two decades, the demand for personal motor vehicles is expected to increase rapidly. In some industrialized countries, per capita motorization levels (number of vehicles per person) already are high. In the developing nations, however, motorization is expected to increase dramatically as economic growth continues and personal incomes rise (Figure 86). Per capita motorization in much of the developing world is projected to more than double between 1997 and 2020 (Figure 87). Still, however, population growth in the developing world is expected to keep national motorization levels

below those in most of the industrialized world. In 2020, the United States is projected to have 797 vehicles per thousand population; but in China, despite a fivefold increase in per capita motorization from 1997 to 2020, the projection for 2020 is only 54 vehicles per thousand.

The economies of developing Asia had been experiencing particularly rapid per capita income expansion until the 1997-1999 economic recession. With the corresponding rise in the standard of living has come a fast-paced increase in the demand for personal means of transport. In many urban centers, such as Bangkok, Manila, Jakarta, Shanghai, and Mumbai, car ownership is among the first symbols of emerging prosperity. Where local infrastructures have not kept pace with expanding motor vehicle fleets, however, street congestion and air pollution have become major problems. Plans for mass transit are being made in some of the developing countries, but the pace of implementation and establishment of the infrastructure is often slow. In Bangkok, for example, discussions of a rapid transit electric train began in the 1970s, but the city's "Skytrain" was not brought into operation until December 1999 [1].

Among the petroleum products used for transportation fuels, the fastest growth is projected for jet fuel

Figure 84. World Oil Consumption for Transportation and Other Uses

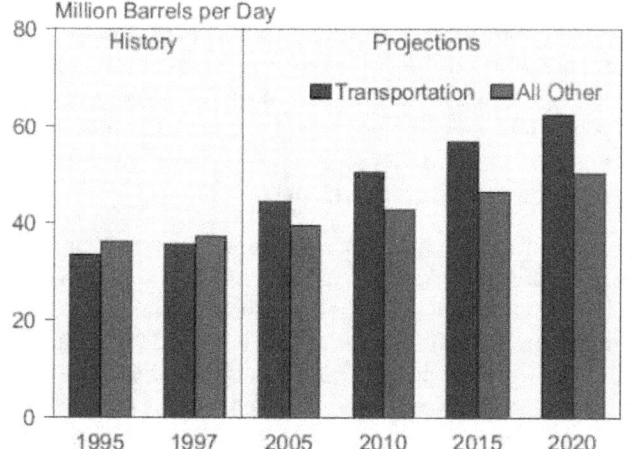

Sources: **History:** Energy Information Administration (EIA), derived from *International Energy Annual 1997*, DOE/EIA-0219(97) (Washington, DC, April 1999). **Projections:** EIA, World Energy Projection System (2000).

Figure 85. World Energy Use for Transportation by Mode, 1980-2020

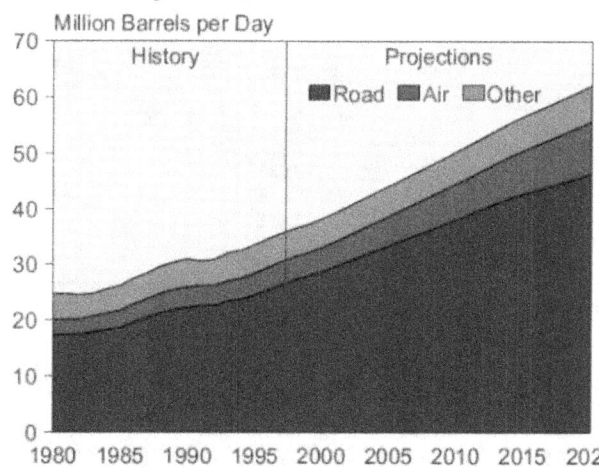

Sources: **History:** Energy Information Administration (EIA), derived from *International Energy Annual 1997*, DOE/EIA-0219(97) (Washington, DC, April 1999). **Projections:** EIA, World Energy Projection System (2000).

Figure 86. Motorization Levels in Selected Countries, 1980-2020

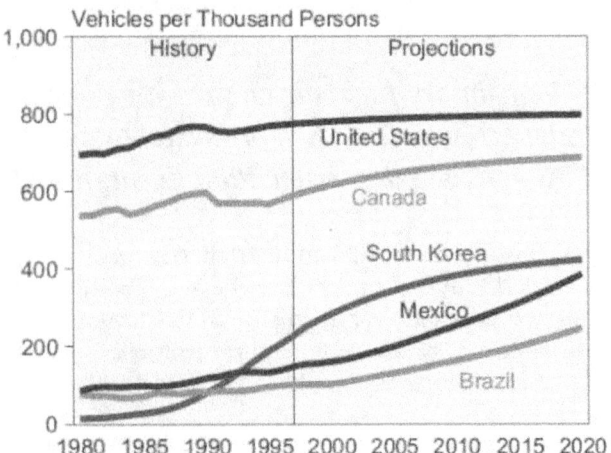

Sources: **History:** Energy Information Administration (EIA), derived from *International Energy Annual 1997*, DOE/EIA-0219(97) (Washington, DC, April 1999). **Projections:** EIA, World Energy Projection System (2000).

Figure 87. Motorization Levels in Selected Countries, 1997 and 2020

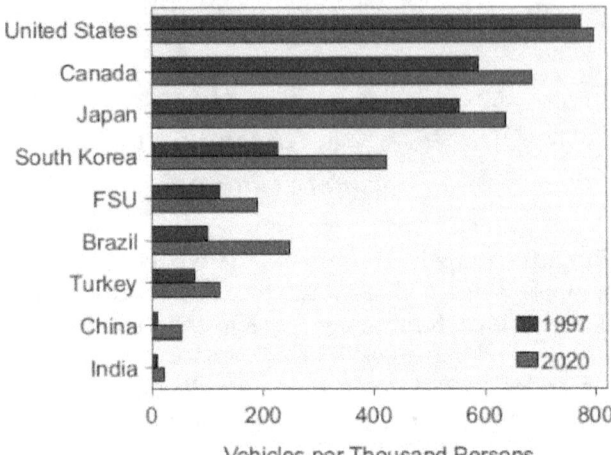

Sources: **1997:** American Automobile Manufacturers Association, *World Motor Vehicle Data* (Detroit, MI, 1997). **2020:** EIA, World Energy Projection System (2000).

consumption. Growth in per capita incomes and the resulting higher standards of living, particularly in the developing countries, are expected to increase the demand for air travel and lead to fast-paced development of the airline industry worldwide (Figure 88). The world's total transportation energy use is projected to grow by 2.5 percent per year between 1997 and 2020, and jet fuel consumption alone is projected to grow by 3.7 percent per year.

In the emerging economies of the developing world, jet fuel use is projected to grow by 5.7 percent annually. Strong economic growth in developing Asia and Central and South America coincide with particularly strong

demand for jet fuel in those regions. In developing Asia (including China, India, and South Korea), jet fuel consumption rises from 0.5 million barrels of oil equivalent per day in 1997 to 2.0 million barrels per day in 2020, and in Central and South America an increase from 0.2 million barrels per day to 0.6 million barrels per day is projected.

Moderate growth in energy use for transportation is expected in the industrialized nations, where transportation infrastructures are well established and motorization levels are already fairly high. As is true worldwide, the fastest growth is expected for jet fuel consumption as a result of increases in personal and business air travel.

Because of its geography and the high tax burden imposed on road fuels throughout the region, lower growth in motor gasoline use is projected for Western Europe than for North America (Figure 89). In addition, a different mix of fuel consumption is expected for road transportation in Western Europe, where a larger share is expected to go to diesel-fueled motor vehicles than to less efficient gasoline engines. In France, Germany, and other European nations, taxes on motor gasoline are higher than those on diesel fuels.

Auto makers are introducing sport utility vehicles into the Western European market, but they will be somewhat smaller than the American version and probably will be fueled by diesel. Between 1998 and 1999, sales of sport utility vehicles in Europe increased by 26 percent, and another 21-percent increase is expected from 1999 to 2000 [2]. In North America, consumer preference for large, less energy-efficient light-duty vehicles (minivans, sport utility vehicles, and small trucks) has contributed to increases in oil consumption over the past several years.

Figure 88. World Transportation Energy Use by Fuel, 1980-2020

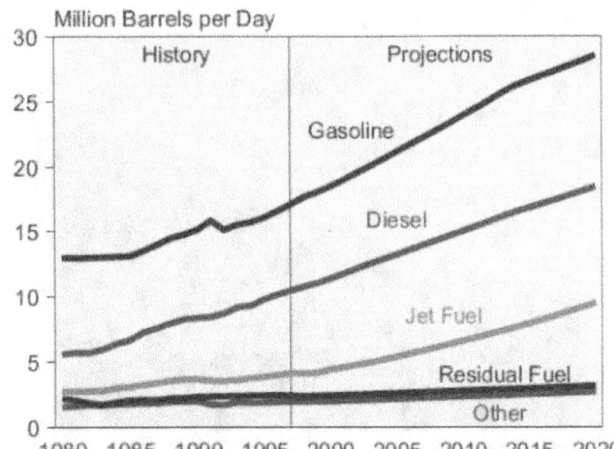

Sources: **History:** Energy Information Administration (EIA), derived from *International Energy Annual 1997*, DOE/EIA-0219(97) (Washington, DC, April 1999). **Projections:** EIA, World Energy Projection System (2000).

Figure 89. Transportation Fuel Mix in the United States and Western Europe, 1997 and 2020

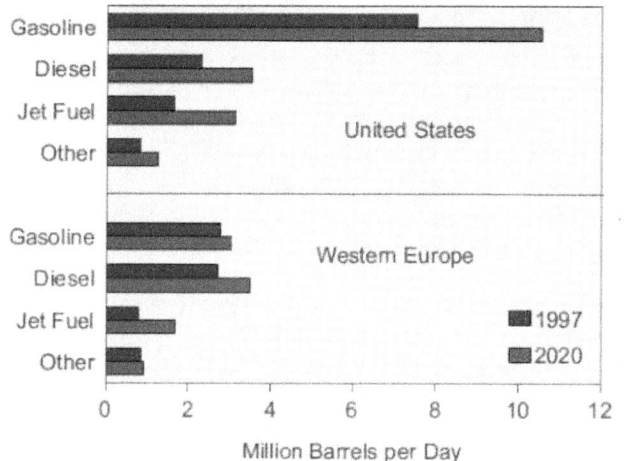

Million Barrels per Day

Sources: **1997:** Energy Information Administration (EIA), derived from *International Energy Annual 1997*, DOE/EIA-0219(97) (Washington, DC, April 1999). **Projections:** EIA, World Energy Projection System (2000).

Regional Activity

North America

Per capita energy use in the North American transportation sector is among the highest in the world. Canada and the United States are large geographic areas, where goods and people often are transported over long distances. Per capita, oil consumption for transportation uses in 1997 averaged about 18 barrels per person per year in the United States and 13 barrels per person in Canada, as compared with about 6 barrels per person per year in Western Europe and industrialized Asia.

United States

In the United States, transportation sector energy consumption is projected to increase by an average of 1.8 percent per year from 1997 to 2020. Growth in U.S. transportation energy demand averaged 2.0 percent per year during the 1970s but slowed in the 1980s as a result of rising fuel prices and new Federal vehicle efficiency standards, which led to an unprecedented 2.1-percent annual increase in average vehicle fuel economy [3]. Over the next two decades, with fuel prices expected to be stable and no new legislative mandates anticipated, fuel economy gains are expected to slow. New car fuel efficiency in the United States is projected to improve from an average of 28.4 miles per gallon in 1999 to 31.6 miles per gallon in 2020.

Petroleum products dominate transportation sector energy use in the United States, where motor gasoline accounts for more than one-half of all transportation fuel use (Figure 90). In the forecast, alternative fuels (compressed natural gas, methane, liquid hydrogen, electricity, and E85—a fuel that is 85 percent ethanol and 15 percent motor gasoline) are projected to contribute about 406,000 barrels of oil equivalent per day by 2020, or about 4 percent of all light-duty vehicle fuel consumption, as a result of current environmental and energy legislation intended to reduce oil use.[22]

Air travel in the United States is also projected to increase. At the same time, however, new aircraft in 2020 are expected to be more than 18 percent more efficient than in 1998 [4]. Ultra-high-bypass engine technology may increase fuel efficiency by as much as 10 percent, and the use of weight-reducing materials may allow for an additional efficiency gain of up to 15 percent.

The United States has taken steps to limit exhaust emissions from its motor vehicle fleet. The Clean Air Act Amendments of 1990 (CAAA90) set "Tier 1" exhaust emission standards for carbon monoxide, hydrocarbons, nitrous oxides, and particular matter for light-duty vehicles and trucks beginning with model year 1994 [5]. CAAA90 also required the U.S. Environmental Protection Agency (EPA) to study more extensive "Tier 2" standards for 2004 model year cars. In July 1998, EPA provided the Congress with a Tier 2 study which concluded that tighter vehicle standards are needed to attain the National Ambient Air Quality Standards for ozone and particulate matter between 2007 and 2010. In May 1999, EPA published a Notice of Proposed

Figure 90. U.S. Transportation Energy Use by Fuel, 1980-2020

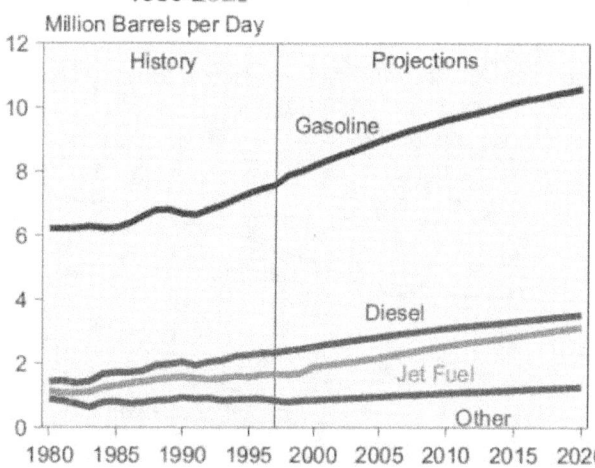

Sources: **History:** Energy Information Administration (EIA), *Short-Term Energy Outlook*, DOE/EIA-0202(99/4Q) (Washington, DC, October 1999), Table A5. **Projections:** EIA, *Annual Energy Outlook 2000*, DOE/EIA-0383(2000) (Washington, DC, December 1999), Table A2.

[22]For example, the Energy Policy Act of 1992 sets new vehicle purchase mandates for vehicle fleet owners, whereby 70 percent of all vehicles must be fueled by alternative fuels by 2006. Also, under the Low Emission Vehicle Program, 10 percent of new vehicle sales in States that agree to participate in the program will be zero-emission vehicles by 2003.

Rulemaking on Tier 2 Emissions Standards for Vehicles and Gasoline Sulfur Standards for Refineries, including standards that would significantly reduce the sulfur content of gasoline throughout the United States to ensure the effectiveness of emission control technologies that will be needed to meet the Tier 2 emissions targets. In 1999, however, the U.S. Circuit Court ruled that the EPA is not authorized to set new standards without indicating their benefits.

In March 1999, California Governor Grey Davis issued an executive order announcing a ban on the use of methyl tertiary butyl ether (MTBE) in gasoline by the end of 2002. MTBE is blended with gasoline to raise its oxygen content, thereby reducing emissions of carbon dioxide and airborne toxic pollutants. The use of MTBE grew in the 1990s, when it was used to meet oxygen requirements for cleaner burning reformulated and oxygenated gasoline. The fuel programs were successfully implemented, but concerns have been raised about the effects of MTBE on water resources. Leaking underground pipes and storage tanks have resulted in the contamination of more than 10,000 groundwater sites in California and of 5 to 10 percent of the water supplies in those areas of the United States required to use reformulated or oxygenated gasoline.

Canada

Canada's transportation sector largely resembles that of the United States (Figure 91). The country is geographically large, and the distances traveled per vehicle are nearly as large as in the United States—substantially larger than in the smaller industrialized nations [6]. As in the United States, oil is the primary fuel used in the transportation sector. A small amount of natural gas is

also used as a pipeline compressor fuel. There are about 21,000 natural gas vehicles operating in Canada at present, but they make up only 0.1 percent of the total vehicle fleet of 18.6 million passenger cars and commercial vehicles [7]. Unless the country enforces new government policies or incentives to increase the penetration of natural gas and propane in the vehicle fleet, their use is expected to remain low throughout the forecast period. Almost three-fourths of the oil consumed in Canada's transportation sector is for road use, and the remainder is used for airplanes, railways, and boats [8].

In Canada, the number of vehicles per thousand persons is about 589, compared with 773 in the United States. The composition of the vehicle fleet closely resembles that of the U.S. fleet, and the NAFTA trade agreement has substantially unified the North American vehicle market [9]. In addition, efficiency programs for new cars and light-duty trucks in Canada are much like the U.S. programs. Both cars and light trucks currently meet voluntary efficiency standards, but light-duty vehicle efficiency improvements have slowed in recent years because of increased sales of light trucks (including vans and sport utility vehicles) for personal transportation.

Mexico

Vehicle ownership is lower in Mexico than in the other countries of North America, estimated at 148 cars per thousand persons, and the transportation infrastructure is less developed. Only 37 percent of Mexico's roads are paved, compared with around 61 percent in the United States [10]. Mexico has about 55,000 miles of paved roads and nearly 2,000 miles of highways [11].

Motor vehicle transportation has contributed to making Mexico City one of the most polluted cities in the world. Indeed, when the city's smog reaches dangerous proportions, the center of the city is closed to traffic and production is shut down in several of the city's factories. To reduce the amount of emissions produced by the estimated 2.5 million vehicles currently operating in Mexico City, the government has issued decals to car owners that ban them from driving one day a week [12].

Mexico has invested at least $5 billion over the past decade in an effort to clean the air in Mexico City [13]. Outdated diesel buses have been replaced, a city oil refinery has been closed down, and some of the hills near the city have been reforested. Nevertheless, ozone levels remain high. In September 1999, two transportation agencies in Mexico—Coordinacion de Transporte de Mexico and "Ruta 89" Union de Taxistas Camesinos Libres Independientes—contracted with IMPCO Technologies to convert 4,100 public transportation vehicles in Mexico City to liquid propane gas systems from gasoline systems [14]. There are plans eventually to convert 70,000 commercial vehicles in the city to liquid propane.

Figure 91. Canadian Transportation Energy Use by Fuel, 1980-2020

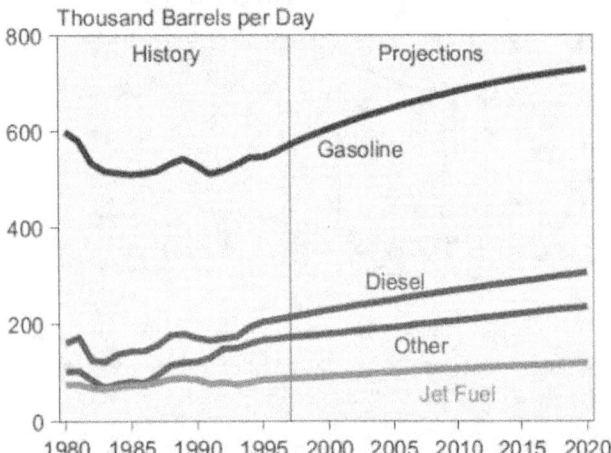

Sources: **History:** Energy Information Administration (EIA), derived from *International Energy Annual 1997*, DOE/EIA-0219(97) (Washington, DC, April 1999). **Projections:** EIA, World Energy Projection System (2000).

Mexico began producing cars with emissions controls in 1991 to mitigate growing concerns about air pollution [15]. The country has also established strict legislation on emissions controls for taxis, trucks, minibuses, and private cars, and the state-owned oil company, Pemex, has been reducing the production of leaded gasoline. In 1997, Pemex increased sales of unleaded gasoline, replaced Nova (a low-quality, highly polluting gasoline) with a higher octane gasoline, and replaced the high-sulfur diesel that was produced at the refineries with the new Pemex Diesel, which contains about 0.05 percent sulfur. Those measures are expected to help curb air pollution somewhat, but increasing levels of car ownership and rising highway use for trade with Central America and the United States suggest that pollution will remain a problem for the country's urban areas.

Western Europe

Most oil in Western Europe is now used in the transportation sector, which is expected to account for most of the growth in the region's oil demand over the next two decades. Most of the increase is expected in the use of diesel, rather than motor gasoline (Figure 92), as a result of lower taxes on diesel in some countries (such as France) and the development of higher grade "city diesel."

Jet fuel use for air travel is also expected to increase in Western Europe. Personal air travel has been rising in the region over the past several years, as air fares have dropped relative to average individual incomes. Rising personal incomes are expected to continue the increase in air travel. Overall, however, the *IEO2000* reference case projects that total energy use for transportation in Western Europe will increase by less than 1 percent per year between 1997 and 2020. In addition, new

technologies, such as fuel cells, may displace some of the demand for oil in the transportation sector in the future.

There is increasing concern in Western Europe about the impact of automobile emissions on the environment. At the end of 1992, the European Commission (EC) initiated a program designed to reduce emissions from road transport [16]. The initiative has evolved into the Auto Oil Programme, jointly conducted by the EC and the European automotive and petroleum industries (see box on pages 138-139). The purpose was to establish a cost-effective program for reducing vehicle emissions and, as a result, improve air quality, particularly in the region's urban areas. The Auto Oil Programme was designed to "provide policy-makers with an objective assessment of the most cost effective measures to reduce emissions from the road-transport sector" [17]. Two directives resulting from the program went into effect on January 1, 2000: (1) that the sale of leaded gasoline in the European Union member countries should be phased out; and (2) that gasoline should have a sulfur content of not more than 150 milligrams per kilogram, a maximum benzene content of 1 percent, and a limit for aromatics of 42 percent by volume of vapor, and diesel should have a sulfur content of not more than 350 milligrams per kilogram.

The EC is now planning the second phase of its Auto Oil Programme, looking at making the fuel specifications more stringent and reducing vehicle emissions even further beginning in 2005. The European motor industry has also committed to reducing carbon dioxide emissions from new vehicles by 25 percent by 2008. There is considerable interest in developing technologies to improve the fuel efficiency of the European automotive fleet, such as direct injection gasoline engines, fuel cells, and hybrid vehicles powered with batteries and conventional engines. Other innovations in vehicle design, such as ultra-light vehicles, are also aimed at reducing fuel consumption and emissions.

As is true for many countries in Western Europe, there has been a shift to the use of diesel fuel in the United Kingdom (UK) [18]. The diesel share of the transportation market has increased from about 22 percent in 1980 to 32 percent in 1997. Much of the increase is attributed to commercial freight vehicles, inasmuch as taxes on diesel fuel for passenger cars rose substantially over the 8-year period. The tax burden (total taxes as a percent of total price) for diesel fuel rose from 53 percent and 59 percent in 1990 for commercial and noncommercial use, respectively, to 83 percent and 86 percent in 1998. The tax burden for motor gasoline ranges from 77 percent to 85 percent, depending on the type of gasoline type.

The diesel share of transportation energy demand has also increased dramatically in France over the past two decades, from 29 percent in 1980 to 41 percent in 1990

Figure 92. Western European Transportation Energy Use by Fuel, 1980-2020

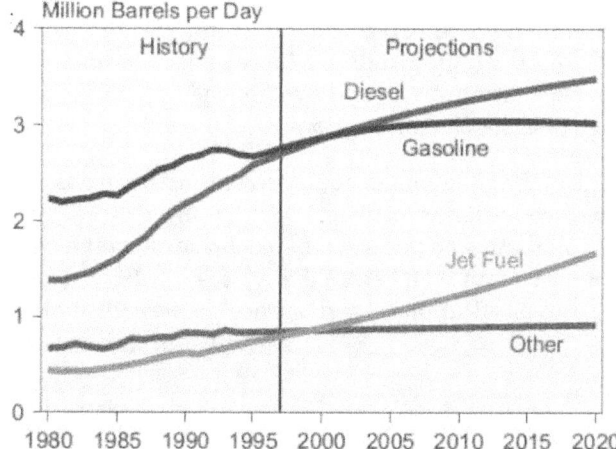

Sources: **History:** Energy Information Administration (EIA), derived from *International Energy Annual 1997*, DOE/EIA-0219(97) (Washington, DC, April 1999). **Projections:** EIA, World Energy Projection System (2000).

Europe's Auto Oil Programme

Industrialized countries have recognized that urban areas are increasingly affected by air pollution from their motor vehicle fleets. The United States, Canada, Japan, and the European Union (EU) member countries, among others, have passed legislation to reduce air pollution from vehicle emissions. In the EU countries, in particular, the impact of legislation passed to improve fuel quality and limit vehicle emissions will become evident in 2000. Member countries must be in compliance with laws passed as a result of the EU's Auto Oil Programme by January 1, 2000. The directives resulting from the first stage of the program, Auto Oil I, will restrict the amounts of benzene and sulfur in vehicle emissions and eliminate lead from motor gasoline. This marks the first step in a long-term commitment by the EU to address air pollution resulting from Western Europe's motor vehicle fleet.

The Auto Oil Programme was established in 1993, and work on the Auto Oil I program was completed in 1996.[a] The EU representatives invited representatives from the automotive and oil industries to participate in developing a cost-effective program for controlling vehicle emissions. A collaboration between different parts of the European Commission, the European oil industry association (EUROPIA), and the European car manufacturer association (ACEA) was formed to work on the Auto Oil Programme, and its results are now being implemented in Auto Oil I.[b]

The collaboration was considered necessary because of the disagreement between oil refinery and automotive industries about who should bear the costs of improving air quality. Emissions are controlled either through tightening the specifications of the fuels used in vehicles or by making the vehicles better equipped to handle pollutants in the fuel. The oil companies wanted the auto companies to pay for new emissions control equipment in cars, and the auto companies wanted the oil companies to pay for tighter fuel standards. Tighter fuel specifications have a rapid impact on the environment, whereas tighter emissions controls take longer—up to 10 years as the new control equipment penetrates the vehicle market.

All the EU member countries are expected to be in compliance with the 1997 European Council directives by 2000. The first of the two directives, COM/96/0163 (COD), originally proposed in June 1996, regulates

emissions from light-duty vehicles. The second, COM/96/0164 (COD), governs the quality of gasoline and diesel fuels.[c] Further directives are planned on emissions from light commercial vehicles, emissions from heavy-duty vehicles, and inspection and maintenance procedures. Under these directives, the sale of leaded fuel was to be phased out by 2000; motor gasoline will have a sulfur content of not more than 150 milligrams per kilogram, a maximum benzene content of 1 percent, and a limit for aromatics of 42 percent by volume of vapor, and diesel fuel will have a sulfur content of not more than 350 milligrams per kilogram.

Auto Oil I was enacted through a compromise reached by the EU Council and the European Parliament in Strasbourg in June 1998. European environment commissioner Ritt Bjerregaard at the time predicted that the legislation would cut exhaust emissions by 70 percent relative to 1990 levels. In addition, the European motor industry has itself committed to reducing carbon dioxide emissions from new vehicles by 25 percent by 2008. To that end, the European automotive industry has been investing heavily in vehicle technology development for the past several years. New engine technologies such as direct injection gasoline, fuel cells, and electric hybrid vehicles, as well as innovations in vehicle design such as ultra-light vehicles, are among the technologies that the industry hopes to use to reduce fuel consumption and emissions.

In addition to the specifications for diesel and gasoline to be implemented by 2000, agreement has been reached on further tightening of fuel specifications, scheduled for implementation by 2005. The new range of specifications is somewhat less inclusive, however, than the specifications included in Auto Oil I. Under Auto Oil I, gasoline qualities in 2005 were to be set at 35 percent by volume of vapor for aromatics and 50 parts per million for sulfur, and diesel sulfur was to be limited to 50 parts per million. More specific details for other specifications are to be established by the second stage of the Auto Oil Programme, Auto Oil II.

Some of the EU member countries are having problems meeting the terms of Auto Oil I. Spain, Greece, and Italy all requested permission to delay implementation of the lead ban well before the August 31, 1999, deadline for applications,[d] and France requested similar

(continued on page 139)

---

[a]T. Neale, DGXVII European Commission, "Euro Synergy Conference, Brussels: New International Investment Projects of European Oil Companies" (October 24, 1997), web site http://europa.eu.int.

[b]Standard & Poor's Platt's, *World Energy Service: European Outlook*, Volume II (Lexington, MA, 1999), p. 5.

[c]Commission of the European Communities, "Guide to the Approximation of European Union Environmental Legislation: The Auto-Oil Programme" (August 25, 1997).

[d]"Energy Alert: EC Delays Decision on Lead," *The Oil Daily*, Vol. 49, No. 217 (November 15, 1999), p. 7.

Europe's Auto Oil Programme (Continued)

exemptions for its overseas territories. Portugal requested a delay for the cut in the sulfur content of diesel and gasoline, and France made the same request for its Réunion territory. In December, after having delayed a decision by nearly 3 months, the European Commission granted permission for Spain, Italy, and Greece to continue marketing leaded gasoline for another 2 years.[e] France has been authorized to sell leaded gasoline until 2005 in Réunion, Martinique, Guadeloupe, and Guyana and is permitted to sell unleaded motor gasoline and diesel with sulfur contents above those permitted by the directives until January 2003. Portugal was granted permission to sell diesel and unleaded gasoline with sulfur contents in excess of the new directive until January 2001 and January 2002, respectively.

Since Auto Oil I was concluded, the European Parliament and the Council have proposed further standards to be implemented by 2005 for both fuels and cars.[f] The standards are to be reviewed by the current Auto Oil II program. In this second stage, the European Commission wishes to tighten fuel specification rules and limit vehicle emissions even further, beginning in 2005. The Auto Oil II program has a wider range of participants than Auto Oil I. It includes all the relevant industries (not only oil and motor vehicles), member states, nongovernment organizations, and research institutes.

The subjects to be dealt with under the Auto Oil II initiative were published in November 1998. Cost-benefit studies were to be considered for rules governing Reid vapor pressure, benzene, olefins, and oxygenates in gasoline, as well as cetane numbers and polyaromatics in diesel. The terms of reference set a deadline of July 1999 for the setting out of scenarios for Auto Oil II, which were to be considered for legislation by the EU by the end of 1999. Two more stages of the Auto Oil Programme (Auto Oil III and IV) are planned, and the Auto Oil II studies are supposed to determine the scope of those future stages.

The costs of the Auto Oil programs to oil refinery companies are not inconsequential. The official estimate from the EU is that costs to the industry may reach $35 billion, around three times the original estimates. The oil industry is not happy with the final Auto Oil targets, arguing that the final proposals moved significantly from the Commission's original June 1996 proposals, which EUROPIA had supported. The cost to the oil industry in the original June 1996 proposal was estimated by the Commission to be about $12 billion (11 billion ECU); however, the original proposal was revised by the European Parliament, adding some 100 amendments and increasing the cost of the plan to oil refiners substantially, to about $60 billion according to EUROPIA. A final compromise agreement between the European Commission and the Parliament, reached in June 1998, left refiners with a cost of approximately $35 billion, $23 billion higher than the original figure, according to oil industry estimates.

[e]European Commission, "Commission Gives Spain, Italy, Greece, Portugal, and France Extra Time To Implement Fuel Directive," Press Release (December 17, 1999), web site http://europa.eu.int.

[f]Standard & Poor's Platt's, *World Energy Service: European Outlook,* Volume I (Lexington, MA, 1999), pp. 5-6.

and almost 50 percent in 1997, while the gasoline share dropped from 51 percent in 1980 to 33 percent by 1997. Tax changes in France since 1990 have increased the tax burden on most oil products, but gasoline has the largest tax share. The current gasoline tax is about 83 percent of the total price, between 14 percent and 17 percent higher than in 1990 (depending on the gasoline type) and 6 percentage points higher than for diesel, which has the second-highest tax burden [19]. The tax burden for diesel has increased from 57 percent for commercial uses and 63 percent for noncommercial uses in 1990 to 73 percent and 78 percent, respectively, in 1999.

The French government has been trying to promote greater energy savings in the transportation sector, primarily because of the environmental impact of vehicle emissions [20]. The country's urban areas have had an increasing number of air pollution alerts over the past several years, which have helped to underscore the growing problem. France's Clean Air Act requires municipalities to replace 20 percent of their vehicle fleets with "clean" vehicles, leading to increased interest in qualifying alternative fuel vehicles, such as those powered by electricity of liquefied petroleum gas.

A number of local governments in France are trying to increase the penetration of electric and other "clean" vehicles by subsidizing sales, expanding the number of recharge points, and offering free parking. Encouraged by these developments, Peugeot and Renault are beginning to market electric vehicles on a large scale. Further, some 350 natural-gas-fueled buses were operating in France at the end of 1999, and another 800 could be on order within 2 years [21]. Around 3,000 natural-gas-fueled private cars and delivery vehicles are also operating on French roads at present.

In Germany, the demand for diesel fuel grew in the 1990s and, as in France and the United Kingdom, displaced a significant amount of motor gasoline use.

During the 1990s gasoline demand declined in response to both slower economic growth and an increasing number of diesel cars [22]. Diesel fuel use grew by 28 percent between 1990 and 1997 in Germany, while motor gasoline use fell by 5 percent between 1990 and 1994 and then increased by 3 percent between 1994 and 1997. Increased trade between eastern and western Germany also contributed to the growth in diesel demand during the 1990s.

In 1989, the car ownership level in eastern Germany was 235 per 1,000 inhabitants, compared with 480 in western Germany. Unification brought an explosion in car ownership in eastern Germany, much of it from the west's used car market. As a result, motorization in eastern Germany now has grown to equal that in western Germany. The 1997 level for Germany as a whole was about 543 vehicles per thousand persons.

Industrialized Asia

In industrialized Asia (Japan, Australia, and New Zealand), the transportation infrastructure is mature and motorization levels are similar to those in other industrialized countries. In the Australasian countries, per capita vehicle ownership in 1997 approached 631 per thousand persons, higher than that of the countries in Western Europe (an average of 509 vehicles per thousand) but lower than the U.S. level (773 vehicles per thousand). Australasian motorization levels are expected to grow steadily over the projection period, closing the gap with the United States.

Japan has a well-established transportation infrastructure with an estimated 719,000 miles of roadway, 15,000 miles of railway, and 170 airports [23]. Japan's motorization levels are somewhat lower than those of the rest of industrialized Asia, mainly because the Japanese rely more heavily on mass transit for passenger commuting [24]. Motorization levels, estimated at around 554 vehicles per thousand currently, are projected to increase to 638 per thousand persons in 2020 in the *IEO2000* reference case. Much of the increase is expected to come from households adding second and third cars.

The Japanese economy has been mired in a long-term recession (the worst in at least 50 years), which has weakened the domestic automotive industry. Although the Japanese government has provided stimulus packages to improve consumer confidence and increase spending, they have been largely unsuccessful, and many analysts are expecting the economy to experience another year of negative economic growth in 2000 [25]. Indeed, Japan's new vehicle sales showed only slight improvement in 1999 over 1998. In November 1999, new vehicle sales increased by 0.2 percent from their November 1998 level, breaking a 31-month sales slump [26]. In the *IEO2000* reference case, Japan's economy is

projected to grow by 1.3 percent annually from 1997 to 2020, and its motorization level is expected to increase by 0.6 percent per year.

Australia, a much larger country geographically than Japan, has 566,000 miles of roads, 24,000 miles of rail, and 408 airports. The country has an extensive and well-developed air transportation system, which has facilitated travel both within and across its borders [27]. Establishing an extensive airline infrastructure has been a key to the success of the tourism industry, which is an important part of the nation's economy. Australia began privatization of its air sector in 1997.

Australia currently has eight international airports. Sydney, the largest, serves almost 21 million passengers each year [28]. Many of the country's international airports have reported substantial gains in international travel, primarily as a result of increased tourism. The strong growth in the air sector has resulted in plans for expanded of improved facilities at many of the airports. Brisbane Airport Corporation, Ltd., announced in December 1999 its intentions to invest $32 million (U.S. dollars) to add a car parking lot and a hotel for business travelers at the Brisbane International Airport [29].

Despite the fact that New Zealand is a small country with only a fraction of the population of its neighbors, Japan and Australia, it is served by 57,000 miles of roads, 2,000 miles of rail, and 111 airports. Most of the transportation energy demand in New Zealand is for road use—primarily, motor gasoline and diesel fuel [30]. Rising incomes are expected to result in an increase in average vehicle usage. Business, trade, and tourism are all expected to contribute to an increase in air travel. The country currently supports three international airports. Each day, more than 21,000 passengers pass through the busiest, at Auckland [31].

Developing Asia

The transportation sector in developing Asia is expected to be among the fastest-growing over the next 23 years in the *IEO2000* reference case (Figure 93). Many of the countries in the region, including China and India, are starting from very low levels of per capita motorization, and although their motorization rates are projected to more than double (and in China to increase fivefold) from 1997 to 2020, they still would be only a fraction of those in the industrialized world. Because the region accounts for a substantial portion of the world's population, however, even the smallest gains in per capita vehicle ownership (and fuel use) will have an enormous impact on world fuel markets. The average motorization level in the region is projected to grow from 18 vehicles per thousand persons in 1997 to 47 in 2020, and total road energy use is projected to expand from 3.6 to 10.2 million barrels per day.

Figure 93. Transportation Energy Use in the Developing World by Region, 1980-2020

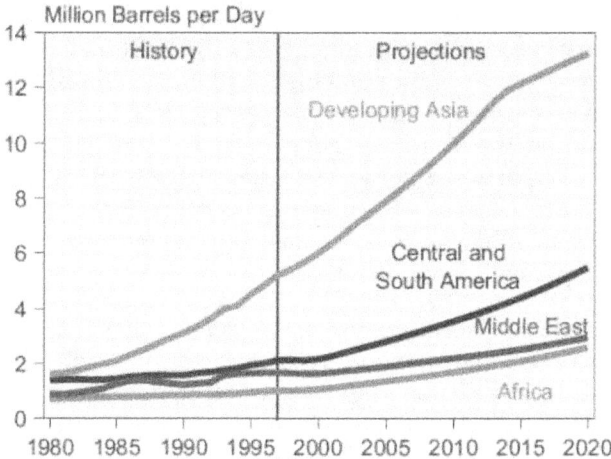

Sources: **History:** Energy Information Administration (EIA), derived from *International Energy Annual 1997*, DOE/EIA-0219(97) (Washington, DC, April 1999). **Projections:** EIA, World Energy Projection System (2000).

Figure 94. Transportation Energy Use in Developing Asia, 1980-2020

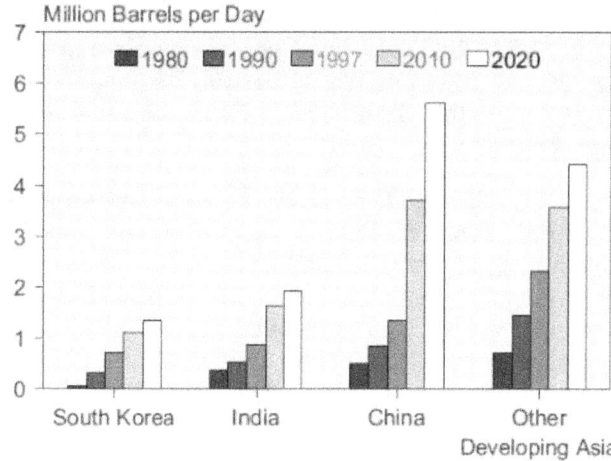

Sources: **History:** Energy Information Administration (EIA), derived from *International Energy Annual 1997*, DOE/EIA-0219(97) (Washington, DC, April 1999). **Projections:** EIA, World Energy Projection System (2000).

Many cities in developing Asia are faced with terrible road congestion because of insufficient infrastructure to handle rapidly growing transportation needs. In Bangkok, urban planners estimate that the average commuter spends 44 days a year in traffic. The government is only now making substantive progress in addressing the problem through mass transit, opening its $1.7 billion "Skytrain" in 1999.

For China, *IEO2000* projects an increment of 4.3 million barrels of oil per day in transportation sector fuel consumption between 1997 and 2020 (Figure 94). The Chinese government has not addressed transportation sector issues in its economic plans for several years, and investment in railways relative to gross domestic product (GDP) has remained at around one-half the levels in other Asian countries, such as South Korea and India [32]. As a result, China has one of the smallest railway networks per capita in the world, and its utilization rate for freight traffic is high.

To redress the lack of planning, China now intends major investment in its transportation infrastructure, including railway, road, and inland waterway projects. Projects are underway to increase railroad mileage from 39,000 miles in 1995 to 42,000 miles in 2000, and to construct 68,000 miles of new road between 1995 and 2000. Personal motor vehicles are still a very small part of China's vehicle fleet, which consists mainly of heavy commercial vehicles. Motorization levels in China remain low, at 10 vehicles per thousand persons, as compared with 228 in South Korea and 240 in Taiwan [33].

Most of the automobiles in China are owned by government and state-owned enterprises or other businesses, which own fleets of cars that may also be used for private purposes. Recent strong economic growth and greater economic liberalization have led to an emerging fleet of privately owned cars. Small businesses are also beginning to use small pickup trucks and vans, and many new taxi companies are operating microvans.

Although privately owned motor vehicles are expected to be the fastest-growing part of China's transportation sector, car density is expected to remain low, and mass transit is expected to remain the primary form of transportation for most people. Truck transport is also expected to expand rapidly. The trucking industry has been progressively deregulated, and since 1983 both collective and private enterprises have been able to license for-hire vehicles.

India is another major country of developing Asia with an underdeveloped transportation infrastructure. Although the country has an estimated 500,000 miles of paved roads and highways and 38,000 miles of railway, they are not sufficient to meet the economic development needs of some parts of the country [34]. Railways, many of which are aging, are the primary means of transport between states. Intrastate roads often suffer from a lack of maintenance. Congestion in the urban centers of the country, as in many other cities of developing Asia, increasingly creates market access problems for the country's industries.

Vehicle ownership in India is estimated at about 9 per thousand persons. Most privately owned automobiles are located in the larger cities of the country. Not all cities are linked by roadways, so that without substantial increases in road construction, growth in vehicle ownership is expected to be constrained in the short-term future.

Air transportation in India also suffers from a lack of development. One of the most important international airports, Mumbai International, has been handling only around 4.5 million passengers per year, compared with Don Muang International Airport in Thailand, which handles more than 15 million. Tourism, however, is an important industry in India.

South Korea has made great strides in the development of its transportation sector in comparison with most of the other countries in developing Asia. Per capita vehicle ownership was estimated at 228 per thousand persons in 1997. The South Korean car market is currently the largest in East Asia, and motorization levels are projected to remain among the highest in developing Asia throughout the projection period, reaching 422 per thousand in 2020—approaching the levels of several West European countries by that time. The government also plans to increase mass transit to reduce congestion and air pollution in South Korea's cities, with plans to install electric railways in the coming decades.

In Indonesia, the government has long recognized the importance of establishing a national transportation network. With more than 13,000 islands covering three time zones, the government has held infrastructure development as a relatively high priority. The country has over 85,000 miles of paved roads, more than 120 miles of highway, and 4,000 miles of railway [35]. Air and sea transportation infrastructures have been upgraded over the past several years in a national effort to improve ports and airports.

Malaysia is another developing Asian country that has recognized the importance of improving its transportation system, which currently includes some 43,000 miles of paved roads, more than 1,200 miles of railways, and approximately 2,000 miles of navigable channels. Road construction—in particular, a highway network on peninsular Malaysia, which provides an efficient north-south link—has been a major focus of the government's infrastructure development plan. Roads are largely financed through build-own-transfer (BOT) schemes. Energy demand in the transportation sector has expanded substantially over the past several decades, at 5.6 percent per year in the 1970s, 8.2 percent in the 1980s, and 7.7 percent in the first 8 years of the 1990s [36]. The development of road infrastructure and a domestic car industry has made this sustained growth possible.

Malaysia has its own car industry (Proton) and a relatively high car density that has risen from less than 100 cars per thousand residents in 1990 to around 150 per thousand in 1998, boosted by government incentives and an increase in per capita incomes. The success of private car ownership has also been linked to the underdeveloped public transport systems. Several large public transportation projects have recently been completed, however, which should restrain the growth in transportation energy demand over the forecast period. New transport infrastructure includes a light rail transit system in Kuala Lumpur and the conversion of the existing rail network in Peninsular Malaysia to double tracks. In addition, the authorities have been improving the country's airports and ports, which have been working over capacity because of a boom in tourism and a rise in exports. Malaysia is a primary destination for tourists within Southeast Asia.

Pakistan's transportation infrastructure development has been somewhat skewed over the past two decades. The size of Pakistan's highway system has doubled since 1980, but growth in the railway infrastructure has stagnated [37]. As a result, the movement of goods has been shifting away from rail, and trucking has become increasingly important. The country has more than 50,000 miles of paved roads, around 200 miles of highways, and about 8,000 miles of railways. Air and maritime transport infrastructures have not been developed to a large extent, although Pakistan's national airport, Karachi International, has been handling close to 5 million passengers annually.

In the Philippines, roads have been the focus of development in the transportation sector, and motor vehicles account for almost 80 percent of the country's total energy use for transportation [38]. The country has 100,000 miles of roads (only a small portion of which are paved), 556 miles of railways, and about 260 airports (only 75 of which have paved runways) [39]. Personal motor vehicle ownership is currently fairly low but is expected to climb rapidly over the next two decades as strong economic growth brings new individual wealth and a rise in personal travel. Many of the roads in the country will require investment for maintenance and expansion to keep up with the expected growth in road travel; however, poor government planning has led to the abandonment of a number of road construction projects because of problems in acquiring the necessary land.

Pollution from motor vehicle emissions is a growing problem, particularly in the country's capital, Manila. The Philippine government has plans to phase out leaded gasoline in Manila at the start of 2000, with further plans to expand the ban to the remainder of the country by 2001.

Air travel slowed substantially in the Philippines during the Asian economic crisis. Philippine Airlines ceased operating in 1998 because it lost money from the lack of passengers. Although other airlines began to fly the Philippine Airlines routes, total air travel dropped substantially. Because many Asian economies are beginning to recover from the recession, the Philippines among them, air travel is expected to begin growing again.

With relatively low motorization levels throughout the Philippines, public transportation accounts for a great deal of the demand for energy in the sector. Manila has a light rail transportation system that began operating in 1958 and currently transports 400,000 commuters each day. The Philippine national railway commuter service handles another 50,000 daily. Mass transit is expected to continue to be an important part of the total transportation infrastructure, and there are already plans to add high-speed mass transit to connect many Philippine cities by 2020.

Taiwan has one of the most sophisticated transportation systems in developing Asia. The country's infrastructure expansions have included the construction of mass transit systems in urban areas, interconnections with other islands, and roads to reach major population and producing sections [40]. Because the country has a fairly high level of motorization (about 240 vehicles per thousand persons), which is expected to increase rapidly over the next decades, many of the transport sector improvements have been implemented in efforts to reduce the effects of vehicle emissions. Currently, Taiwan has around 12,000 miles of paved roads and 3,000 miles of railways. The country has well-developed ports and airports that are able to handle large commercial shipments.

Thailand has, more than any other country in southeast Asia, epitomized the problems that occur when rapid increases in personal automobile demand outpace development of the transportation infrastructure. There are notorious congestion and traffic jams in the country's largest city, Bangkok. The liberalization of the Thai car market—achieved by the government reduction of import duties—resulted in a rapid expansion of personal transport. Unfortunately, the expansion brought severe congestion and pollution problems to Bangkok.

The economic crisis that struck Thailand in mid-1997 served to alleviate some of the problems caused by the rapid development of personal vehicle ownership. City streets became less clogged when unemployment rose, the volume of freight transport declined, and many people lost their cars to bank repossessions. However, the economic crisis also meant that most major infrastructure projects that had been planned were delayed. The elevated mass transit system of Hopewell, the subway project of the Metropolitan Rapid Transit Authority, and the elevated electric train system of the Bangkok Transit System all faced delay. As the country begins to come out of its recession, there are some indications that mass transit construction is also coming back. The 16-mile, $1.7 billion elevated rail system, Skytrain, completed in December 1999 [41], is only a small portion of Bangkok's proposed 180-mile mass transit master plan. The recession also accelerated plans to privatize the country's roads, railways, airports, and ports. As a result, it may be easier to attract new private investment to begin infrastructure improvements.

The economic boom of the early 1990s resulted in a well-established Thai air transportation network [42]. The importance of the tourism industry in Thailand also helped to attract investment in the air sector. The country's main airport, Don Muang International, typically handles more than 15 million passengers per year. The economic downturn in Japan has dampened local demand for jet fuel, however, because Japanese tourists constitute one of the largest visitor groups.

Central and South America

Dynamic growth is expected for the transportation sector in Central and South America over the *IEO2000* forecast period. Motorization rates are projected to expand by 3.6 percent per year between 1997 and 2020, rising from 86 vehicles per thousand persons to 194 by the end of the projection period. Growth in Brazil's vehicle fleet is projected to be even more robust, resulting in a 4.0-percent average annual increase in motorization rates over the forecast (Figure 95).

Many of the Central and South American economies found themselves in recession in 1999, partly as a result of Asian economic crisis and the January 1999 devaluation of the real in Brazil, the region's largest economy. Despite signs of economic recovery, which were particularly encouraging in Brazil, the recession had a negative impact on the region's automotive industry in 1999. Standard's and Poor's Platt's has estimated that demand for automobiles in South America will have declined so much by 2000 that all the growth achieved in the region over the past 5 years will be reversed [43].

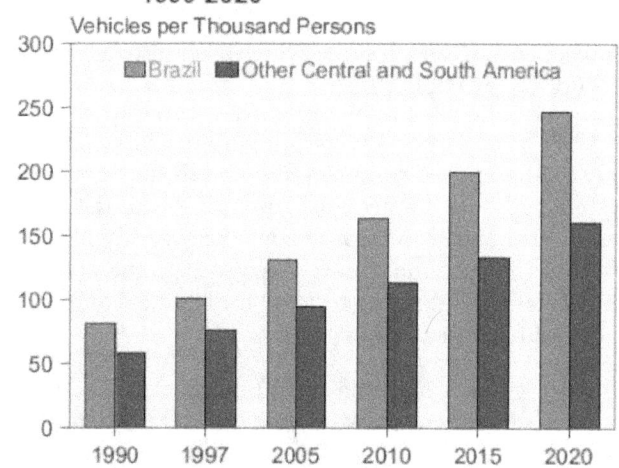

Figure 95. Motorization in Levels in Brazil and Other Central and South America, 1990-2020

Sources: **History:** American Automobile Manufacturers Association, *World Motor Vehicle Data* (Detroit, MI, 1997). **Projections:** EIA, World Energy Projection System (2000).

Brazil's economy actually performed better than many analysts expected at the beginning of 1999, after the devaluation of the country's currency sent its stock markets into a downward spiral [44]. In an atmosphere of tight monetary and fiscal policies, the economy experienced a period of restricted investment and limited domestic demand, which meant that economic activity slowed in the second half of 1998 and at the beginning of 1999. The country's industrial activity declined sharply, with some sectors reporting 50-percent reductions in production. Temporary fiscal and monetary measures were introduced during 1999 to help sustain the government deficit, and the slowdown in economic activity has helped to check the rate of inflation. In the end, the recession was not as severe in Brazil as in some of its neighboring countries, and positive economic growth is expected in 2000.

Brazil has some 139,440 miles of paved roads, and the Trans-Amazon Highway [45], which connects the road system of northeastern Brazil to those of Colombia and Peru [46], accounts for another 3,109 miles. Brazil's automotive sector has a unique fuel mix, in that the country's vehicle fleet uses gasoline, diesel, and, to a lesser extent, alcohol and natural gas [47] (see also page 98 in "Hydroelectricity and Other Renewable Resources"). Alcohol fuel use is a legacy from the 1979 Proácool program, in which the government tried to encourage the consumption of ethanol to ease Brazil's dependency on foreign oil, allowing the country to expand oil production and reserves. In addition, diesel fuels are allowed only in commercial vehicles, and natural gas and liquefied petroleum gas are limited to commercial fleets.

Although it includes 18,721 miles of railways, the Brazilian rail system has become increasingly inefficient over the past several years because of a lack of investment and declining technological expertise [48]. The main railways have now been privatized, however, and an influx of private investment should permit improvements in the rail system and increased competition between trains and trucks, which currently transport most of Brazil's bulk cargo.

In 1998, 3,265 airports were in operation in Brazil, but only 514 had paved runways [49]. The air transportation system provides connections among most of the country's states and also with major cities worldwide. Air transport is becoming more important to the country, both in terms of passenger travel for business and pleasure and for transporting goods.

Argentina has more than 40,000 miles of roads, more than 20,000 miles of railways, and around 6,800 miles of navigable channels [50]. Motorization levels in the country, currently estimated at 179 cars per thousand people, are fairly high relative to those in much of the rest of the region. The high economic growth enjoyed by Argentina for much of the 1990s resulted in higher per capita incomes and strong growth in demand for personal motor vehicles. New automotive plants were built by some of the world's major car and truck manufacturers. In 1998, however, total car and truck production began to decrease and sales fell as the government and international creditors restricted access to credit, and demand from the rest of the Mercosur trading block members declined.[23]

Argentina was particularly hard hit by the Brazilian economic downturn. The devaluation of Brazil's real meant a substantial loss of competitive power for Argentine industry and a loss of exports to Brazil, including automobiles. This has already caused some strain within the Mercosur, particularly in the relationship between Argentina and Brazil. Industrial production in Argentina declined by about 12.5 percent in the first half of 1999 relative to the same time period a year earlier, and unemployment increased to 14.5 percent.

The economic troubles in the region meant a delay in progress on the Mercosur Automotive Agreement. Argentina and Brazil had signed a letter of intent at the end of 1998 that outlined the general terms of an agreement on the subject, but in the atmosphere of economic recession little progress was made in 1999, and the "Common Regime for the Automotive Industry" did not come into effect on January 1, 2000, as originally agreed [51]. The agreement would have included free intrazone trade, common external tariffs, vehicle import rules, environmental protection rules, and user safety standards, among others.

Argentina has begun to improve its airport infrastructure, and the improvements have resulted in an increase in air transportation for passengers and freight. The total number of passengers traveling through the country's airports increased by 1 million, from 12 million in 1997 to 13 million in 1998, and airborne freight movement increased from 169,000 tons in 1997 to 180,000 tons in 1998.

There has been increasing interest from foreign air services in participating in Argentina's air transportation sector. In 1998, United Parcel Service, a U.S. company, acquired several Argentine firms and started a local express mail service. A growing market for mail services, along with the rising number of international firms settling in the country, should help sustain demand for commercial air transport over the next several years.

---

[23]The Mercosur trading block is made up of Argentina, Brazil, Paraguay, and Uruguay. Chile and Bolivia are Associate Members.

In Colombia the political situation worsened in 1999, with government forces involved in an escalating armed conflict with members of the Colombian Armed Revolutionary Forces (FARC) since mid-1999. More than one-third of the country was placed under martial law, and more than 200 government soldiers were killed by the rebel forces in June alone. Representatives from the Colombian government and FARC began discussing negotiations for a peace settlement in January 1999, but talks repeatedly broke down throughout the year. Even when negotiations resumed in December, there seemed little hope that an agreement would be reached, given the FARC demands that the government overhaul nearly every aspect of national life and its resistance to government demands that it give up kidnaping and drug trafficking, its main sources of income [52].

The continuing armed conflict between government and the FARC could keep private and international investment out of Colombia, which would certainly not help the development of its transportation sector. The country does remain an important leader in road transport in the Andean region and is gaining market share in air and transportation [53]. Along with the political problems facing the country, there are several factors that might prevent the transportation sector from achieving its full potential. For one, only a small portion of the country's roads are paved—approximately 20 percent of the total of 68,000 miles of roads. Thus, many of the roads can only be used by vehicles that can handle the terrain, and other vehicles are limited to those roads less affected by rain.

Chile has around 7,690 miles of paved roads and is connected to neighboring Peru by 2,146 miles of the Inter-American Highway [54]. Vehicle ownership is estimated at 114 per thousand people. In addition, Chile possesses about 450 miles of navigable channels and 5,000 miles of railways. Air transportation is well developed in the country, and its main airport, Comodoro Arturo Merino Benítez (in Santiago), handles an average of 2 million passengers per year.

Peru's transportation sector is among the smallest in South America. Mountainous regions, dense jungles, unpaved highways, and earthquakes have all worked to constrain development. Although the government has been easing import restrictions on both old and new cars, vehicle ownership remains quite low at 40 vehicles per thousand persons. The country is connected to Colombia and Chile by 1,550 miles of the Inter-American Highway system, and it has approximately 4,740 miles of paved roads. There are also 1,491 miles of railways and 5,344 miles of navigable channels [55].

Venezuela has around 19,000 miles of paved roads and is interconnected to Guyana and to Colombia by 802 miles of the Inter-American Highway system [56]. There are 338 miles of railways, which are mainly used for transporting freight, and 4,412 miles of navigable channels. In addition, Venezuela has an extensive airway system, which connects most of the country. Its primary international airport, Simón Bolívar International, handles more than 6 million passengers a year. Car ownership is estimated at 103 cars per thousand persons, higher than its neighbors, Colombia and Guyana. Venezuela is the third largest producer of cars in Latin America, after Brazil and Argentina.

Most of the growth projected for Venezuela's transportation sector comes from expected higher levels of domestic commercial transportation and leisure driving as economic conditions improve. An increase in jet fuel consumption results from projected moderate growth in tourism and air freight transportation, encouraged by the establishment of foreign firms in the country.

Eastern Europe and the Former Soviet Union

The transportation sector in Eastern Europe and the former Soviet Union (EE/FSU), like every other sector in the region, was severely affected by the collapse of the Soviet Union. In 1990, total energy consumption in the FSU transportation sector was 2.7 million barrels per day; but by 1997, the region's transportation energy use had fallen dramatically to 1.3 million barrels per day (Figure 96). In Eastern Europe, transportation energy use fell less sharply in the early 1990s, and by 1997 it had recovered to its 1990 level of 0.6 million barrels per day. In the long term, with expected economic recovery in the region as a whole, energy consumption for transportation in the FSU is projected to reach its 1990 level by 2020, and in Eastern Europe it is expected to nearly doubles the 1997 level by 2020.

Figure 96. Transportation Energy Use in Eastern Europe and the Former Soviet Union, 1990-2020

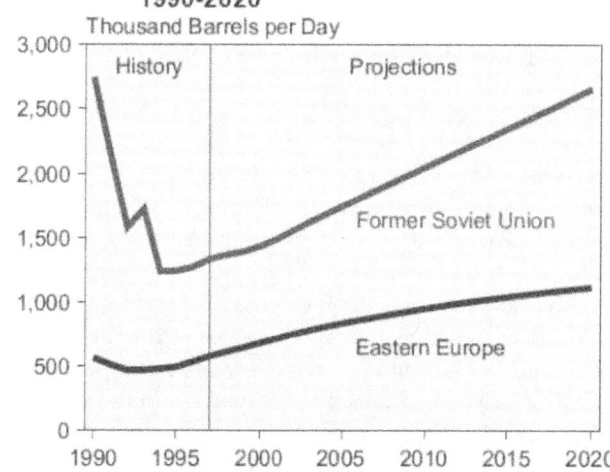

Sources: **History:** Energy Information Administration (EIA), derived from *International Energy Annual 1997*, DOE/EIA-0219(97) (Washington, DC, April 1999). **Projections:** EIA, World Energy Projection System (2000).

The transportation infrastructure in parts of Eastern Europe and the FSU has suffered from the economic decline of the past decade. The World Bank, among other institutions, has made extensive loans available to various countries in the region to improve the transportation network. In Russia, the World Bank's International Bank for Reconstruction and Development (IBRD) approved $400 million in 1999 to improve the selected high-priority road network on the federal road system, including Siberia and the Far East [57]. Loans from the IBRD have also been approved to rehabilitate and maintain key roads in Kazakhstan and to modernize and improve the railway systems of Croatia and Macedonia. In December 1999, another $14 million was approved for emergency road repair and rehabilitation in Albania for restoring roads heavily damaged by refugee convoys and NATO vehicles during the Kosovo crisis earlier in the year [58]. These loans amounted to about $652 million in 1999.

The economic recovery in the countries of Eastern Europe has been more sustained than in the former Soviet Union, and there is renewed interest from foreign automobile manufacturers in the growth potential of Eastern Europe. The economies of Poland and Hungary have been particularly strong in recent years, and demand for well-built, inexpensive cars and small trucks is expected to increase as a result [59]. In the decade since the fall of the Berlin Wall, the demand for personal vehicles has increased by more than 35 percent in Eastern Europe, to 1.7 million vehicles in 1999. Automotive industry analysts expect the demand to increase by another 26 percent, to 2.1 million vehicles, by 2004. General Motors Corporation has established an auto plant in Gliwice, building small cars for the European car market. Italy's Fiat S.p.A. and Volkswagen AG also have invested in the East European market. Volkswagen owns about 70 percent of Skoda AS, the Czech Republic-based automaker that is the fastest-growing brand in Germany. France's Renault SA holds the majority share of the Romanian Automobiles Dacia.

The outlook for automobile sales in the FSU is not as promising as in Eastern Europe. Political instability and economic contraction, along with bureaucratic delays, have reduced sales of personal motor vehicles in the region and made it more difficult for auto manufacturers to penetrate the market. That said, some automotive industry experts believe there may be demand for as many as 1.5 million cars in the FSU within the next 5 years, and the region is increasingly attracting the interest of the world's automotive companies. In the summer of 1998, Ford Motor Company finalized a deal with Russia's Bankers House of St. Petersburg to build 25,000 cars in the country by the end of 2001. General Motors Corporation is also discussing a joint venture with Russian

car manufacturer Avtovaz to build Opel cars in Togliatti, located 600 miles west of Moscow.

The Russian upper house of parliament, Council of Federation, has approved a law regulating the use of natural gas as a motor fuel [60]. According to Russian experts, for every 100,000 vehicles being converted from oil fuel to natural gas, 2.5 million metric tons of fuel annually can be saved, and the volume of exhaust gases can be reduced by 100,000 tons. Today, there are about 200 gas-fuel filling stations in Russia, with the capacity to serve about 30,000 vehicles daily. In Moscow alone there are more than 2 million cars and other motor vehicles.

Africa

The transportation sector has remained largely undeveloped in the economies of Africa, with limited roads that are not, in general, well maintained. Low per capita incomes have kept the number of vehicles per person among the lowest in the world [61]. For example, in Nigeria—Africa's most populous country—there are only an average 12 vehicles per thousand persons; and even in South Africa—the region's most developed economy—there are only about 139 vehicles per thousand persons [62]. In much of the region railways are used primarily to transport goods to market, but the locomotives are outdated, and railway lines are in disrepair. Improvements in economic activity in Africa would make an expansion of trade possible, which in turn would likely stimulate investment in the region's transportation infrastructure.

In Nigeria, the oil boom of the 1970s resulted in construction of a large transportation network connecting oil-producing centers to marketing sectors. Today, Nigeria has more than 20,000 miles of paved roads and approximately 2,000 miles of railways [63], but political instability, government corruption, and low economic growth (despite its rich oil and gas resources) has made it difficult for the country to implement repairs in the transportation sector. As a result, the infrastructure has deteriorated over the past two decades. In November 1999, the government announced it had approved 45 new road construction projects worth an estimated $625 million (U.S.) [64].

Airlines in Nigeria have also fallen into disrepair, and airline travel has declined substantially over the past several years. The situation could change, however, as a result of plans to privatize the state-run airline, Nigeria Airlines Limited (NAL). The Nigerian government and the World Bank signed a pact to restructure NAL in October 1999 [65]. One positive development is that, in 1999, after a 2-year struggle by Nigeria's private airlines to liberalize the Bilateral Air Services Agreement, 28 African countries signed the Yamasoukro Agreement, which will allow Africa's private airlines to fly all air routes in Africa beginning on January 1, 2000 [66].

Nigeria's rail transportation system is poorly developed, and most of its locomotives are in a state of disrepair. There are plans to rehabilitate the national railway network at an estimated cost of $528 million (U.S.), but those plans, along with the proposed Lagos-Abuja high speed railway system, have not yet advanced beyond the planning stages [67].

Another African country with a fairly extensive transportation network is South Africa. Although there are good interconnections among the country's industrial production centers, the transportation infrastructure has not sufficiently kept pace with national needs. South Africa has more than 35,000 miles of paved roads, including highways, and more than 13,000 miles of railways [68]. The government allocated about $472 million (U.S.) in fiscal year 1993, $489 million in fiscal year 1994, and another $500 million in 1995 for road maintenance and repair. However, South Africa's Department of Transportation estimated in 1995 that $1,028 million would be needed to complete all the necessary repairs on the nations's roads [69]. There are plans to construct several roads, including the $200 million DeBeers Pass Road and the $75 million Heidelberg-Villers Road, but both are still only in the planning stages [70].

There is a substantial railway network in South Africa, serving the mining and heavy industries of the country along with those of neighboring countries. Spoornet is the largest heavy hauler and transporter of general freight in South Africa [71]. It was created in 1990 when the South African government decided to commercialize its transportation sector business interests and deregulate the nation's transportation industry. The railway system seems to be suffering from the poor economic conditions of the past 2 years. Spoornet went from a company making a profit of $91 million (U.S.) in 1998 to a loss of $22 million in 1999 [72]. South Africa's ports are among the most important in the world, serving industry from South Africa, Zimbabwe, Zambia, Botswana, Swaziland, and Lesotho [73].

The airline infrastructure in South Africa needs substantial improvement to be able to accommodate the expected growth in tourism and business travel over the next decade. Already, the country's main international airport, Johannesburg International (formerly Jan Smuts International), handles around 5 million passengers each year [74]. Airports Company South Africa (ACSA) is the country's main airline services company, handling 90 percent of all airline passengers traveling throughout South Africa [75].

ACSA, which was privatized in 1998, operates nine of South Africa's major airports: Johannesburg International, Capetown International, Duban International, Kimberley Airport, Port Elizabeth Airport, Bloemfontein Airport, George Airport, East London Airport, and Upington Airport. In 1999 it acquired Pilanesberg International Airport near Sun City. The company expects the number of international travelers through the nation's airports to double within the next 8 years, as tourism and business travel continue to increase. To that end, the company has been working to improve the international airports with several major additions and refurbishments. ACSA has committed $3.25 million (U.S.) to infrastructure improvements at Pilanesberg International [76].

Morocco has an extensive road network relative to those of other African countries. It is a natural transit point between Europe and Africa, and there are plans to expand its transportation infrastructure further in the near future [77]. In terms of the road network, the north-south axis of the country is well established, with nearly 40,000 miles of roads [78]. Highways link Casablanca to Tangier (in the north) and to Agadir (in the south). The country is currently developing its east-west axis. The Moroccan government has committed to connecting all of the country's major cities by paved roads by 2002 [79].

Rail transportation in Morocco, under the control of the National Office of Railways, is relatively well established with about 1,200 miles of rail lines [80]. A rapid commuter service is in operation linking Rabat, Casablanca, El Jadida, Marrakesh, and Agadir. Morocco has six international airports, and strong growth is expected for the nation's airline industry in the future [81]. The government plans to privatize the state-run airline, Royal Air Maroc (RAM), and an international tender to evaluate the privatization was issued in November 1999 [82]. Air France and Spain's Iberia airline currently own minor shares of the company. RAM purchased seven Boeing 737 aircraft in February 1999 and announced in September that it would launch an international tender to buy an additional 20 new aircraft at an estimated total cost of $1.5 billion between 2002 and 2012.

Algeria has more than 30,000 miles of paved highways and around 2,500 miles of railways [83]. Because most goods in the country are transported by rail, the government does not plan to allow this important segment of the transportation sector to be privatized [84]. Modernization of Algeria's railways is considered to be a priority. In October 1999, the government reached a $2 billion (U.S.) agreement with the French companies Spie Enertrance, RailTech, and Cogifer Travaux Ferroviaires on a 10-year contract for maintenance of the existing rail lines and expansion of the system. The state-owned Societe Nationale du Transport Ferovier owns 200 trains, acquired in the 1970s, that are due for modernization. There are also plans to purchase two locomotives to link Algiers and Oran. The consortium of French companies is also in negotiation with the Algerian government to complete the underground subway system in Algiers.

Middle East

In the Middle East, as in Africa, transportation infrastructure has not been extensively developed and motorization levels remain fairly low. Motorization is expected to increase slowly among the Middle Eastern countries, in part because many Middle Eastern countries actively discourage women from driving, ultimately limiting the part of the population able to own automobiles [85]. Nevertheless, the *IEO2000* reference case projects that motorization rates in the region will increase by 2.3 percent per year between 1997 and 2020, reaching 91 vehicles per thousand persons by the end of the forecast period, substantially lower than today's motorization rates in the industrialized world (Figure 97).

Since the end of the Iran-Iraq War in 1988, transportation energy use in Iran has grown substantially, by about 6 percent per year, as reconstruction of the oil refinery network has allowed the easing and eventual removal of restrictions on the sale of transportation fuels [86]. The country's Roads and Transport Ministry has also begun construction on a high-speed railway line between Tehran and Isfahan and plans to construct a 300-mile extension of the Mashad-Bafq rail line.

Another Middle Eastern country that has seen fast-paced growth in its transportation sector is Israel. The number of cars in Israel has increased rapidly since 1985, a result of increasing economic prosperity, and motor gasoline use has grown by an estimated 5 percent per year [87]. Currently, per capita vehicle ownership is estimated at 250 vehicles per thousand persons. Israel's Ben Gurion International Airport serves about 7.4 million passengers each year [88]. Tourism and business travel are expected to continue increasing, and the government has estimated that the airport will handle as many as 16 million international travelers a year by 2010 [89]. With that in mind, the Israeli Transport Ministry has invested some $500 million in improving the infrastructure of the Ben Gurion Airport and expects to invest $330 million in improvements before 2010.

Several road and airport projects are planned for transportation sector development in the United Arab Emirates (UAE) [90]. The Ministry of Public Works and Housing started construction on 15 highway projects in 1997, as well as several maintenance projects. The country also has plans to construct a major inter-Emirate highway that would, upon completion in 2005, link all seven emirates, from Abu Dhabi to Ras al-Khaimah and across to Fujairah. The government also plans to upgrade and expand the UAE's six international airports. Abu Dhabi International Airport will add a new runway and satellite terminal by 2002, and facilities at Dubai International Airport are also being expanded.

References

1. S. Mydans, "Bangkok Opens Skytrain, But Will It Ease Car Traffic?" *The New York Times* (December 6, 1999), p. A3.

2. "Translating S.U.V. Into European: Re-engineering the Old World," *The New York Times* (December 14, 1999), pp. C1 and C14.

3. Energy Information Administration, *Annual Energy Outlook 2000*, DOE/EIA-0383(2000) (Washington, DC, December 1999).

4. Energy Information Administration, *Annual Energy Outlook 2000*, DOE/EIA-0383(2000) (Washington, DC, December 1999), p. 60.

5. Energy Information Administration, *Annual Energy Outlook 2000*, DOE/EIA-0383(2000) (Washington, DC, December 1999), p. 13.

6. Standard & Poor's Platt's, *World Energy Service: Canadian Energy Outlook* (Lexington, MA, Fall-Winter 1998-1999), p. 18.

7. Natural Resources Canada, "Backgrounder: Natural Gas for Vehicles," Vol. 99/18(d) (1999), web site http://nrn1.nrcan.gc.ca; and Ward's/pembertons, *World Vehicle Population*, 1998 Edition (Southfield, MI, 1998), p. 116.

8. Standard & Poor's Platt's, *World Energy Service: Canadian Energy Outlook* (Lexington, MA, Fall-Winter 1998-1999), p. 18.

9. Ward's/pembertons, *World Auto Atlas and Directory*, 1998 Edition (Southfield, MI, 1998), pp. 52 and 204.

Figure 97. Motorization in Levels in the Middle East and Africa, 1990-2020

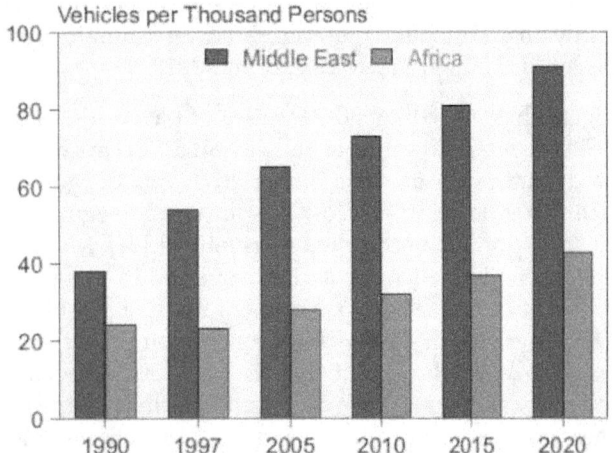

Sources: **History:** American Automobile Manufacturers Association, *World Motor Vehicle Data* (Detroit, MI, 1997). **Projections:** EIA, World Energy Projection System (2000).

10. Ward's/pembertons, *World Auto Atlas and Directory*, 1998 Edition (Southfield, MI, 1998), p. 122.

11. Standard & Poor's Platt's, *World Energy Service: Latin American Outlook, 1999* (Lexington, MA, 1999), pp. 158-159.

12. G. Mohan, "Giant Antennas Answer to D.F. Smog Problem?" elan News Release (January 31, 1999), web site http://earthsystems.org/list/elan.

13. ABCNEWS.com, "Mexico City Air Hurting Kids" (March 11, 1999), web site http://abcnews.go.com.

14. E-Wire Press Release, "Mexico City and IMPCO To Convert Thousands of Vehicles in a Committed Effort To Clean the City's Air" (September 14, 1999), web site http://ens.lycos.com/e-wire.

15. Standard & Poor's Platt's, *World Energy Service: Latin American Outlook, 1999* (Lexington, MA, 1999), pp. 158-159.

16. Standard & Poor's Platt's, *World Energy Service: European Outlook*, Vol. II (Lexington, MA, 1999), p. 5.

17. Standard & Poor's Platt's, *World Energy Service: European Outlook*, Vol. II (Lexington, MA, 1999), pp. 5-6.

18. Standard & Poor's Platt's, *World Energy Service: European Outlook*, Vol. II (Lexington, MA, 1999), p. 189.

19. Standard & Poor's Platt's, *World Energy Service: European Outlook*, Vol. II (Lexington, MA, 1999), p. 63.

20. Standard & Poor's Platt's, *World Energy Service: European Outlook*, Vol. II (Lexington, MA, 1999), pp. 60-61.

21. "Gas Vehicle Pact Signed," *Financial Times: International Gas Report*, No. 387 (November 26, 1999), p. 22.

22. Standard & Poor's Platt's, *World Energy Service: European Outlook*, Vol. II (Lexington, MA, 1999), pp. 86-87.

23. U.S. Central Intelligence Agency, *World Factbook 1999*, web site www.odci.gov/cia/publications/factbook/mo.html#trans.

24. Standard & Poor's Platt's, *World Energy Service: Asia Pacific Outlook* (Lexington, MA, 1999), pp. 137-138.

25. Standard & Poor's DRI, *Asian Automotive Industry Forecast* (Lexington, MA, July 1999), p. 149.

26. "Global Automotive Report: Japan New Vehicle Sales Increase 0.2 Percent," *The Detroit News* (December 2, 1999), web site http://detnews.com.

27. Standard & Poor's Platt's, *World Energy Service: Asia Pacific Outlook* (Lexington, MA, 1999), pp. 25-26.

28. "Facts at a Glance," Unofficial Sydney International Airport web site (August 11, 1999), www.acay.com.au/~willt/yssy/welcome.html.

29. Brisbane Airport Corporation, Ltd., "Airport Starts Public Comment Phase for Proposed $51 Million Development," Media Release (December 8, 1999).

30. Standard & Poor's Platt's, *World Energy Service: Asia Pacific Outlook* (Lexington, MA, 1999), pp. 212-213.

31. "Record Results," *Auckland Airport Touchdown Newsletter* (1999), web site www.auckland-airport.co.nz.

32. Standard & Poor's Platt's, *World Energy Service: Asia Pacific Outlook* (Lexington, MA, 1999), pp. 54-55.

33. Ward's/pembertons, *World Auto Atlas and Directory*, 1998 Edition (Southfield, MI, 1998), p. 186.

34. Standard & Poor's Platt's, *World Energy Service: Asia Pacific Outlook* (Lexington, MA, 1999), p. 84.

35. Standard & Poor's Platt's, *World Energy Service: Asia Pacific Outlook* (Lexington, MA, 1999), pp. 111-112.

36. Standard & Poor's Platt's, *World Energy Service: Asia Pacific Outlook* (Lexington, MA, 1999), pp. 189-190.

37. Standard & Poor's Platt's, *World Energy Service: Asia Pacific Outlook* (Lexington, MA, 1999), pp. 235-236.

38. Standard & Poor's Platt's, *World Energy Service: Asia Pacific Outlook* (Lexington, MA, 1999), pp. 261-262.

39. U.S. Central Intelligence Agency, *World Factbook 1999*, web site www.odci.gov/cia/publications/factbook/index.html.

40. Standard & Poor's Platt's, *World Energy Service: Asia Pacific Outlook* (Lexington, MA, 1999), p. 285.

41. S. Mydans, "Bangkok Opens Skytrain, But Will It Ease Car Traffic?" *The New York Times* (December 6, 1999), p. A3.

42. Standard & Poor's Platt's *World Energy Service: Asia Pacific Outlook* (Lexington, MA, 1999), pp. 309-310

43. D.P. Mazal, "Overview: A Change of Perspective," *South American Automotive Industry Forecast Report*, Second Quarter 1999 (Lexington, MA: Standard & Poor's DRI, 1999), pp. 3-5.

44. Standard & Poor's Platt's, *World Energy Service: Latin American Outlook, 1999* (Lexington, MA, 1999), p. 1.

45. Standard & Poor's Platt's, *World Energy Service: Latin American Outlook, 1999* (Lexington, MA, 1999), pp. 75-76.

46. "Amazon River," *Compton's Encyclopedia Online*, web site www.optonline.com/comptons/ceo/00157_A.html).

47. Standard & Poor's Platt's, *World Energy Service: Latin American Outlook, 1999* (Lexington, MA, 1999), pp. 75-76.

48. Standard & Poor's Platt's, *World Energy Service: Latin American Outlook, 1999* (Lexington, MA, 1999), pp. 75-76.

49. U.S. Central Intelligence Agency, *World Factbook 1999*, web site www.odci.gov/cia/publications/factbook/br.html.

50. Standard & Poor's Platt's, *World Energy Service: Latin American Outlook, 1999* (Lexington, MA, 1999), pp. 22-23.

51. "What is MERCOSUR," *Mercosur Trade and Investment Report*, web site www.mercosurinvestment.com (not dated).

52. "Peace Talks Face Deadlock," *Latin American Monitor: Andean Group*, Vol. 16, No. 12 (December 1999), p. 5.

53. Standard & Poor's Platt's, *World Energy Service: Latin American Outlook, 1999* (Lexington, MA, 1999), p. 130.

54. Standard & Poor's Platt's, *World Energy Service: Latin American Outlook, 1999* (Lexington, MA, 1999), pp. 102-103.

55. Standard & Poor's Platt's, *World Energy Service: Latin American Outlook, 1999* (Lexington, MA, 1999), pp. 183-184.

56. Standard & Poor's Platt's, *World Energy Service: Latin American Outlook, 1999* (Lexington, MA, 1999), p. 212.

57. World Bank, *World Bank Annual Report 1999*, web site www.worldbank.org.

58. World Bank, "Emergency Loan To Help Restore Albania's Damaged National Road Network," News Release No. 2000/122/ECA (December 7, 1999).

59. D. Howes, "Eastern Europe Holds a Special Market Allure for U.S. Automakers," *The Detroit News* (November 9, 1999), web site http://detnews.com.

60. "Russia Approves 'Gas-Fired Cars' Law," *Financial Times: Gas Markets Week: Europe*, No. 29 (November 22, 1999), p. 12.

61. Standard & Poor's Platt's, *World Energy Service: Africa/Middle East* (Lexington, MA, 1999), pp. 6-7.

62. Ward's/pembertons, *World Auto Atlas and Directory*, 1998 Edition (Southfield, MI, 1998), pp. 138 and 168.

63. Standard & Poor's Platt's *World Energy Service: Africa/Middle East* (Lexington, MA, 1999), p. 85.

64. Mbendi News Release, "Nigerian Government Approves Funds for Road Projects" (November 5, 1999), web site www.mbendi.co.za.

65. Mbendi News Release (October 24, 1999).

66. T. Ojudun, "African Private Airlines To Have Open Skies in the Continent," *The News (Lagos)* (December 6, 1999), web site http://africanews.org.

67. Mbendi Company Project List, "Transport and Storage Projects" (November 17, 1999), web site www.mbendi.co.za.

68. Standard & Poor's Platt's, *World Energy Service: Africa/Middle East* (Lexington, MA, 1999), p. 111.

69. South African Embassy, "Business & Economy: Rad Expenditure," web site www.southafrica.net (not dated).

70. Mbendi Company Project List, "Transport and Storage Projects" (November 17, 1999), web site www.mbendi.co.za.

71. Spoornet, "Company Profile" (1998), web site www.spoornet.co.za.

72. "Transformation News: To Survive, Spoornet Must Take the Pain Now," *The Sunday Times* (July 18, 1999), web site www.spoornet.co.za.

73. "South Africa Demographics" (February 16, 1998), web site http://tcol.co.uk/southafr/sou2.htm.

74. Standard & Poor's Platt's, *World Energy Service: Africa/Middle East* (Lexington, MA, 1999), p. 111.

75. Airports Company South Africa, "Company Profile," web site www.airports.co.za/ (not dated).

76. Airports Company South Africa, "More Regional Airports Takeovers," Press Release (November 10, 1999), web site www.airports.co.za.

77. Standard & Poor's Platt's, *World Energy Service: Africa/Middle East* (Lexington, MA, 1999), p. 68.

78. U.S. Central Intelligence Agency, *World Factbook 1999*, web site www.odci.gov/cia/publications/factbook/mo.html#trans.

79. "Morocco: The Country," Morocco Today, web site www.morocco-today.com (not dated).

80. "Morocco, Land of Opportunity," *Morocco & Africa Today: International Edition* (1995), web site www.morocco-today.com.

81. Standard & Poor's Platt's, *World Energy Service: Africa/Middle East* (Lexington, MA, 1999), p. 68.

82. "Morocco Moves on Privatization of State-Owned Airline Company RAM," *North Africa Journal* (November 12, 1999), web site www.africanews.com.

83. Standard & Poor's Platt's, *World Energy Service: Africa/Middle East* (Lexington, MA, 1999), p. 26.

84. "Algeria Business News," *North Africa Journal* (October 22, 1999), web site www.africanews.com.

85. Standard & Poor's DRI, *World Energy Service: Africa/Middle East Outlook* (Lexington, MA, 1998), p. 399.

86. Standard & Poor's Platt's, *World Energy Service: Africa/Middle East* (Lexington, MA, 1999), p. 151.

87. Standard & Poor's Platt's, *World Energy Service: Africa/Middle East* (Lexington, MA, 1999), p. 175.

88. "Airport Ranking (Passengers)," web site www. airconnex.com (April 7, 1999).

89. D. Ben-Tal, "Israel Braces for Huge Influx of International Travelers," *Jewish Bulletin of Northern California* (November 17, 1995), web site http:// jewishsf.com.

90. Standard & Poor's Platt's, *World Energy Service: Africa/Middle East* (Lexington, MA, 1999), p. 241.

Environmental Issues and World Energy Use

In the coming decades, global environmental issues could significantly affect patterns of energy use around the world. Any future efforts to limit carbon emissions are likely to alter the composition of total energy-related carbon emissions by energy source.

The importance of carbon dioxide emissions as an environmental issue of international concern has grown substantially since 1992, when the United Nations Framework Convention on Climate Change was adopted because of increasing concern over rising atmospheric concentrations of greenhouse gases and their possible adverse effects on the global climate system. World energy use has emerged at the center of the issue.

The two major anthropogenic (human-caused) sources of carbon dioxide emissions worldwide are the combustion of fossil fuels and land-use changes (primarily, deforestation). Currently, approximately three-quarters of global carbon dioxide emissions result from fossil fuel combustion, although the share varies by region. Net carbon releases from these two anthropogenic sources are believed to be largely responsible for the rapid rise in atmospheric concentrations of carbon dioxide since pre-industrial times [1].

Since 1970, global carbon emissions from the combustion of fossil fuels have gradually increased, reaching 5.8 billion metric tons in 1990 and 6.2 billion metric tons in 1997 (Figure 98). In the *International Energy Outlook 2000* (*IEO2000*) reference case, annual energy-related carbon emissions are projected to rise to 10 billion metric tons by 2020—an increase of 62 percent over 1997 emissions.

During the 1990s, growth in global energy-related carbon emissions was dampened by several circumstances. Starting early in the decade, economic and political upheaval in Eastern Europe and the former Soviet Union (EE/FSU) brought a considerable decline in emissions from the countries in the region. In 1991, the economic recession in the United States induced a drop in U.S. carbon emissions. Declining energy intensities and carbon intensities across Western Europe, resulting mainly from a reduction in coal use, led to a substantial drop in the region's emissions between 1990 and 1994; and while energy-related carbon emissions from Western Europe have since increased, carbon emissions in 1997 were still below their 1990 level. More recently, continued economic problems in Russia and recession in Japan and several East Asian economies have slowed the growth of worldwide carbon emissions. As the new century begins, however, those economies are expected to enter a period of recovery, and emissions are projected

to rise at rates closer to those before the economic downturn.

Based on expectations of regional economic growth and energy demand, global carbon emissions over the first two decades of the 21st century are projected to increase at an average annual rate of 2.1 percent, more than twice the rate of increase from 1990 to 1997. The growth in world emissions anticipated toward the end of the forecast period, between 2015 and 2020, is largely a result of rapidly increasing energy demand in developing countries and the EE/FSU and reductions in nuclear power generation across North America and Western Europe (to be replaced primarily by fossil fuel energy sources). The projected increase between 2015 and 2020 amounts to 1.1 billion metric tons of carbon, which is more than three times the increase in global emissions between 1990 and 1997.

Developing countries as a group account for the majority (70 percent) of the projected growth in global carbon emissions. Carbon emissions from China, which are expected to rise from 0.8 billion metric tons in 1997 to 2.1 billion metric tons in 2020, constitute about 33 percent of the projected global increase. Carbon emissions from the industrialized world are expected to rise by 0.9 billion

Figure 98. World Carbon Emissions by Region, 1970-2020

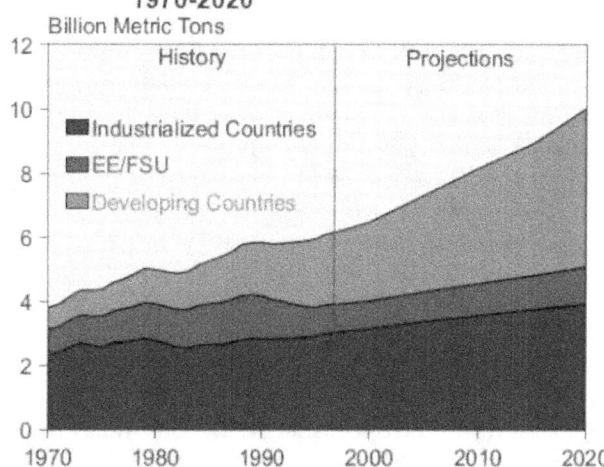

Sources: **History:** Energy Information Administration (EIA), Office of Energy Markets and End Use, International Statistics Database and *International Energy Annual 1997*, DOE/EIA-0219(97) (Washington, DC, April 1999). **Projections:** EIA, World Energy Projection System (2000).

metric tons between 1997 and 2020, led by an increase of 0.6 billion metric tons in emissions from North America. In total, the industrialized nations account for 23 percent of the projected increase in global carbon emissions.

As economic recovery takes hold in the transitional economies of the EE/FSU, an increase in emissions from this region is projected to make up the remaining 7 percent of the 1997-2020 growth in world carbon emissions. As a result of reduced economic activity, carbon emissions from the EE/FSU declined by 34 percent between 1990 and 1997. Despite the projected upswing, EE/FSU carbon emissions in 2020 are expected to be about 0.2 billion metric tons lower than in 1990.

By 2020, developing countries are expected to surpass the industrialized countries in both energy demand and carbon emissions, given expectations of rapid population growth, rising personal incomes, improved standards of living, and greater industrialization (Figure 99). Carbon emissions from the developing countries are expected to make up 46 percent of the world total in 2015 and 49 percent of the total in 2020, as compared with 42 percent and 39 percent, respectively, for the industrialized countries. On a per capita basis, however, carbon emissions from the industrialized nations are expected to remain far higher than those from most of the developing countries (Figure 100). Only South Korea's per capita emissions, projected at 3.6 metric tons per person in 2020, rival those of the industrialized nations.

Future levels of energy-related carbon emissions are likely to differ significantly from the *IEO2000* projections if measures to stabilize atmospheric concentrations of global greenhouse gases are enacted, such as those outlined in the Kyoto Protocol of the Framework Convention on Climate Change [2]. The Protocol, which calls for carbon emission reductions and limits to emissions growth from Annex I countries between 2008 and 2012, could have profound effects on future energy use worldwide. As of January 13, 2000, 22 of 84 signatories had ratified the Protocol; however, none of those 22 countries would be required to reduce emissions under the terms of the treaty.[24] Consequently, the *IEO2000* projections do not reflect the potential effects of the Kyoto Protocol or other possible climate change policy measures.

Although the Kyoto Protocol would require large cuts in carbon emissions in the Annex I countries, it would not result in worldwide stabilization of emissions. In the *IEO2000* reference case, world carbon emissions are projected to grow from 5.8 billion metric tons in 1990 to 8.1

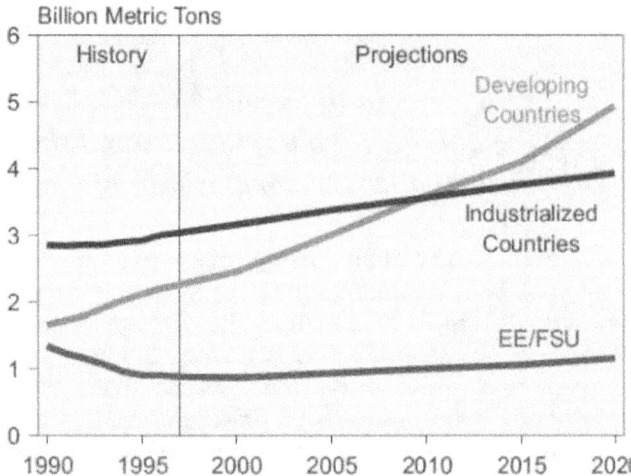

Figure 99. World Carbon Emissions by Region, 1990-2020

Sources: **History:** Energy Information Administration (EIA), Office of Energy Markets and End Use, International Statistics Database and *International Energy Annual 1997*, DOE/EIA-0219(97) (Washington, DC, April 1999). **Projections:** EIA, World Energy Projection System (2000).

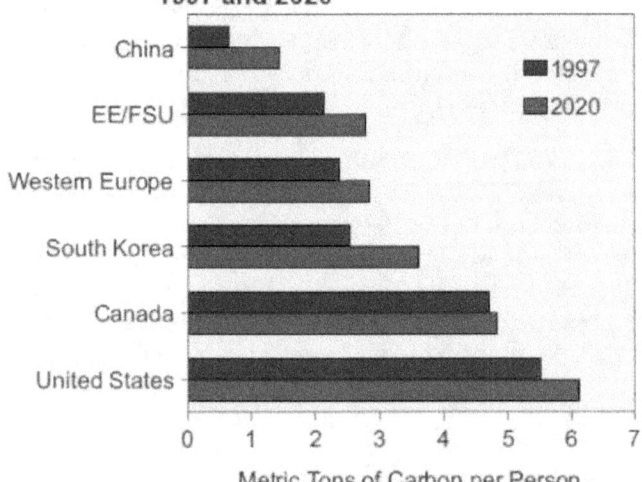

Figure 100. Per Capita Carbon Emissions in Selected Regions and Countries, 1997 and 2020

Sources: **1997:** Energy Information Administration (EIA), Office of Energy Markets and End Use, International Statistics Database and *International Energy Annual 1997*, DOE/EIA-0219(97) (Washington, DC, April 1999). **2020:** EIA, World Energy Projection System (2000).

billion metric tons in 2010, a 40-percent increase (Figure 101). This projection does not differ significantly from last year's carbon forecast for carbon emissions. Even if the Kyoto agreement did go into effect, worldwide carbon emissions still would increase by 31 percent from

[24] The Kyoto Protocol would enter into force 90 days after ratification by at least 55 Parties to the United Nations Framework Convention on Climate Change, including developed (Annex I) countries representing at least 55 percent of the total 1990 carbon dioxide emissions from the Annex I group. The Protocol opened for signature on March 15-16, 1999. By that time, seven countries (Antigua and Barbuda, El Salvador, Fiji, Maldives, Panama, Trinidad and Tobago, and Tuvalu) had already ratified it. As of January 13, 2000, the Protocol had also been ratified by the Bahamas, Cyprus, Georgia, Guatemala, Jamaica, Micronesia, Nicaragua, Niue, Paraguay, Turkmenistan, Uzbekistan, Bolivia, Ecuador, Mongolia, and Palau.

1990 to 2010, because emissions from the developing world would continue to increase. In the long term, the participation of developing nations will be needed to achieve global stabilization of carbon emissions.

By fuel, the *IEO2000* reference case projections indicate that the combustion of coal will contribute 1.1 billion metric tons to the total increase in worldwide emissions between 1997 and 2020, natural gas 1.3 billion metric tons, and oil 1.5 billion metric tons. Again, however, any future efforts to limit carbon emissions are also likely to alter the composition of total energy-related carbon emissions by energy source.

Considering the differences in regional patterns of fuel use, projections for carbon emissions by energy source differ among regions. In the industrialized world, increased emissions from the consumption of oil and natural gas constitute the majority of the rise in energy-related carbon emissions between 1997 and 2020 (Figure 102). Because of its importance in the transportation sector, oil is expected to remain the primary source of carbon emissions in industrialized nations, accounting for roughly 50 percent of their total in 2020.

Natural-gas-related emissions in the industrialized countries are projected to increase from 0.6 billion metric tons in 1997 to just over 1 billion metric tons in 2020—a larger increase than is expected for emissions from oil or coal. By 2020, natural gas is projected to account for roughly one-fourth of overall carbon emissions in the industrialized world, equal to the coal share; however, the amount of energy derived from natural gas in 2020 is projected at 70.6 quadrillion British thermal units (Btu), compared with only 40.5 quadrillion Btu from coal.

For the EE/FSU, the projected increase in natural-gas-related emissions over the forecast period is even more pronounced than that for the industrialized world (Figure 103). With natural gas consumption in the EE/FSU region projected to rise from 22.4 quadrillion Btu in 1997 to 41.2 quadrillion Btu in 2020, related emissions are expected to increase by 84 percent. Carbon

Figure 101. Carbon Emissions in Annex I and Non-Annex I Nations in Three Cases, 1990, 2010, and 2020

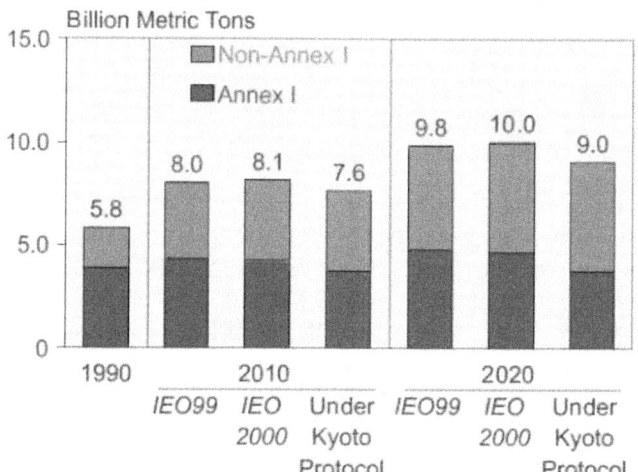

Sources: **1990**: Energy Information Administration (EIA), Office of Energy Markets and End Use, International Statistics Database and *International Energy Annual 1997*, DOE/EIA-0219(97) (Washington, DC, April 1999). *IEO99*: EIA, International Energy Outlook 1999, DOE/EIA-0484(99) (Washington, DC, March 1999). *IEO2000*: World Energy Projection System (2000).

Figure 102. Carbon Emissions in the Industrialized World by Fuel Type, 1990-2020

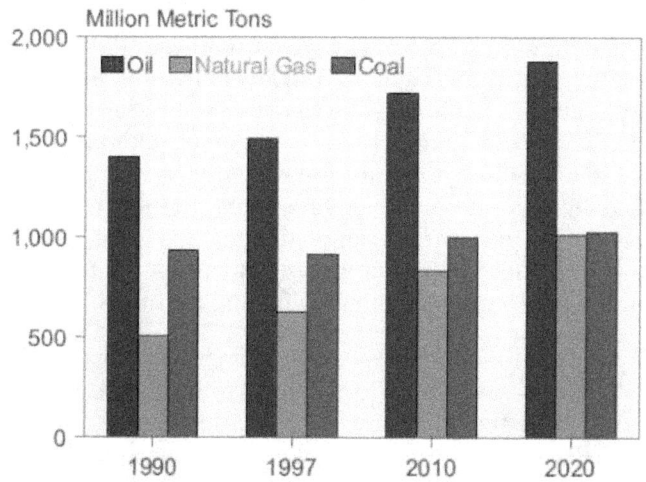

Sources: **History:** Energy Information Administration (EIA), Office of Energy Markets and End Use, International Statistics Database and *International Energy Annual 1997*, DOE/EIA-0219(97) (Washington, DC, April 1999). **Projections:** EIA, World Energy Projection System (2000).

Figure 103. Carbon Emissions in Eastern Europe and the Former Soviet Union by Fuel Type, 1990-2020

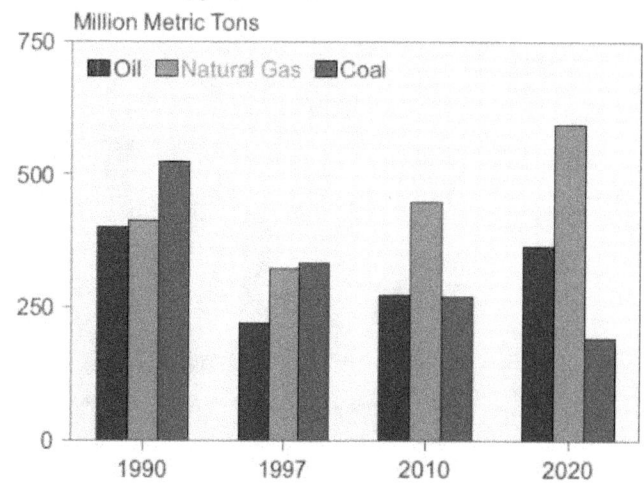

Sources: **History:** Energy Information Administration (EIA), Office of Energy Markets and End Use, International Statistics Database and *International Energy Annual 1997*, DOE/EIA-0219(97) (Washington, DC, April 1999). **Projections:** EIA, World Energy Projection System (2000).

emissions from the combustion of coal in the region are expected to drop by 43 percent over the same period. In 1997, coal accounted for 38 percent of total EE/FSU carbon emissions and natural gas 37 percent; their respective shares in 2020 are projected to be 17 percent and 52 percent.

In the developing world, carbon emissions from the combustion of all fossil fuels are projected to increase, although emissions from coal and oil use grow more slowly than those from natural gas (Figure 104). Despite a 3-percent drop in its share of total energy consumption between 1997 and 2020, coal remains the largest source of energy-related carbon emissions in the developing world—most notably, in China and India, where it remains the primary fuel for electricity generation. Coal's share of total carbon emissions from the developing nations is projected to decline from 48 percent in 1997 to 44 percent in 2020 and oil's from 41 percent to 38 percent, while the natural gas share rises from 11 percent to 18 percent.

The remainder of this chapter examines the influences of economic growth, energy demand, and fuel mix on trends in energy-related carbon emissions and discusses some of the issues in the international debate on climate change policy and the mitigation of global greenhouse gas emissions. The following are highlights of the *IEO2000* projections in this chapter:

- The projected increases in carbon emissions are the result of both economic growth and population growth, despite projected declines in the energy intensity of economic activity and the carbon intensity of energy supply.

Figure 104. Carbon Emissions in the Developing World by Fuel Type, 1990-2020

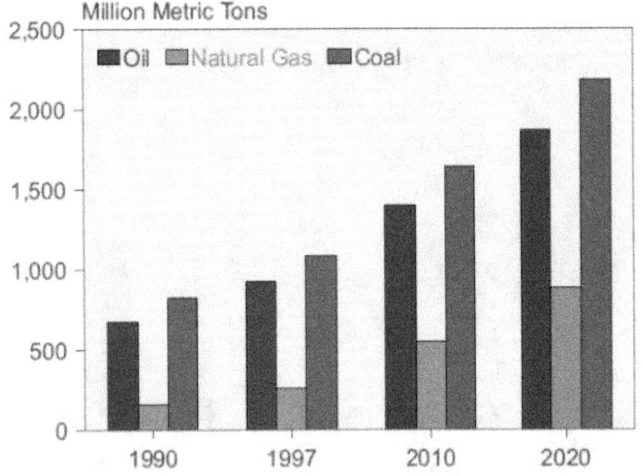

Sources: **History:** Energy Information Administration (EIA), Office of Energy Markets and End Use, International Statistics Database and *International Energy Annual 1997*, DOE/EIA-0219(97) (Washington, DC, April 1999). **Projections:** EIA, World Energy Projection System (2000).

- In the reference case, the expected rate of increase in carbon emissions for the developing world is three times that for the industrialized world.

- Economic recovery in the transitional economies of the EE/FSU is expected to reverse the region's trend of declining carbon emissions during the 1990s.

Factors Influencing Trends in Energy-Related Carbon Emissions

A degree of uncertainty surrounds any forecast of energy-related carbon emissions. Major sources of uncertainty include estimates of primary energy consumption in total and by fuel source, which are used, in conjunction with fuel-specific factors for the amount of carbon emitted per unit of energy consumed, to derive regional emissions projections.

The *carbon intensity of energy supply* is a measure of the amount of carbon associated with each unit of energy produced. It directly links changes in carbon emissions levels with changes in energy usage. Carbon emissions vary by energy source, with coal being the most carbon-intensive fuel, followed by oil, then natural gas. Nuclear power and some renewable energy sources (i.e., solar, wind, and hydropower) do not generate carbon emissions. As changes in the fuel mix alter the share of total energy demand met by more carbon-intensive fuels relative to less carbon-intensive or "carbon-free" energy sources, overall carbon intensity changes. Over time, declining carbon intensity can offset increasing energy consumption to some extent. If energy consumption increased and carbon intensity declined at the same rate, carbon emissions would remain constant.

The *energy intensity of economic activity* is a measure of energy consumption per unit of economic activity. It relates changes in economic activity to changes in energy consumption. As a country's energy intensity changes, so does the influence of a given level of economic activity on carbon emissions. Increased energy use and economic growth generally occur together, although the degree to which they are linked varies across regions and stages of economic development. In industrialized countries, growth in energy demand has historically lagged behind economic growth, whereas the two are more closely correlated in developing countries.

As with carbon intensity, regional energy intensities do not necessarily remain constant over time. The rate at which the energy efficiency of an economy's capital stock (vehicles, appliances, manufacturing equipment, etc.) improves affects trends in energy intensity. New stock is often more energy efficient than the older equipment it replaces. In addition to the availability of more energy-efficient technologies, however, the rate of efficiency improvement is determined by the rate of capital stock turnover, the dynamics between energy and non-energy prices, investment in research and development, and the makeup of the existing capital stock. Changes in the energy efficiency of capital stocks in individual economies can produce changes in regional energy intensities, with corresponding effects on expectations for future levels of energy consumption, fuel mix, and carbon emissions.

Structural shifts in national economies can also lead to changes in energy intensity, when the shares of economic output attributable to energy-intensive and non-energy-intensive industries change. For example, iron, steel, and cement are among the most energy-intensive industries, and countries whose economies rely on production from such energy-intensive industries tend to have high energy intensities. When their economies shift toward less energy-intensive activities, their national energy intensities may decline. Other influences on regional energy intensity trends include changes in consumer tastes and preferences, climate, taxation, the availability of energy supply, government regulations and standards, and the structure of energy markets themselves.

The *Kaya Identity* is a mathematical expression that is used to describe the relationship among the factors that influence trends in energy-related carbon emissions:

$$C = (C / E) \times (E / GDP) \times (GDP / POP) \times POP \ .$$

The formula links total energy-related carbon emissions (C) to energy (E), the level of economic activity as measured by gross domestic product (GDP), and population size (POP) [3]. The first two components on the right-hand side represent the carbon intensity of energy supply (C/E) and the energy intensity of economic activity (E/GDP), as discussed above. Economic growth is viewed from the perspective of changes in output per capita (GDP/POP). At any point in time, the level of energy-related carbon emissions can be seen as the product of the four Kaya Identity components—energy intensity, carbon intensity, output per capita, and population size.

Regional Trends

The Kaya Identity provides an intuitive approach to the interpretation of historical trends and future projections of energy-related carbon emissions. Essentially, it illustrates how the percentage rate of change in carbon emission levels over time approximates the percentage rate of change across the four Kaya components.[25] Between 1970 and 1997, reductions in energy intensity and/or carbon intensity did not produce a sustained downward trend in energy-related carbon emissions from the industrialized world, the developing world, or the transitional economies of the EE/FSU taken as a group (Table 22).[26] In the EE/FSU, lower annual carbon emissions during the 1990s resulted primarily from reductions in economic output (and energy consumption) per capita and only secondarily from a slight decline in the region's carbon intensity.

In the *IEO2000* reference case projections for regional carbon emissions, economic growth and population growth continue to overshadow expected reductions in energy intensity and carbon intensity, particularly in the developing world. Accordingly, any future reductions in carbon emissions would require accelerated declines in energy intensity and/or carbon intensity. Increasing the share of future energy demand met by low-carbon or carbon-free energy sources, however, may require significant changes in existing energy infrastructure. The Kaya Identity components do not provide a framework for estimating costs associated with efforts to reduce either carbon intensity or energy intensity.

The oil crises of the 1970s, among other circumstances, triggered reductions in energy intensity in the United States and other industrialized countries; however, there has been little incentive in the past to reduce carbon intensity. Energy-related carbon dioxide emissions have become an issue of international concern only in recent years. Any historical reductions in carbon intensity have been the unintended results of other actions or developments, such as the introduction of nuclear power and the emergence of natural gas as a competitive fuel for electricity generation.

Industrialized Countries

For the industrialized countries as a group, economic growth has been the driving force behind growth in energy-related carbon emissions in the 1990s, and it is expected to remain so over the forecast period (Figure 105). Both the average energy intensity and the average carbon intensity of the industrialized countries declined by approximately 4 percent between 1990 and 1997. The *IEO2000* projections indicate further reductions in energy intensity, whereas carbon intensity is expected to increase slightly over the forecast period, mainly as a result of reductions in nuclear power generation.

Current energy intensities differ among the industrialized regions. For example, North America has a higher

---

[25]In terms of rates of changes, the Kaya Identity can be expressed as [$d(\ln C) / dt = d(\ln C / E) / dt + d(\ln E / GDP) / dt + d(\ln GDP / POP) / dt + d(\ln POP) / dt$], which shows that, over time, the rate of change in carbon emissions is equal to the sum of the rate of change across the four Kaya components (i.e. the rate of change in carbon intensity, plus the rate of change in energy intensity, plus the rate of change in output per capita, plus the rate of change in population).

[26]In contrast to the industrialized countries taken as a group, declines in energy intensity and carbon intensity in Western Europe during the early 1990s did lead to emissions reductions.

Table 22. Average Annual Percentage Change in Carbon Emissions and the Kaya Identity Components by Region, 1970-2020

Parameter	History			Projections	
	1970-1980	1980-1990	1990-1997	1997-2010	2010-2020
Industrialized World					
Carbon Intensity	-0.4%	-0.5%	-0.6%	0.0%	0.1%
Energy Intensity	-1.1%	-1.9%	-0.5%	-1.0%	-1.2%
Output per Capita	2.4%	2.1%	1.4%	1.8%	1.7%
Population	0.9%	0.7%	0.7%	0.5%	0.4%
Carbon Emissions	1.7%	0.4%	0.9%	1.2%	1.0%
Developing World					
Carbon Intensity	-0.7%	-0.2%	-0.3%	-0.1%	0.0%
Energy Intensity	-0.1%	1.0%	-0.7%	-0.8%	-1.3%
Output per Capita	3.2%	1.6%	3.8%	3.2%	3.3%
Population	2.2%	2.1%	1.9%	1.4%	1.2%
Carbon Emissions	4.7%	4.5%	4.6%	3.6%	3.2%
Transitional Economies					
Carbon Intensity	-0.7%	-0.7%	-0.9%	-0.3%	-0.4%
Energy Intensity	1.4%	0.2%	0.2%	-2.0%	-2.9%
Output per Capita	2.4%	0.9%	-5.2%	3.4%	4.9%
Population	0.9%	0.7%	0.0%	0.0%	0.0%
Carbon Emissions	4.0%	1.2%	-5.8%	0.9%	1.5%

Note: Using an average annual rate of change in carbon emissions between any two years mathematically approximates the actual combined effect on emission levels from changes in the four Kaya Identity components. Across years where there were large changes in either carbon emission levels or the Kaya Identity components themselves, comparisons based on an average annual rate of change measure may yield round-off differences.

Sources: **History:** Energy Information Administration (EIA), Office of Energy Markets and End Use, International Statistics Database and *International Energy Annual 1997*, DOE/ EIA-0219(97) (Washington, DC, April 1999). **Projections:** EIA, World Energy Projection System (2000).

Figure 105. Changes in Carbon Emissions and the Kaya Identity Components in the Industrialized World, 1990-2020

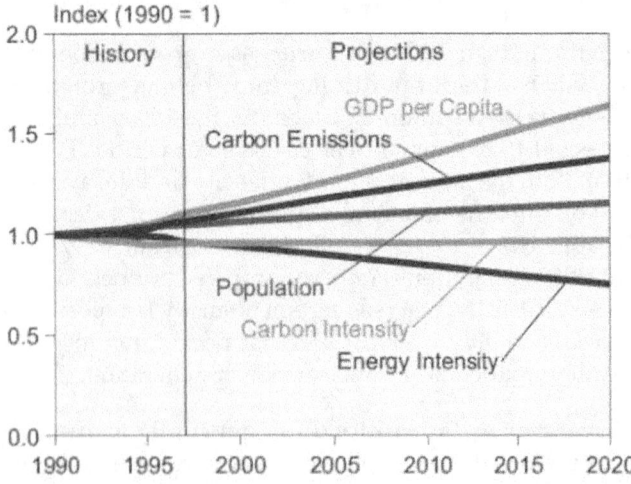

Sources: **History:** Energy Information Administration (EIA), Office of Energy Markets and End Use, International Statistics Database and *International Energy Annual 1997*, DOE/ EIA-0219(97) (Washington, DC, April 1999). **Projections:** EIA, World Energy Projection System (2000).

average energy intensity than either Western Europe or industrialized Asia (Figure 106). There are also marked differences within North America, with the energy intensity of the United States less than that of either Mexico or Canada. From 1997 to 2020, energy intensities are projected to drop by between 23 percent and 27 percent in North America and Western Europe, where per capita energy use is expected to remain fairly stable and energy consumption is projected to rise less rapidly than GDP. A 24-percent decline in energy intensity is projected for Australasia, compared with only a 12-percent drop for Japan, where a continued period of slow economic growth is expected.

Since 1990, Western Europe has shown a slightly greater decline in carbon intensity than either North America or industrialized Asia (Figure 107)—a direct result of the sharp decrease in Western Europe's coal consumption in favor of natural gas and nuclear power. After the reunification of its eastern states, Germany's lignite production was curtailed, and the use of lignite for home heating was replaced by natural gas and other fuels. In the United Kingdom, hard coal production declined

Figure 106. Energy Intensity in the Industrialized and EE/FSU Regions, 1990-2020

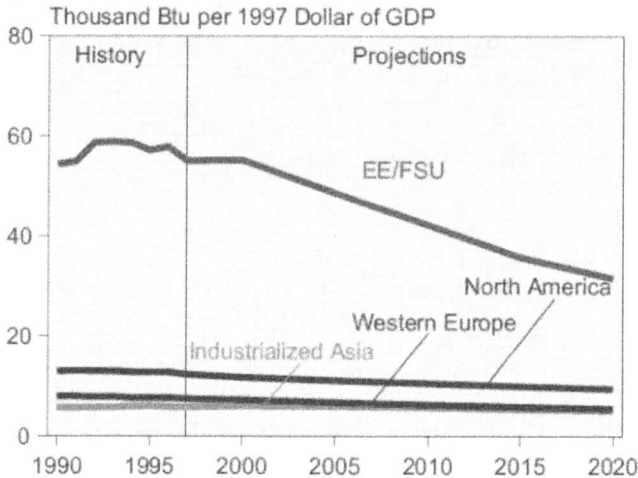

Sources: **History:** Energy Information Administration (EIA), Office of Energy Markets and End Use, International Statistics Database and *International Energy Annual 1997*, DOE/EIA-0219(97) (Washington, DC, April 1999). **Projections:** EIA, World Energy Projection System (2000).

Figure 107. Carbon Intensity in the Industrialized and EE/FSU Regions, 1990-2020

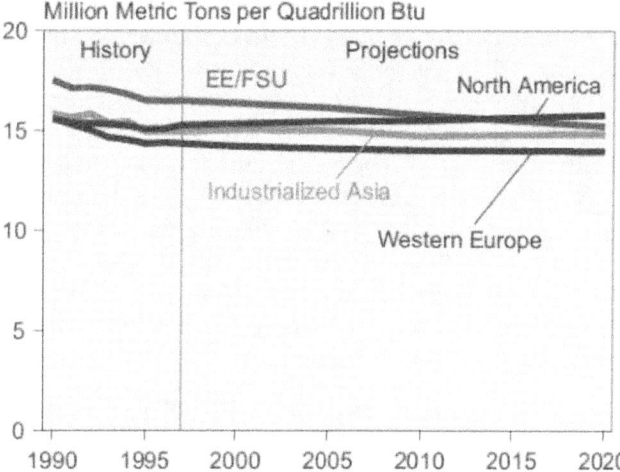

Sources: **History:** Energy Information Administration (EIA), Office of Energy Markets and End Use, International Statistics Database and *International Energy Annual 1997*, DOE/EIA-0219(97) (Washington, DC, April 1999). **Projections:** EIA, World Energy Projection System (2000).

significantly as a result of the elimination of subsidies and privatization in the electricity sector, which led to a rapid increase in gas-fired electricity generation at the expense of coal [4]. Hard coal production also declined in France, while nuclear capacity increased. Annual carbon emissions for Western Europe declined by nearly 2 percent between 1990 and 1997 (equivalent to 15 million metric tons of carbon) as a result of the drop in carbon intensity in conjunction with reduced energy intensity. Carbon emissions rose in all other industrialized regions.

Over the next 20 years, coal use is expected to continue declining in Western Europe, as natural gas consumption continues to increase. Renewable energy use is also expected to increase, largely as a result of several country-level programs aimed at increasing electricity generation from wind energy. On the other hand, projected decreases in the region's total nuclear capacity are expected to slow the decline in carbon intensity over the forecast period.

Carbon intensity in industrialized Asia was 15 million metric tons of carbon per quadrillion Btu of energy consumed in 1997, slightly higher than in Western Europe. In the *IEO2000* projections, carbon intensity in Japan remains virtually unchanged through 2020. The main source of non-carbon-emitting energy in Japan is nuclear power, and plans for the construction of new facilities and life extensions for existing nuclear units are uncertain. The reference case projections show a slight increase in the share of Japan's energy consumption coming from nuclear power and natural gas. In contrast to Western Europe, coal consumption for electricity generation is expected to increase in Japan.

In 1997, energy-related carbon emissions from the United States accounted for 86 percent of North America's carbon emissions. A slight decline in that share is projected by 2020 as a result of the expected rapid growth in economic output, energy use, and carbon emissions in Mexico. For North America as a whole, increased energy consumption is expected to be coupled with rising carbon intensity over the forecast period, largely as a consequence of the expected retirements of older nuclear units in Canada and the United States and the replacement of nuclear power generation with fossil-fuel-based generation.

In many of the industrialized countries, with the exceptions of France and Japan, environmental and safety concerns and economic pressures have prevented the addition of new nuclear facilities in recent years. In Germany, the government has promised a complete phaseout of nuclear power, although no schedule has been established [5]. It remains to be seen how climate change policy will affect prospects for the nuclear industry worldwide. Building new nuclear capacity is generally more expensive than building new coal- or gas-fired generators, but if carbon restrictions are enacted, new nuclear construction and life extensions for existing nuclear units may be economically justified.

Eastern Europe and the Former Soviet Union

The economic and political problems in the EE/FSU that brought about declines in both energy consumption and economic growth during the 1990s also produced a substantial decline in carbon emissions. Currently, the FSU is the most energy-intensive region in the world, although its carbon intensity is comparable to those of the industrialized nations (Figures 106 and 107). Over the projection period, a forecast for strong economic growth and the replacement of older capital stock from the Soviet era with more efficient equipment contribute to declines in energy intensity for both Eastern Europe and the FSU. Despite these expectations, however, energy intensity in the EE/FSU is still projected to be more than four times those in most of the industrialized countries in 2020. On the other hand, *IEO2000* projects a 19-percent decline in Eastern Europe's carbon intensity, which is significantly greater than the reduction projected for the industrialized world. The decline in Eastern Europe results in part from the expected replacement of coal-fired electricity generation with natural gas throughout the region.

The transitional economies of the EE/FSU region have become central to the debate on climate change policy. Based on the emissions targets outlined in the Kyoto Protocol, several EE/FSU countries may be able to generate emissions "credits" by virtue of declines in their emission levels during the 1990s. Such credits could potentially be used to offset reductions required in other Annex I countries. At present, however, rules for the creation and exchange of credits remain to be worked out.

Developing Countries

Decreases in both energy intensity and carbon intensity since 1990 have had a negligible influence on carbon emissions in the developing world (Figure 108). In the *IEO2000* reference case, a threefold increase in carbon emissions between 1990 and 2020 results from the strong economic growth and significant population growth expected for the region. Over the forecast period, average per capita income in the region is expected to more than double, and the region's total population is projected to increase by 35 percent. As a result, significant increases in electricity consumption and transportation energy use are projected for many of the developing countries, only partially offset by declines in energy intensity and carbon intensity of 21 percent and 1 percent, respectively, from 1997 to 2020.

Among the developing nations and on a worldwide basis, the strongest economic growth and most rapid increases in energy consumption between 1997 and 2020 are expected for developing Asia. At the same time, however, more rapid declines in energy intensity and carbon intensity are expected for developing Asia than for the other developing regions (Figures 109 and 110). A 3-percent drop in carbon intensity is projected for developing Asia as a whole, with greater declines in some countries. In India, increased use of natural gas, hydropower, and nuclear energy for electricity generation produces a projected 11-percent decline in carbon intensity from 1997 to 2020, and a similar decline is

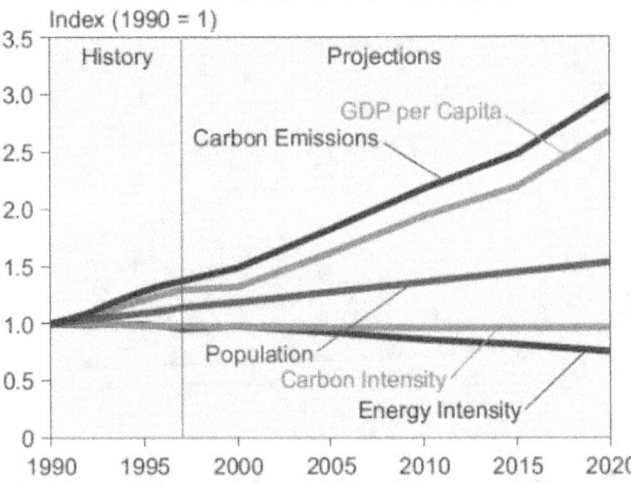

Figure 108. Changes in Carbon Emissions and the Kaya Identity Components in the Developing World, 1990-2020

Sources: **History:** Energy Information Administration (EIA), Office of Energy Markets and End Use, International Statistics Database and *International Energy Annual 1997*, DOE/EIA-0219(97) (Washington, DC, April 1999). **Projections:** EIA, World Energy Projection System (2000).

projected for South Korea, primarily because of increased natural gas use.

In China, rapidly increasing energy consumption over the forecast period is accompanied by a 35-percent decrease in energy intensity and a 4-percent decrease in carbon intensity. China's carbon intensity still is expected to be the highest in the world, however, and its annual carbon emissions are projected to increase from 822 million metric tons in 1997 to 2,091 million metric tons in 2020. Coal continues to be the primary fuel for China's electricity and industrial sectors, given the nation's abundant coal reserves and limited access to alternative sources of energy. To meet the projected rise in electricity demand, increases in both nuclear power and hydroelectric generation are expected.

Central and South America is the only developing region where both energy intensity and carbon intensity are projected to increase in the forecast—by 1 percent and 12 percent, respectively, from 1997 to 2020. A 29-percent increase in carbon intensity is projected for Brazil alone. Currently, many South American nations depend on hydropower for electricity generation, which can prove problematic for maintaining electricity supply during times of drought. Consequently, significant increases in natural-gas-fired capacity are expected. In Brazil, the natural gas share of energy consumption is projected to rise from 3 percent in 1997 to 18 percent in 2020, and the oil share is also expected to increase as a result of strong growth in transportation energy demand.

Figure 109. Energy Intensity in the Developing World by Region, 1990-2020

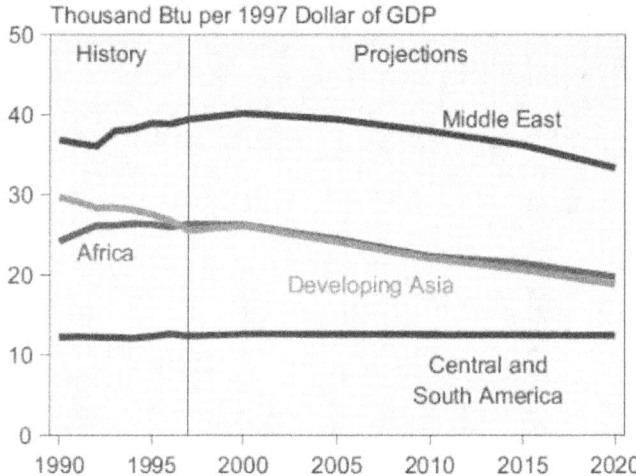

Thousand Btu per 1997 Dollar of GDP

Sources: **History:** Energy Information Administration (EIA), Office of Energy Markets and End Use, International Statistics Database and *International Energy Annual 1997*, DOE/EIA-0219(97) (Washington, DC, April 1999). **Projections:** EIA, World Energy Projection System (2000).

Figure 110. Carbon Intensity in the Developing World by Region, 1990-2020

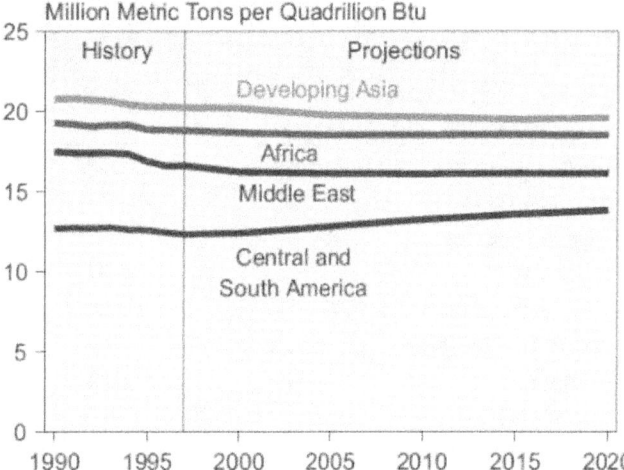

Million Metric Tons per Quadrillion Btu

Sources: **History:** Energy Information Administration (EIA), Office of Energy Markets and End Use, International Statistics Database and *International Energy Annual 1997*, DOE/EIA-0219(97) (Washington, DC, April 1999). **Projections:** EIA, World Energy Projection System (2000).

The *IEO2000* projections show slight declines in carbon intensity for the Middle East and Africa and declines in energy intensity of approximately 15 percent and 25 percent, respectively. Nevertheless, Africa is expected to remain one of the most carbon-intensive regions worldwide (second only to developing Asia), and the Middle East remains one of the most energy-intensive regions, given the dominance of petroleum-based industries in most of the Middle Eastern economies. Within the developing world, the slowest growth rates in carbon emissions are projected for Africa and the Middle East, with overall increases of 77 percent and 86 percent, respectively, from 1997 to 2020. Those increases still are larger than the increases projected for any of the industrialized countries, with the exception of Mexico.

Historically, biomass—including wood, charcoal, and agriculture residues—has played an important role in meeting the energy needs of the developing world. Biomass is generally considered a carbon-neutral energy source, in that carbon emissions from the combustion of biomass are largely balanced by transfers of carbon from the atmosphere back into biomass as part of the global carbon cycle. If the use of biomass resources is not sustainable, however, it can contribute to a net rise in carbon emissions. The issue of sustainability, which has received much attention in the context of woodfuel use in developing countries, is discussed in the box on page 163.

Issues in Climate Change Policy

The Framework Convention on Climate Change

The United Nations Framework Convention on Climate Change (UNFCCC), which was developed and endorsed in Rio de Janeiro, Brazil, in 1992, continues to be the center of international debate on the environment. The most ambitious proposal coming out of the subsequent conferences of the parties has been the Kyoto Protocol, which was developed by the third conference of the parties (COP-3) in Kyoto, Japan, in December 1997.

Under the terms of the Kyoto Protocol, Annex I industrialized countries would agree to reduce their carbon emissions by 7 percent below 1990 levels, corresponding to a 24-percent reduction from the *IEO2000* baseline projection for 2010 (Table 23).[27] On a percentage basis, the cuts vary by region. For Western Europe, with a target of 8 percent below 1990, a 15-percent reduction from the 2010 baseline would be required. Europe has been replacing coal use with natural gas in recent years and has seen a decline in regional energy intensity, leading to a 1.6-percent decrease in carbon emissions between 1990 and 1997. The *IEO2000* reference case, which incorporates no carbon emissions reduction efforts, projects a baseline level of emissions in 2010 that would be 8.8 percent higher than in 1990. The United States, whose Kyoto Protocol emissions target is 7 percent below 1990 levels, would have to reduce emissions 30 percent below its

---

[27] The reduction goals are calculated as the average over the 2008 to 2012 time frame; 2010, the midpoint, is commonly used as the reference year for calculating emissions reductions.

Table 23. Projected Effects of the Kyoto Protocol on Carbon Emissions in Annex I Countries, 2010

Region and Country	Carbon Emissions (Million Metric Tons)			Change From *IEO2000* Reference Case, 2010 (Million Metric Tons)	Change From 1990 Emissions (Percent)	Change From *IEO2000* Reference Case, 2010 (Percent)
	1990	2010, *IEO2000* Reference Case	2010, Kyoto Protocol Target			
Annex I Industrialized Countries						
North America	1,472	1,947	1,370	-577	-7	-30
United States	1,345	1,787	1,251	-536	-7	-30
Canada	127	160	119	-41	-6	-26
Western Europe	934	1,016	860	-156	-8	-15
Industrialized Asia	364	457	354	-103	-3	-23
Japan	274	331	257	-74	-6	-22
Australasia	90	126	97	-29	7	-23
Total Annex I Industrialized	2,769	3,420	2,584	-836	-7	-24
Annex I Transitional (EE/FSU)						
Former Soviet Union	854	591	853	261	0	44
Eastern Europe	281	244	300	56	7	23
Total Annex I EE/FSU	1,135	835	1,153	318	2	38
Total	3,904	4,255	3,737	-519	-4	-12

Sources: **1990:** Energy Information Administration (EIA), *International Energy Annual 1997*, DOE/EIA-0219(97) (Washington, DC, April 1999). **Projections:** EIA, World Energy Projection System (2000).

projected baseline, which in the absence of reduction efforts would be 33 percent higher than the 1990 level.

For the Annex I countries as a group—including those in the EE/FSU—the required reduction under the Kyoto Protocol would be 4 percent below 1990 or 12 percent below the 2010 baseline. In the *IEO2000* projections, the Annex I EE/FSU countries are separated from the non-Annex I EE/FSU countries. In the 1999 *IEO*, 374 million metric tons of carbon were projected to be available from all EE/FSU countries in 2010 for possible trading; in *IEO2000* it is estimated that 318 million metric tons will be available from the Annex I EE/FSU countries. Slightly higher projected rates of economic growth (and carbon emissions) in the transitional economies also contribute to the lower estimate of available carbon credits in *IEO2000*.

The details of the Kyoto Protocol have been the subject of international negotiations since the end of 1997 when COP-3 took place. COP-4, held in Buenos Aires, Argentina, in November 1998, was intended to provide details for some of the general proposals of the Kyoto Protocol. While COP-4 did provide some general guidance, in the form of a Buenos Aires Plan of Action, most of the details were deferred to future meetings. The Buenos Aires Plan itself is a pledge on behalf of the parties to strengthen the implementation of the Framework Convention and prepare for the future entry into force of the Kyoto Protocol. Under the Plan, the parties resolved to demonstrate progress on financial mechanisms and transfer of technology.

COP-5, held in Bonn, Germany, in November 1999, developed a timetable for implementation of the Buenos Aires Plan of Action. The goal of the timetable is to adopt key decisions mandated by the Plan at COP-6, which will take place in The Hague, Netherlands, in November 2000. The COP-6 meeting will be preceded by a week of informal talks and workshops and two 1-week rounds of negotiations (June 12-16 and September 11-15).

Several issues remain to be resolved before the key decisions governing the Kyoto Protocol are adopted. The issues include the allowability of and rules for trading of emissions credits among industrialized countries, rules on carbon sequestration projects, and other land-use and forestry issues.

Approaches to Achieving Reductions in Energy-Related Greenhouse Gas Emissions

A variety of approaches to reducing energy-related emissions of carbon dioxide and other greenhouse gases are possible. One generic approach would be the adoption of a carbon tax or fee. Such a tax would add a premium to the price of any activity that generates carbon emissions. The higher the tax, the less would be the demand for those activities that generate the carbon. If a carbon tax were the only approach used to meet the Kyoto Protocol reduction targets, the taxing authority would have to estimate the tax level that would be required to achieve a specified reduction in emissions. The tax might or might not achieve the desired reduction, depending on the demand response to price changes.

The Role of Biomass in Energy-Related Carbon Emissions Estimates

Biomass, largely in the form of firewood, agricultural residues, and charcoal, plays a significantly larger role in meeting the energy needs of many developing countries than in most industrialized nations. While the supply and application of biomass energy varies widely by region, its use is most prevalent in households of developing Asia and Sub-Saharan Africa (excluding South Africa), where other fuels may be either unavailable or unaffordable. Because data on biomass use in developing countries are often sparse or inadequate (largely due to its decentralized and non-marketed use in many regions), *IEO2000* does not explicitly include projections for biomass energy consumption or for carbon emissions related to its production and use.

Containing approximately 45 percent carbon by weight, biomass energy fuels can emit more carbon dioxide than either oil or natural gas per unit of energy produced. Unlike the combustion of fossil fuels, however, the burning of biomass for energy is not generally considered to contribute to the buildup of carbon dioxide in the atmosphere, because corresponding amounts of carbon are absorbed during the growth of biomass crops as a part of the natural carbon cycle.[a] Only under circumstances of unsustainable use, where the harvest of biomass for fuel use is not balanced by regrowth of tree or plant resources, would carbon dioxide emissions from biofuels result in net additions to atmospheric concentrations.

Although the use of fuelwood in developing countries may contribute to localized degradation of forests in some regions, it is not considered to be a significant cause of large-scale deforestation, which is a major anthropogenic source of net carbon dioxide emissions in the developing world.[b] In developing countries, economic, social, and demographic pressures have led to the clearing of forested areas for agriculture, pasture land, or housing. For the most part, the gathering of fuelwood from areas being deforested is, rather, a byproduct of the process of land conversion.

Although biomass is generally considered a "carbon-neutral" renewable energy source, its use is not necessarily environmentally benign. Incomplete combustion of biomass can produce, in addition to carbon dioxide, significant levels of other gases such as carbon monoxide, methane, nonmethane hydrocarbons, nitrous oxide, and various particulates. Moreover, unlike carbon dioxide emissions from biomass combustion, emissions of methane and nitrous oxide produce net increases in their concentrations in the atmosphere. In addition, serious health risks may be posed by prolonged exposure to pollutants from biomass burning. Environment and health issues are particularly relevant to biomass energy use in developing countries, where low-quality biomass fuels are widely used in inefficient stoves for cooking and heating, with poor ventilation.

The viability of biomass energy as an option for mitigating the increase in atmospheric concentrations of carbon dioxide and other greenhouse gases depends not only on the conditions of its harvest (i.e. resource sustainability) but also on the technologies used to convert biofuels into useful energy. Carbon sequestration by standing forests may in fact yield a greater overall net reduction in anthropogenic carbon emissions than the harvesting of trees for a sustainable flow of fuelwood to displace fossil fuel use if the wood is used inefficiently.[c]

[a]R.A. Houghton, "Converting Terrestrial Ecosystems From Sources to Sinks of Carbon," *Ambio*, Vol. 25, No. 4 (1996), pp. 267-272.
[b]D.O. Hall and F. Rosillo-Calle, "Evaluating Environmental Effects and Carbon Sources and Sinks Resulting From Biomass Production and Use in Developing Countries," in *Proceedings from IEA Workshop on Biomass Energy: Data, Analysis and Trends* (Paris, France, 1998), pp. 293-314.
[c]G. Marland and B. Schlamadinger, "Forests for Carbon Sequestration or Fossil Fuel Substitution? A Sensitivity Analysis," *Biomass and Bioenergy*, Vol. 13, No. 6 (1997), pp. 331-420.

Another approach would be to specify limits on carbon emissions in order to assure the achievement of overall reduction goals. For large sources, such as electric utilities, the cost of monitoring could be relatively low, because data on fuel consumption are readily available. In addition, large emitters might be more inclined to engage in credit trading, which could reduce the cost of meeting the target. For emissions sources that are small, mobile, or dispersed, such as personal vehicles, direct measurement of fuel consumption is likely to be impractical, however. Standards programs, such as the automotive Corporate Average Fuel Economy (CAFE)

standards in the United States, would be more manageable, but if the standards increased the cost of new equipment they might be counterproductive, causing consumers to postpone purchases of new equipment and maintain older, less efficient equipment for longer periods of time.

Other possible approaches to reducing emissions include appliance efficiency standards, which would reduce energy consumption, and renewable portfolio standards, which would lower the amount of carbon emitted per kilowatthour of electricity consumed. Tax

Other Principal Greenhouse Gases Included in the Kyoto Protocol

Although carbon dioxide, primarily from the combustion of fossil fuels, is the most important greenhouse gas, the Kyoto Protocol also addresses reductions in five other greenhouse gases emitted from a wider range of sources. The principal greenhouse gases other than carbon dioxide are methane and nitrous oxide from both energy and non-energy sources. Also included are hydrofluorocarbons, perfluorocarbons, and sulfur hexafluoride, which have relatively high global warming potentials but are emitted in small volumes and, for the most part, are not energy-related.[a]

In 1990, according to emissions inventories submitted to the UNFCCC,[b] anthropogenic emissions of methane from the 37 Annex I countries totaled 631 million metric tons carbon equivalent (see table below).[c] Methane constitutes 12.8 percent of total greenhouse gas emissions in the 1990 baseline for the Annex I countries. In the United States, methane emissions were about 10.5 percent of the 1990 total but only 9 percent of total emissions in 1998, largely as a result of increased methane recovery for energy use at landfill sites.[d]

The sources of methane emissions differ from one region to the next. For example, "fugitive" emissions from sources such as leaking valves made up an estimated 69 percent of total methane emissions from FSU nations in 1990, as compared with 35 percent of the total for North America and only 16 percent for Australasia (where domesticated livestock accounted for 62 percent of total methane emissions in 1990). High

(continued on page 165)

Methane Emissions in Annex I Countries by End Use, 1990
(Million Metric Tons Carbon Equivalent)

Region and Country	Energy Fuel Use	Energy Fugitive Fuel	Agriculture Livestock	Agriculture Other	Waste	Other	Total
Annex I Industrialized Countries							
North America	5.25 (2.8%)	65.70 (34.6%)	52.97 (27.9%)	2.72 (1.4%)	62.42 (32.9%)	0.73 (0.4%)	189.90 (30.1%)
United States	3.76 (2.2%)	58.25 (34.3%)	47.53 (28.0%)	2.72 (1.6%)	57.27 (33.7%)	0.33 (0.2%)	169.86 (26.9%)
Canada.	1.49 (7.4%)	7.44 (37.1%)	5.44 (27.1%)	0.00 (0.0%)	5.15 (25.7%)	0.40 (2.0%)	20.04 (3.2%)
Western Europe	4.94 (3.8%)	26.01 (19.9%)	51.67 (39.5%)	2.70 (2.1%)	43.76 (33.4%)	1.88 (1.4%)	130.95 (20.8%)
Industrialized Asia	1.20 (2.4%)	7.44 (15.2%)	27.63 (56.4%)	4.06 (8.3%)	7.10 (14.5%)	1.60 (3.3%)	49.03 (7.8%)
Japan.	0.51 (5.8%)	0.95 (10.8%)	2.66 (30.1%)	2.16 (24.5%)	2.26 (25.5%)	0.29 (3.3%)	8.84 (1.4%)
Australasia	0.69 (1.7%)	6.49 (16.1%)	24.97 (62.1%)	1.90 (4.7%)	4.84 (12.0%)	1.31 (3.2%)	40.20 (6.4%)
Total Annex I Industrialized. .	11.38 (3.1%)	99.14 (26.8%)	132.27 (35.8%)	9.48 (2.6%)	113.28 (30.6%)	4.21 (1.1%)	369.89 (58.7%)
Annex I Transitional (EE/FSU)							
Former Soviet Union.	1.41 (0.7%)	144.31 (68.9%)	43.16 (20.6%)	0.74 (0.4%)	17.52 (8.4%)	2.29 (1.1%)	209.43 (33.2%)
Eastern Europe	1.44 (2.8%)	22.87 (44.7%)	13.75 (26.9%)	0.27 (0.5%)	12.51 (24.4%)	0.36 (0.7%)	51.19 (8.1%)
Total Annex I EE/FSU	2.85 (1.1%)	167.18 (64.1%)	56.91 (21.8%)	1.01 (0.4%)	30.03 (11.5%)	2.65 (1.0%)	260.62 (41.3%)
Total	14.24 (2.3%)	266.32 (42.2%)	189.18 (30.0%)	10.50 (1.7%)	143.31 (22.7%)	6.85 (1.1%)	630.51 (100.0%)

Source: United Nations Framework Convention on Climate Change, *Country Inventories*, web site www.unfcc.de/index.html.

[a]The exception is sulfur hexafluoride used as insulation for electrical breakers and switches.
[b]United Nations Framework Convention on Climate Change, *National Communications from Parties Included in Annex I to the Convention, Greenhouse Gas Inventory Data, 1990-1997* (Bonn, Germany, September 29, 1999). The only year for which a complete inventory is available for all Annex I countries is 1990.
[c]Methane emissions are converted to carbon dioxide equivalents by multiplying the amount of methane by weight (110 million metric tons) times 21 (the 100-year global warming potential for methane). To convert from carbon dioxide to carbon, the amount of carbon dioxide (2,312 million metric tons) is divided by 44/12 or 3.667. The amount of carbon equivalent is therefore 631 million metric tons.
[d]Energy Information Administration, *Emissions of Greenhouse Gases in the United States 1998*, DOE/EIA-0573(98) (Washington DC, October 1999).

Other Principal Greenhouse Gases Included in the Kyoto Protocol (Continued)

levels of fugitive methane emissions are an indication of old, poorly maintained natural gas infrastructure.

The sources of methane emissions can influence the cost of reducing them. For example, according to a report by the U.S. Environmental Protection Agency, about 30 percent of the projected emissions for the years 2000, 2010, and 2020 from fugitive emissions of natural gas systems in the United States could be avoided, based on the value of the saved methane.[e] If the same analysis applies to the FSU, substantial reductions could be possible at a relatively low cost.

In the UNFCCC inventory for 1990, anthropogenic emissions of nitrous oxide in the Annex I countries totaled 280 million metric tons carbon equivalent (see table below).[f] Nitrous oxide constitutes 5.7 percent of

total greenhouse gas emissions in the 1990 baseline for the Annex I countries and 6 percent in the 1990 U.S. baseline.

Nitrous oxide emissions are the most difficult to estimate of the three major gases. Variables affecting emissions levels include the penetration of catalytic converters, which emit nitrous oxide, in vehicle fleets. Catalytic converters were widely used on automobiles in the United States and Japan in 1990, and transportation energy sources accounted for 14 percent and 22 percent of their nitrous oxide totals, respectively—compared with only 0.1 percent of the FSU total. In 1990, 41 percent of the world's anthropogenic methane emissions were attributable to the Annex I transitional economies, compared with only 16 percent of nitrous oxide emissions.

Nitrous Oxide Emissions in Annex I Countries by End Use, 1990
(Million Metric Tons Carbon Equivalent)

Region and Country	Energy Transporta-tion	Energy Other	Industry	Agriculture	Waste	Other	Total
Annex I Industrialized Countries							
North America	5.39 (13.8%)	4.57 (4.1%)	11.24 (10.1%)	77.35 (69.5%)	2.37 (2.1%)	0.42 (0.4%)	111.34 (39.7%)
United States	13.61 (14.2%)	3.89 (4.1%)	8.12 (8.5%)	68.05 (71.1%)	2.11 (2.2%)	0.00 (0.0%)	95.78 (34.2%)
Canada	1.78 (11.4%)	0.68 (4.3%)	3.13 (20.1%)	9.30 (59.8%)	0.25 (1.6%)	0.42 (2.7%)	15.55 (5.6%)
Western Europe	3.27 (3.0%)	11.42 (10.5%)	30.64 (28.3%)	57.79 (53.3%)	1.10 (1.0%)	4.13 (3.8%)	108.35 (38.7%)
Industrialized Asia	1.56 (10.9%)	0.79 (5.5%)	2.16 (15.0%)	9.18 (63.7%)	0.43 (3.0%)	0.28 (1.9%)	14.41 (5.1%)
Japan	1.09 (22.1%)	0.54 (11.0%)	2.02 (41.0%)	0.79 (16.0%)	0.41 (8.4%)	0.08 (1.5%)	4.93 (1.8%)
Australasia	0.47 (5.0%)	0.25 (2.7%)	0.14 (1.4%)	8.39 (88.6%)	0.02 (0.2%)	0.20 (2.1%)	9.48 (3.4%)
Total Annex I Industrialized . .	20.22 (8.6%)	16.78 (7.2%)	44.04 (18.8%)	144.32 (61.7%)	3.90 (1.7%)	4.83 (2.1%)	234.09 (83.6%)
Annex I Transitional (EE/FSU)							
Former Soviet Union	0.03 (0.1%)	2.24 (8.2%)	2.31 (8.5%)	22.12 (81.3%)	0.06 (0.2%)	0.44 (1.6%)	27.20 (9.7%)
Eastern Europe	0.27 (1.4%)	4.81 (25.5%)	5.09 (26.9%)	8.68 (46.0%)	0.03 (0.1%)	0.02 (0.1%)	18.89 (6.7%)
Total Annex I EE/FSU	0.30 (0.6%)	7.05 (15.3%)	7.40 (16.0%)	30.81 (66.8%)	0.08 (0.2%)	0.46 (1.0%)	46.09 (16.4%)
Total	20.52 (7.3%)	23.83 (8.5%)	51.44 (18.4%)	175.13 (62.5%)	3.98 (1.4%)	5.28 (1.9%)	280.18 (100.0%)

Source: United Nations Framework Convention on Climate Change, *Country Inventories*, web site www.unfcc.de/index.html.

[e]U.S. Environmental Protection Agency, *U.S. Methane Emissions 1990-2020: Inventories, Projections, and Opportunities for Reductions*, EPA 430-R-99-013 (Washington, DC, September 1999).
[f]Nitrous oxide is converted to carbon dioxide equivalent by multiplying the amount of nitrous oxide by weight (3.3 million metric tons) times 310 (the 100-year global warming potential for nitrous oxide). To convert from carbon dioxide to carbon, the amount of carbon dioxide (1,027 million metric tons) is divided by 44/12 or 3.667. The amount of carbon equivalent is therefore 280 million metric tons.

incentives for purchases of energy-efficient or carbon-free equipment, such as solar photovoltaic units at homes or businesses, could also be an option.

Estimating Costs

A primary goal in the analysis of carbon mitigation policy is to estimate a carbon permit price for the trading of emissions credits. So-called "abatement curves" for carbon dioxide and other gases pair the total amount of emissions reduced domestically with a cost per unit of carbon emitted (usually per ton) to achieve that reduction. Carbon abatement curves can be used to estimate the prices that countries (or emitting entities within countries) would be willing to pay for carbon permits. In general, an emitting entity would not pay more for a permit than the internal cost of achieving an equivalent reduction. The trading of carbon permits can influence the cost estimates for achieving the overall goals of the Kyoto Protocol by encouraging the least costly emissions reductions to be made wherever they are possible. Other factors that could influence the cost of achieving carbon policy goals include the following:

- **Ancillary Benefits.** These are reductions in other gases (or potentially other pollutants) that would result from efforts to reduce greenhouse gases. Such an accounting of costs and benefits would attempt to include all pollution reductions resulting from actions taken to reduce greenhouse gases. As discussed below, however, not all the reductions of other pollutants would yield lower cost estimates from a global warming perspective.

- **Offsetting Effects.** The opposite of an ancillary benefit is the offsetting effect created when one pollutant is reduced that may be beneficial to other aspects of the environment. One example is the cooling effect of sulfur dioxide (SO_2) that results from the reflective properties of sulfate aerosols. While reducing SO_2 is vital to limiting acid deposition, its reduction leads to warmer temperatures. If the sulfur reduction is accomplished through investment in scrubbers at coal-fired power plants, there is a disincentive to retire that capital equipment early in order to reduce carbon dioxide emissions. Thus, the order in which policies are addressed could have a significant influence on the cost and likelihood of achieving reduction goals [6]. One of the implications of offsetting effects is that ancillary benefits should be calculated as a joint optimization problem. This greatly complicates the task, increasing the amount of resources needed to engage in such an analysis.

- **Carbon Sinks.** Evidence points to enhancement of carbon sinks, principally, the sequestration of carbon in biomass and soils, as a cost-effective approach to mitigating greenhouse gas emissions. The mechanisms of carbon sequestration are not yet well understood, however, and it is not clear what sorts of projects should qualify under international agreements.

- **Clean Development Mechanism.** The clean development mechanism (CDM) is a proposed program under the Kyoto Protocol program that would allow industrialized countries to finance emissions reduction or avoidance projects in developing countries and credit some or all of the reductions achieved against their own emissions limitation targets. The rules governing the CDM have not yet been determined. Depending on how the rules are structured, the CDM could be a cost-effective way for industrialized Annex I countries to meet their Kyoto Protocol goals.

The climate change issue is both technically complex and politically difficult to assess. Many factors interact to influence the literal as well as the policy climate. In addition, the costs and benefits of reducing greenhouse gas emissions, at least in the short to medium term, would not be borne equally by all countries. Whether or not the Kyoto Protocol is ultimately ratified by enough countries to become binding, the issue of global warming is not likely to disappear. If the baseline emissions growth paths projected in *IEO2000* are realized, in the absence of evidence conclusively refuting current theories on climate change, the pressure to take action during the first quarter of the 21st century is almost certain to increase.

References

1. Intergovernmental Panel on Climate Change, *Climate Change 1995: The Science of Climate Change* (Cambridge, UK: Cambridge University Press, 1996).

2. United Nations, *Kyoto Protocol to the United Nations Framework Convention on Climate Change*, web site www.unfccc.de/index.html.

3. J.P. Bruce, L. Hoesung, and E.F. Haitus, Editors, *Climate Change 1995: Economic and Social Dimensions of Climate Change* (Melbourne, Australia: Cambridge University Press, 1996), p. 27.

4. International Energy Agency, *Coal Information 1997* (Paris, France, 1998).

5. M. Hibbs, "German Utilities Won't Negotiate Without Three Major Concessions," *Nucleonics Week*, Vol. 40 (August 5, 1999), p. 5.

6. A. McDonald, "Combating Acid Deposition and Climate Change: Priorities for Asia," *Environment*, Vol. 41, No. 3 (April 1999), p. 4.

Appendix A

Reference Case Projections:

- World Energy Consumption
- Gross Domestic Product
- Carbon Emissions
- Nuclear Power Capacity
- World Population

Table A1. World Total Energy Consumption by Region, Reference Case, 1990-2020
(Quadrillion Btu)

Region/Country	History			Projections				Average Annual Percent Change, 1997-2020
	1990	1996	1997	2005	2010	2015	2020	
Industrialized Countries								
North America	99.9	112.0	112.5	126.9	135.0	142.4	148.5	1.2
United States[a]	84.0	93.9	94.2	105.3	111.3	116.7	120.9	1.1
Canada.	10.9	12.6	12.5	14.2	15.0	15.8	16.3	1.1
Mexico	5.0	5.5	5.8	7.4	8.7	10.0	11.3	3.0
Western Europe	59.9	63.3	64.0	69.6	72.6	75.4	78.4	0.9
United Kingdom	9.3	9.7	9.9	10.9	11.4	12.1	12.6	1.1
France	9.3	10.5	10.4	11.3	11.8	12.2	12.6	0.8
Germany	14.8	14.2	14.2	15.3	15.9	16.6	17.2	0.8
Italy.	6.7	7.3	7.3	7.8	8.2	8.4	8.8	0.8
Netherlands	3.3	3.7	3.7	4.1	4.3	4.4	4.6	0.9
Other Western Europe	16.6	17.9	18.5	20.3	21.0	21.7	22.6	0.9
Industrialized Asia.	23.0	26.9	27.1	29.2	31.1	32.2	33.1	0.9
Japan	18.1	21.4	21.3	22.6	24.1	24.8	25.4	0.8
Australasia	4.9	5.4	5.8	6.6	7.0	7.3	7.7	1.2
Total Industrialized	182.8	202.1	203.7	225.7	238.7	250.0	259.9	1.1
EE/FSU								
Former Soviet Union	61.0	41.8	40.8	43.1	47.3	51.0	57.3	1.5
Eastern Europe	15.3	12.7	12.5	14.4	15.6	16.7	18.3	1.7
Total EE/FSU	76.4	54.5	53.3	57.5	63.0	67.7	75.7	1.5
Developing Countries								
Developing Asia	51.4	73.5	75.3	105.0	126.4	144.3	172.6	3.7
China.	27.0	35.8	36.7	55.0	68.1	79.2	97.3	4.3
India	7.8	11.6	11.8	17.0	20.4	23.1	27.3	3.7
South Korea	3.7	7.1	7.5	9.3	10.7	11.9	13.4	2.6
Other Asia	13.0	19.0	19.3	23.7	27.2	30.1	34.7	2.6
Middle East.	13.1	17.1	17.9	22.5	26.2	29.3	34.3	2.9
Turkey	2.0	2.6	2.7	3.3	3.9	4.3	5.1	2.8
Other Middle East	11.1	14.5	15.2	19.1	22.4	25.0	29.2	2.9
Africa	9.3	11.1	11.4	14.1	15.8	17.8	20.6	2.6
Central and South America . . .	13.7	17.8	18.3	24.2	30.1	35.3	44.7	4.0
Brazil.	5.4	6.9	7.2	8.9	10.8	12.5	15.5	3.4
Other Central/South America . .	8.3	10.9	11.1	15.4	19.3	22.8	29.1	4.3
Total Developing	87.6	119.4	122.9	165.8	198.5	226.7	272.1	3.5
Total World	346.7	376.0	379.9	449.0	500.2	544.4	607.7	2.1
Annex I								
Industrialized	177.8	196.6	197.9	218.3	230.0	240.0	248.7	1.0
EE/FSU	64.9	46.8	45.5	48.5	52.9	56.8	63.3	1.4
Total Annex I	242.6	243.4	243.4	266.8	282.9	296.9	312.0	1.1

[a]Includes the 50 States and the District of Columbia. U.S. Territories are included in Australasia.

Notes: EE/FSU = Eastern Europe/Former Soviet Union. Energy totals include net imports of coal coke and electricity generated from biomass in the United States. Totals may not equal sum of components due to independent rounding. The electricity portion of the national fuel consumption values consists of generation for domestic use plus an adjustment for electricity trade based on a fuel's share of total generation in the exporting country.

Sources: **History:** Energy Information Administration (EIA), *International Energy Annual 1997*, DOE/EIA-0219(97) (Washington, DC, April 1999). **Projections:** EIA, *Annual Energy Outlook 2000*, DOE/EIA-0383(2000) (Washington, DC, December 1999), Table A1; and World Energy Projection System (2000).

Table A2. World Total Energy Consumption by Region and Fuel, Reference Case, 1990-2020
(Quadrillion Btu)

Region/Country	History			Projections				Average Annual Percent Change, 1997-2020
	1990	1996	1997	2005	2010	2015	2020	
Industrialized Countries								
North America								
Oil	40.4	43.0	43.8	49.9	53.9	57.7	61.4	1.5
Natural Gas	22.7	26.8	26.9	30.0	33.5	36.9	39.0	1.6
Coal	20.5	22.5	23.0	26.3	26.7	27.5	28.4	0.9
Nuclear.	7.0	8.2	7.7	8.1	7.6	6.3	5.3	-1.6
Other	9.2	11.4	11.2	12.6	13.4	13.9	14.4	1.1
Total	**99.9**	**112.0**	**112.5**	**126.9**	**135.0**	**142.4**	**148.5**	**1.2**
Western Europe								
Oil	25.8	28.2	28.5	30.2	30.8	31.2	31.5	0.4
Natural Gas	10.0	13.5	13.4	17.4	19.8	22.7	25.9	2.9
Coal	12.4	8.4	8.7	7.6	7.3	6.9	6.4	-1.3
Nuclear.	7.4	8.7	8.8	8.9	8.6	8.0	7.3	-0.8
Other	4.4	4.5	4.7	5.5	6.0	6.6	7.2	1.9
Total	**59.9**	**63.3**	**64.0**	**69.6**	**72.6**	**75.4**	**78.4**	**0.9**
Industrialized Asia								
Oil	12.5	14.3	14.1	14.6	15.2	15.8	16.2	0.6
Natural Gas	2.9	3.6	3.5	4.4	4.8	5.2	5.6	2.0
Coal	4.2	4.7	5.0	5.5	5.6	5.7	5.8	0.6
Nuclear.	2.0	2.9	3.1	3.1	3.8	3.7	3.7	0.7
Other	1.4	1.4	1.4	1.7	1.7	1.8	1.8	1.2
Total	**23.0**	**26.9**	**27.1**	**29.2**	**31.1**	**32.2**	**33.1**	**0.9**
Total Industrialized								
Oil	78.7	85.5	86.4	94.6	99.8	104.7	109.0	1.0
Natural Gas	35.6	43.9	43.8	51.9	58.2	64.8	70.6	2.1
Coal	37.2	35.6	36.6	39.4	39.6	40.2	40.5	0.4
Nuclear.	16.3	19.9	19.6	20.1	20.0	18.0	16.2	-0.8
Other	15.0	17.3	17.2	19.8	21.1	22.3	23.5	1.4
Total	**182.8**	**202.1**	**203.7**	**225.7**	**238.7**	**250.0**	**259.9**	**1.1**
EE/FSU								
Oil	21.0	11.4	11.9	12.7	14.7	17.0	19.7	2.2
Natural Gas	28.8	23.9	22.4	26.6	31.2	34.7	41.2	2.7
Coal	20.8	13.6	13.3	12.3	10.7	9.4	7.6	-2.4
Nuclear.	2.9	2.8	2.8	2.9	3.0	3.1	2.7	-0.2
Other	2.8	2.9	2.8	3.0	3.3	3.6	4.5	2.0
Total	**76.4**	**54.5**	**53.3**	**57.5**	**63.0**	**67.7**	**75.7**	**1.5**
Developing Countries								
Developing Asia								
Oil	16.0	25.1	26.3	34.4	41.0	47.5	51.5	3.0
Natural Gas	3.2	5.5	6.0	11.4	16.0	20.6	28.8	7.0
Coal	28.1	37.5	37.7	50.6	59.3	65.2	80.1	3.3
Nuclear.	0.9	1.3	1.3	2.1	2.6	3.0	3.2	3.9
Other	3.2	4.0	3.9	6.5	7.4	8.0	9.1	3.7
Total	**51.4**	**73.5**	**75.3**	**105.0**	**126.4**	**144.3**	**172.6**	**3.7**

See notes at end of table.

Table A2. World Total Energy Consumption by Region and Fuel, Reference Case, 1990-2020 (Continued)
(Quadrillion Btu)

Region/Country	History			Projections				Average Annual Percent Change, 1997-2020
	1990	1996	1997	2005	2010	2015	2020	
Developing Countries (Continued)								
Middle East								
Oil	8.1	9.9	10.1	11.4	13.6	16.0	18.9	2.8
Natural Gas	3.9	5.8	6.3	9.0	10.2	10.8	12.5	3.0
Coal	0.8	0.9	0.9	1.0	1.1	1.2	1.2	0.9
Nuclear.	0.0	0.0	0.0	0.0	0.1	0.2	0.2	--
Other.	0.4	0.6	0.6	1.0	1.2	1.3	1.5	4.3
Total	**13.1**	**17.1**	**17.9**	**22.5**	**26.2**	**29.3**	**34.3**	**2.9**
Africa								
Oil	4.2	4.9	5.1	6.7	8.2	10.1	12.3	3.9
Natural Gas	1.5	1.9	2.0	2.5	2.6	2.7	3.0	1.8
Coal	3.0	3.5	3.5	3.9	3.9	4.0	4.1	0.7
Nuclear.	0.1	0.1	0.1	0.1	0.1	0.1	0.1	-0.5
Other.	0.6	0.6	0.6	0.9	0.9	0.9	1.0	2.0
Total	**9.3**	**11.1**	**11.4**	**14.1**	**15.8**	**17.8**	**20.6**	**2.6**
Central and South America								
Oil	7.0	8.7	8.9	11.0	13.3	15.9	19.1	3.4
Natural Gas	2.2	3.1	3.2	6.3	9.5	11.7	17.1	7.5
Coal	0.6	0.7	0.6	0.7	0.8	0.8	0.9	1.8
Nuclear.	0.1	0.1	0.1	0.2	0.2	0.2	0.1	0.8
Other.	3.9	5.2	5.4	6.0	6.4	6.7	7.4	1.4
Total	**13.7**	**17.8**	**18.3**	**24.2**	**30.1**	**35.3**	**44.7**	**4.0**
Total Developing Countries								
Oil	35.2	48.5	50.4	63.6	76.2	89.4	101.7	3.1
Natural Gas	10.8	16.4	17.6	29.2	38.3	45.8	61.5	5.6
Coal	32.4	42.6	42.9	56.2	65.1	71.1	86.4	3.1
Nuclear.	1.1	1.5	1.6	2.4	3.0	3.5	3.6	3.7
Other.	8.1	10.4	10.5	14.4	15.9	16.9	19.0	2.6
Total	**87.6**	**119.4**	**122.9**	**165.8**	**198.5**	**226.7**	**272.1**	**3.5**
Total World								
Oil	134.9	145.4	148.7	170.9	190.7	211.1	230.4	1.9
Natural Gas	75.1	84.1	83.9	107.7	127.7	145.3	173.3	3.2
Coal	90.5	91.8	92.8	107.9	115.4	120.7	134.5	1.6
Nuclear.	20.4	24.2	24.0	25.4	26.0	24.6	22.5	-0.3
Other.	25.9	30.5	30.6	37.2	40.4	42.8	47.0	1.9
Total	**346.7**	**376.0**	**379.9**	**449.0**	**500.2**	**544.4**	**607.7**	**2.1**

Notes: EE/FSU = Eastern Europe/Former Soviet Union. Energy totals include net imports of coal coke and electricity generated from biomass in the United States. Totals may not equal sum of components due to independent rounding. The electricity portion of the national fuel consumption values consists of generation for domestic use plus an adjustment for electricity trade based on a fuel's share of total generation in the exporting country.

Sources: **History:** Energy Information Administration (EIA), *International Energy Annual 1997*, DOE/EIA-0219(97) (Washington, DC, April 1999). **Projections:** EIA, *Annual Energy Outlook 2000*, DOE/EIA-0383(2000) (Washington, DC, December 1999), Table A1; and World Energy Projection System (2000).

Table A3. World Gross Domestic Product (GDP) by Region, Reference Case, 1990-2020
(Billion 1997 Dollars)

Region/Country	History			Projections				Average Annual Percent Change, 1997-2020
	1990	1996	1997	2005	2010	2015	2020	
Industrialized Countries								
North America	**7,726**	**8,784**	**9,145**	**11,470**	**12,848**	**14,343**	**15,775**	**2.4**
United States[a]	6,846	7,804	8,111	10,104	11,217	12,437	13,588	2.3
Canada.	547	603	631	794	902	1,000	1,069	2.3
Mexico	333	377	403	571	729	906	1,119	4.5
Western Europe	**7,569**	**8,284**	**8,570**	**10,345**	**11,595**	**12,928**	**14,338**	**2.3**
United Kingdom	1,143	1,268	1,311	1,577	1,780	2,038	2,299	2.5
France	1,269	1,361	1,412	1,717	1,928	2,111	2,295	2.1
Germany	1,839	2,045	2,122	2,512	2,796	3,099	3,428	2.1
Italy.	1,060	1,128	1,159	1,373	1,540	1,715	1,901	2.2
Netherlands	306	351	363	444	498	559	625	2.4
Other Western Europe	1,951	2,131	2,202	2,721	3,053	3,406	3,790	2.4
Industrialized Asia.	**4,048**	**4,595**	**4,669**	**4,925**	**5,438**	**6,004**	**6,517**	**1.5**
Japan	3,667	4,142	4,199	4,334	4,779	5,270	5,708	1.3
Australasia	381	453	469	591	659	734	808	2.4
Total Industrialized	**19,343**	**21,663**	**22,384**	**26,739**	**29,881**	**33,275**	**36,630**	**2.2**
EE/FSU								
Former Soviet Union	1,049	595	600	674	866	1,121	1,458	3.9
Eastern Europe	358	347	368	510	629	776	944	4.2
Total EE/FSU	**1,407**	**942**	**968**	**1,183**	**1,494**	**1,897**	**2,402**	**4.0**
Developing Countries								
Developing Asia	**1,735**	**2,749**	**2,951**	**4,354**	**5,734**	**6,997**	**9,197**	**5.1**
China.	440	851	926	1,599	2,193	2,751	3,761	6.3
India	266	367	382	602	793	967	1,266	5.4
South Korea	273	419	476	660	870	1,063	1,378	4.7
Other Asia	755	1,111	1,166	1,494	1,878	2,216	2,793	3.9
Middle East.	**356**	**441**	**454**	**570**	**692**	**811**	**1,029**	**3.6**
Turkey	140	176	189	240	298	355	463	4.0
Other Middle East	216	266	265	330	394	456	567	3.4
Africa.	**386**	**426**	**434**	**576**	**708**	**828**	**1,043**	**3.9**
Central and South America . . .	**1,127**	**1,409**	**1,481**	**1,925**	**2,396**	**2,817**	**3,591**	**3.9**
Brazil.	660	778	803	1,033	1,295	1,530	1,959	4.0
Other Central/South America . .	467	631	678	892	1,101	1,287	1,632	3.9
Total Developing	**3,604**	**5,026**	**5,320**	**7,426**	**9,530**	**11,453**	**14,861**	**4.6**
Total World	**24,353**	**27,631**	**28,672**	**35,348**	**40,905**	**46,625**	**53,892**	**2.8**
Annex I								
Industrialized	19,010	21,286	21,981	26,168	29,153	32,368	35,512	2.1
EE/FSU	1,255	856	878	1,060	1,335	1,617	2,147	4.0
Total Annex I	**20,266**	**22,142**	**22,859**	**27,228**	**30,488**	**33,986**	**37,659**	**2.2**

[a]Includes the 50 States and the District of Columbia. U.S. Territories are included in Australasia.

Notes: EE/FSU = Eastern Europe/Former Soviet Union. Totals may not equal sum of components due to independent rounding.

Sources: **History:** The WEFA Group, *World Economic Outlook: 20-Year Extension* (Eddystone, PA, April 1997). **Projections:** Standard & Poor's DRI, *World Economic Outlook*, Vol. 1 (Lexington, MA, 3rd Quarter 1999); Energy Information Administration (EIA), *Annual Energy Outlook 2000*, DOE/EIA-0383(2000) (Washington, DC, December 1999), Table A20; and EIA, World Energy Projection System (2000).

Table A4. World Oil Consumption by Region, Reference Case, 1990-2020
(Million Barrels per Day)

Region/Country	History			Projections				Average Annual Percent Change, 1997-2020
	1990	1996	1997	2005	2010	2015	2020	
Industrialized Countries								
North America	**20.4**	**21.9**	**22.3**	**25.5**	**27.5**	**29.4**	**31.2**	**1.5**
United States[a]	17.0	18.3	18.6	21.1	22.5	23.9	25.1	1.3
Canada.	1.7	1.8	1.9	2.0	2.1	2.2	2.2	0.7
Mexico	1.7	1.8	1.9	2.3	2.8	3.3	4.0	3.3
Western Europe	**12.5**	**13.7**	**13.8**	**14.6**	**14.9**	**15.1**	**15.3**	**0.4**
United Kingdom	1.8	1.8	1.8	2.0	2.0	2.1	2.2	0.8
France	1.8	1.9	2.0	2.1	2.2	2.2	2.2	0.5
Germany	2.7	2.9	2.9	3.1	3.1	3.2	3.2	0.4
Italy.	1.9	2.1	2.0	2.2	2.2	2.2	2.3	0.4
Netherlands	0.7	0.8	0.8	0.8	0.9	0.9	0.9	0.6
Other Western Europe	3.6	4.1	4.3	4.4	4.5	4.5	4.5	0.2
Industrialized Asia	**6.2**	**7.1**	**6.9**	**7.2**	**7.5**	**7.8**	**8.0**	**0.6**
Japan	5.1	5.9	5.7	5.8	6.0	6.2	6.3	0.4
Australasia	1.0	1.2	1.2	1.4	1.5	1.6	1.7	1.4
Total Industrialized	**39.0**	**42.6**	**43.1**	**47.3**	**49.9**	**52.3**	**54.5**	**1.0**
EE/FSU								
Former Soviet Union	8.4	4.0	4.3	4.4	5.3	6.3	7.6	2.5
Eastern Europe	1.6	1.4	1.4	1.6	1.7	1.7	1.8	1.0
Total EE/FSU	**10.0**	**5.4**	**5.7**	**6.0**	**7.0**	**8.1**	**9.4**	**2.2**
Developing Countries								
Developing Asia	**7.6**	**12.0**	**12.6**	**16.5**	**19.7**	**22.8**	**24.7**	**3.0**
China.	2.3	3.5	3.8	5.4	7.1	8.8	9.5	4.1
India	1.2	1.7	1.8	2.7	3.2	3.7	4.1	3.7
South Korea	1.0	2.2	2.3	2.7	3.1	3.4	3.6	2.0
Other Asia	3.1	4.6	4.8	5.6	6.4	7.0	7.6	2.0
Middle East.	**3.9**	**4.7**	**4.8**	**5.4**	**6.5**	**7.6**	**9.0**	**2.8**
Turkey	0.5	0.6	0.6	0.8	1.0	1.1	1.3	3.1
Other Middle East	3.4	4.1	4.2	4.6	5.5	6.5	7.7	2.7
Africa	**2.1**	**2.4**	**2.5**	**3.2**	**4.0**	**4.8**	**5.9**	**3.9**
Central and South America . . .	**3.4**	**4.2**	**4.4**	**5.4**	**6.5**	**7.8**	**9.3**	**3.4**
Brazil.	1.3	1.7	1.8	2.3	2.9	3.5	4.4	4.0
Other Central/South America . .	2.1	2.5	2.6	3.1	3.7	4.2	4.9	2.8
Total Developing	**17.0**	**23.3**	**24.2**	**30.6**	**36.6**	**43.0**	**49.0**	**3.1**
Total World	**66.0**	**71.3**	**73.0**	**83.9**	**93.5**	**103.4**	**112.8**	**1.9**
Annex I								
Industrialized	37.3	40.8	41.2	44.9	47.1	49.0	50.6	0.9
EE/FSU	8.1	4.4	4.6	4.9	5.7	6.6	7.6	2.2
Total Annex I	**45.4**	**45.2**	**45.9**	**49.9**	**52.8**	**55.6**	**58.1**	**1.0**

[a]Includes the 50 States and the District of Columbia. U.S. Territories are included in Australasia.

Notes: EE/FSU = Eastern Europe/Former Soviet Union. Energy totals include net imports of coal coke and electricity generated from biomass in the United States. Totals may not equal sum of components due to independent rounding. The electricity portion of the national fuel consumption values consists of generation for domestic use plus an adjustment for electricity trade based on a fuel's share of total generation in the exporting country.

Sources: **History:** Energy Information Administration (EIA), *International Energy Annual 1997*, DOE/EIA-0219(97) (Washington, DC, April 1999). **Projections:** EIA, *Annual Energy Outlook 2000*, DOE/EIA-0383(2000) (Washington, DC, December 1999), Table A21; and World Energy Projection System (2000).

Table A5. World Natural Gas Consumption by Region, Reference Case, 1990-2020
(Trillion Cubic Feet)

Region/Country	History			Projections				Average Annual Percent Change, 1997-2020
	1990	1996	1997	2005	2010	2015	2020	
Industrialized Countries								
North America	**22.0**	**26.1**	**26.1**	**29.1**	**32.5**	**35.8**	**37.9**	**1.6**
United States[a]	18.7	22.0	22.0	23.9	27.0	29.9	31.5	1.6
Canada.	2.4	3.0	3.0	3.4	3.7	4.0	4.3	1.6
Mexico	0.9	1.1	1.2	1.8	1.9	2.0	2.1	2.4
Western Europe	**10.1**	**13.7**	**13.5**	**17.5**	**19.9**	**22.8**	**25.9**	**2.9**
United Kingdom	2.1	3.2	3.2	4.1	4.6	5.2	5.8	2.6
France	1.0	1.3	1.3	1.7	2.0	2.3	2.8	3.5
Germany	2.7	3.6	3.4	4.3	5.0	5.9	6.6	2.9
Italy.	1.7	2.0	2.0	2.2	2.4	2.6	2.8	1.4
Netherlands	1.5	1.9	1.8	2.1	2.2	2.3	2.4	1.4
Other Western Europe	1.2	1.8	1.8	3.1	3.6	4.5	5.4	4.8
Industrialized Asia.	**2.6**	**3.3**	**3.2**	**4.0**	**4.4**	**4.7**	**5.2**	**2.0**
Japan	1.9	2.4	2.3	2.9	3.2	3.4	3.7	2.0
Australasia	0.8	0.9	0.9	1.1	1.3	1.4	1.5	2.1
Total Industrialized	**34.8**	**43.0**	**42.9**	**50.7**	**56.9**	**63.3**	**69.0**	**2.1**
EE/FSU								
Former Soviet Union	25.0	20.8	19.7	22.4	25.2	27.3	32.0	2.1
Eastern Europe	3.1	2.9	2.6	4.2	5.9	7.3	9.2	5.6
Total EE/FSU	**28.1**	**23.7**	**22.3**	**26.5**	**31.1**	**34.6**	**41.2**	**2.7**
Developing Countries								
Developing Asia	**3.0**	**5.2**	**5.7**	**10.5**	**14.7**	**18.9**	**26.3**	**6.9**
China.	0.5	0.7	0.7	2.4	3.9	5.8	8.6	11.2
India	0.4	0.7	0.8	1.7	2.7	3.4	4.8	7.9
South Korea	0.1	0.4	0.5	0.8	1.1	1.5	2.4	6.8
Other Asia	1.9	3.4	3.6	5.5	7.0	8.2	10.6	4.9
Middle East.	**3.7**	**5.6**	**6.0**	**8.6**	**9.8**	**10.3**	**12.0**	**3.0**
Turkey	0.1	0.3	0.3	0.5	0.6	0.7	1.0	4.8
Other Middle East	3.6	5.3	5.7	8.1	9.1	9.6	11.0	2.9
Africa.	**1.4**	**1.8**	**1.8**	**2.3**	**2.4**	**2.5**	**2.8**	**1.8**
Central and South America . . .	**2.0**	**2.8**	**2.9**	**5.7**	**8.5**	**10.5**	**15.3**	**7.5**
Brazil.	0.1	0.2	0.2	0.8	1.4	1.6	2.5	11.6
Other Central/South America . .	1.9	2.6	2.7	4.9	7.1	9.0	12.9	7.0
Total Developing	**10.1**	**15.3**	**16.4**	**27.1**	**35.4**	**42.2**	**56.4**	**5.5**
Total World	**73.0**	**82.1**	**81.6**	**104.2**	**123.3**	**140.1**	**166.5**	**3.1**
Annex I								
Industrialized	33.9	41.9	41.7	48.9	54.9	61.3	66.9	2.1
EE/FSU	24.2	20.4	18.9	22.2	25.9	29.0	34.6	2.7
Total Annex I	**58.1**	**62.3**	**60.6**	**71.1**	**80.9**	**90.3**	**101.5**	**2.3**

[a]Includes the 50 States and the District of Columbia. U.S. Territories are included in Australasia.

Notes: EE/FSU = Eastern Europe/Former Soviet Union. Energy totals include net imports of coal coke and electricity generated from biomass in the United States. Totals may not equal sum of components due to independent rounding. The electricity portion of the national fuel consumption values consists of generation for domestic use plus an adjustment for electricity trade based on a fuel's share of total generation in the exporting country.

Sources: **History:** Energy Information Administration (EIA), *International Energy Annual 1997*, DOE/EIA-0219(97) (Washington, DC, April 1999). **Projections:** EIA, *Annual Energy Outlook 2000*, DOE/EIA-0383(2000) (Washington, DC, December 1999), Table A13; and World Energy Projection System (2000).

Table A6. World Coal Consumption by Region, Reference Case, 1990-2020
(Million Short Tons)

Region/Country	History			Projections				Average Annual Percent Change, 1997-2020
	1990	1996	1997	2005	2010	2015	2020	
Industrialized Countries								
North America	**959**	**1,076**	**1,102**	**1,250**	**1,271**	**1,313**	**1,364**	**0.9**
United States[a]	896	1,006	1,028	1,175	1,195	1,232	1,279	1.0
Canada.	55	59	62	60	59	64	68	0.4
Mexico	9	12	12	15	17	17	17	1.4
Western Europe	**896**	**579**	**583**	**521**	**501**	**479**	**450**	**-1.1**
United Kingdom	119	70	78	63	59	53	42	-2.7
France	36	25	23	17	11	11	9	-3.7
Germany	528	279	277	261	254	247	238	-0.7
Italy.	25	19	17	17	17	16	15	-0.4
Netherlands	15	15	20	18	16	14	11	-2.4
Other Western Europe	173	170	168	145	144	138	133	-1.0
Industrialized Asia	**233**	**265**	**281**	**306**	**311**	**317**	**321**	**0.6**
Japan	125	145	143	164	168	173	175	0.9
Australasia	108	119	138	142	143	144	145	0.2
Total Industrialized	**2,088**	**1,920**	**1,966**	**2,076**	**2,083**	**2,109**	**2,134**	**0.4**
EE/FSU								
Former Soviet Union	848	471	453	401	358	316	262	-2.4
Eastern Europe	524	438	424	418	355	304	241	-2.4
Total EE/FSU	**1,373**	**909**	**877**	**819**	**713**	**620**	**502**	**-2.4**
Developing Countries								
Developing Asia	**1,552**	**2,117**	**2,126**	**2,866**	**3,368**	**3,708**	**4,571**	**3.4**
China.	1,124	1,514	1,532	2,161	2,584	2,882	3,658	3.9
India	242	352	342	438	492	517	565	2.2
South Korea	42	59	65	71	81	87	89	1.4
Other Asia	145	191	188	196	211	222	259	1.4
Middle East.	**66**	**77**	**81**	**88**	**97**	**99**	**101**	**0.9**
Turkey	60	67	70	73	81	83	85	0.9
Other Middle East	6	10	11	15	16	16	16	1.4
Africa.	**152**	**179**	**184**	**203**	**204**	**206**	**213**	**0.7**
Central and South America . . .	**27**	**39**	**35**	**35**	**37**	**38**	**43**	**0.9**
Brazil.	17	21	21	22	23	24	29	1.4
Other Central/South America . .	10	19	14	14	14	14	14	0.0
Total Developing	**1,797**	**2,412**	**2,426**	**3,192**	**3,707**	**4,051**	**4,928**	**3.1**
Total World	**5,258**	**5,240**	**5,269**	**6,087**	**6,503**	**6,781**	**7,564**	**1.6**
Annex I								
Industrialized	2,080	1,908	1,954	2,061	2,067	2,092	2,118	0.3
EE/FSU	1,162	781	756	697	612	531	434	-2.4
Total Annex I	**3,242**	**2,690**	**2,710**	**2,759**	**2,679**	**2,624**	**2,551**	**-0.3**

[a]Includes the 50 States and the District of Columbia. U.S. Territories are included in Australasia.

Notes: EE/FSU = Eastern Europe/Former Soviet Union. Totals may not equal sum of components due to independent rounding. The electricity portion of the national fuel consumption values consists of generation for domestic use plus an adjustment for electricity trade based on a fuel's share of total generation in the exporting country. To convert short tons to metric tons, divide each number in the table by 1.102.

Sources: **History:** Energy Information Administration (EIA), *International Energy Annual 1997*, DOE/EIA-0219(97) (Washington, DC, April 1999). **Projections:** EIA, *Annual Energy Outlook 2000*, DOE/EIA-0383(2000) (Washington, DC, December 1999), Table A16; and World Energy Projection System (2000).

Table A7. World Nuclear Energy Consumption by Region, Reference Case, 1990-2020
(Billion Kilowatthours)

Region/Country	History			Projections				Average Annual Percent Change, 1997-2020
	1990	1996	1997	2005	2010	2015	2020	
Industrialized Countries								
North America	**649**	**770**	**717**	**754**	**707**	**590**	**491**	**-1.6**
United States[a]	577	675	629	674	627	511	427	-1.7
Canada.	69	88	78	72	72	72	56	-1.4
Mexico	3	7	10	8	8	8	8	-0.9
Western Europe	**703**	**826**	**835**	**844**	**821**	**757**	**693**	**-0.8**
United Kingdom	59	86	89	82	80	76	70	-1.1
France	298	377	374	401	409	411	395	0.2
Germany	145	152	162	151	132	106	106	-1.8
Italy.	0	0	0	0	0	0	0	0.0
Netherlands	3	4	2	3	3	0	0	-100.0
Other Western Europe	198	207	207	207	198	165	122	-2.3
Industrialized Asia.	**192**	**287**	**306**	**305**	**368**	**363**	**358**	**0.7**
Japan	192	287	306	305	368	363	358	0.7
Australasia	0	0	0	0	0	0	0	0.0
Total Industrialized	**1,544**	**1,883**	**1,859**	**1,904**	**1,896**	**1,711**	**1,541**	**-0.8**
EE/FSU								
Former Soviet Union	201	194	193	194	202	213	182	-0.2
Eastern Europe	54	60	63	73	70	68	61	-0.2
Total EE/FSU	**256**	**254**	**256**	**267**	**272**	**281**	**243**	**-0.2**
Developing Countries								
Developing Asia	**88**	**128**	**130**	**205**	**258**	**296**	**312**	**3.9**
China.	0	14	11	38	69	88	112	10.5
India	6	7	10	14	22	33	43	6.3
South Korea	50	70	73	100	107	116	106	1.6
Other Asia	32	37	35	53	59	60	51	1.7
Middle East.	**0**	**0**	**0**	**0**	**10**	**17**	**17**	**0.0**
Turkey	0	0	0	0	0	6	6	0.0
Other Middle East	0	0	0	0	10	10	10	0.0
Africa	**8**	**12**	**13**	**11**	**11**	**11**	**11**	**-0.5**
Central and South America . . .	**9**	**9**	**10**	**15**	**17**	**17**	**13**	**0.8**
Brazil.	2	2	3	8	9	9	9	4.7
Other Central/South America . .	7	7	7	7	8	8	4	-2.8
Total Developing	**105**	**149**	**153**	**232**	**296**	**341**	**353**	**3.7**
Total World	**1,905**	**2,286**	**2,268**	**2,402**	**2,464**	**2,333**	**2,136**	**-0.3**
Annex I								
Industrialized	1,541	1,876	1,849	1,896	1,888	1,703	1,533	-0.8
EE/FSU	255	252	254	263	267	278	240	-0.2
Total Annex I	**1,797**	**2,128**	**2,103**	**2,158**	**2,155**	**1,981**	**1,773**	**-0.7**

[a]Includes the 50 States and the District of Columbia. U.S. Territories are included in Australasia.

Notes: EE/FSU = Eastern Europe/Former Soviet Union. Totals may not equal sum of components due to independent rounding. The electricity portion of the national fuel consumption values consists of generation for domestic use plus an adjustment for electricity trade based on a fuel's share of total generation in the exporting country.

Sources: **History:** Energy Information Administration (EIA), *International Energy Annual 1997*, DOE/EIA-0219(97) (Washington, DC, April 1999). **Projections:** EIA, *Annual Energy Outlook 2000*, DOE/EIA-0383(2000) (Washington, DC, December 1999), Table A16; and World Energy Projection System (2000).

Table A8. World Consumption of Hydroelectricity and Other Renewable Energy by Region, Reference Case, 1990-2020
(Quadrillion Btu)

Region/Country	History			Projections				Average Annual Percent Change, 1997-2020
	1990	1996	1997	2005	2010	2015	2020	
Industrialized Countries								
North America	**9.2**	**11.4**	**11.2**	**12.6**	**13.4**	**13.9**	**14.4**	**1.1**
United States[a]	5.8	7.3	7.2	7.6	7.8	8.0	8.4	0.7
Canada.	3.1	3.7	3.6	4.6	5.0	5.2	5.4	1.8
Mexico	0.3	0.4	0.4	0.5	0.6	0.6	0.6	2.3
Western Europe	**4.4**	**4.5**	**4.7**	**5.5**	**6.0**	**6.6**	**7.2**	**1.9**
United Kingdom	0.1	0.0	0.0	0.2	0.2	0.2	0.3	8.3
France	0.6	0.7	0.6	0.6	0.7	0.7	0.8	0.8
Germany	0.2	0.2	0.2	0.4	0.6	0.8	0.8	5.7
Italy.	0.4	0.5	0.5	0.6	0.6	0.7	0.8	1.9
Netherlands	0.0	0.0	0.0	0.1	0.1	0.1	0.2	15.5
Other Western Europe	3.2	3.1	3.2	3.7	3.8	4.0	4.3	1.3
Industrialized Asia	**1.4**	**1.4**	**1.4**	**1.7**	**1.7**	**1.8**	**1.8**	**1.2**
Japan	1.0	0.9	1.0	1.1	1.2	1.2	1.2	1.0
Australasia	0.4	0.5	0.4	0.5	0.6	0.6	0.6	1.5
Total Industrialized	**15.0**	**17.3**	**17.2**	**19.8**	**21.1**	**22.3**	**23.5**	**1.4**
EE/FSU								
Former Soviet Union	2.4	2.3	2.2	2.4	2.4	2.5	2.6	0.7
Eastern Europe	0.4	0.6	0.6	0.6	0.9	1.1	1.8	4.9
Total EE/FSU	**2.8**	**2.9**	**2.8**	**3.0**	**3.3**	**3.6**	**4.5**	**2.0**
Developing Countries								
Developing Asia	**3.2**	**4.0**	**3.9**	**6.5**	**7.4**	**8.0**	**9.1**	**3.7**
China.	1.3	1.9	1.8	3.8	4.2	4.3	4.5	4.0
India	0.7	0.7	0.7	1.2	1.4	1.7	2.4	5.6
South Korea	0.1	0.1	0.1	0.1	0.2	0.2	0.3	7.6
Other Asia	1.1	1.3	1.3	1.5	1.6	1.8	1.9	1.6
Middle East.	**0.4**	**0.6**	**0.6**	**1.0**	**1.2**	**1.3**	**1.5**	**4.3**
Turkey	0.2	0.4	0.4	0.4	0.5	0.5	0.6	1.4
Other Middle East	0.1	0.2	0.2	0.6	0.7	0.8	0.9	7.8
Africa	**0.6**	**0.6**	**0.6**	**0.9**	**0.9**	**0.9**	**1.0**	**2.0**
Central and South America . . .	**3.9**	**5.2**	**5.4**	**6.0**	**6.4**	**6.7**	**7.4**	**1.4**
Brazil.	2.2	2.8	3.0	3.0	3.1	3.1	3.2	0.4
Other Central/South America . .	1.7	2.4	2.4	3.0	3.3	3.6	4.2	2.4
Total Developing	**8.1**	**10.4**	**10.5**	**14.4**	**15.9**	**16.9**	**19.0**	**2.6**
Total World	**25.9**	**30.5**	**30.6**	**37.2**	**40.4**	**42.8**	**47.0**	**1.9**
Annex I								
Industrialized	14.7	16.9	16.9	19.3	20.6	21.7	22.9	1.3
EE/FSU	2.2	2.1	2.1	2.2	2.5	2.7	3.4	2.1
Total Annex I	**16.8**	**19.0**	**19.0**	**21.5**	**23.0**	**24.4**	**26.3**	**1.4**

[a]Includes the 50 States and the District of Columbia. U.S. Territories are included in Australasia.
Notes: EE/FSU = Eastern Europe/Former Soviet Union. Totals may not equal sum of components due to independent rounding. The electricity portion of the national fuel consumption values consists of generation for domestic use plus an adjustment for electricity trade based on a fuel's share of total generation in the exporting country. U.S. totals include net electricity imports, methanol, and liquid hydrogen.
Sources: **History:** Energy Information Administration (EIA), *International Energy Annual 1997*, DOE/EIA-0219(97) (Washington, DC, April 1999). **Projections:** EIA, *Annual Energy Outlook 2000*, DOE/EIA-0383(2000) (Washington, DC, December 1999), Table A1; and World Energy Projection System (2000).

Table A9. World Net Electricity Consumption by Region, Reference Case, 1990-2020
(Billion Kilowatthours)

Region/Country	History			Projections				Average Annual Percent Change, 1997-2020
	1990	1996	1997	2005	2010	2015	2020	
Industrialized Countries								
North America	**3,359**	**3,869**	**3,908**	**4,373**	**4,726**	**5,066**	**5,357**	**1.4**
United States[a]	2,817	3,247	3,279	3,647	3,909	4,155	4,350	1.2
Canada.	435	479	475	524	563	601	632	1.2
Mexico	107	144	154	202	254	310	375	3.9
Western Europe	**2,064**	**2,266**	**2,262**	**2,573**	**2,819**	**3,075**	**3,343**	**1.7**
United Kingdom	286	317	310	341	372	405	437	1.5
France	324	382	376	433	475	516	554	1.7
Germany	485	479	477	514	567	623	679	1.5
Italy.	222	249	257	305	345	383	425	2.2
Netherlands	71	85	88	104	114	126	137	2.0
Other Western Europe	675	755	754	876	946	1,023	1,110	1.7
Industrialized Asia.	**930**	**1,090**	**1,117**	**1,305**	**1,416**	**1,486**	**1,555**	**1.5**
Japan	750	884	905	1,021	1,113	1,167	1,217	1.3
Australasia	180	207	212	285	302	319	338	2.0
Total Industrialized	**6,353**	**7,226**	**7,287**	**8,252**	**8,960**	**9,628**	**10,255**	**1.5**
EE/FSU								
Former Soviet Union	1,488	1,106	1,081	1,122	1,233	1,330	1,494	1.4
Eastern Europe	418	399	403	428	487	543	621	1.4
Total EE/FSU	**1,906**	**1,505**	**1,484**	**1,550**	**1,720**	**1,873**	**2,115**	**1.6**
Developing Countries								
Developing Asia	**1,260**	**2,006**	**2,103**	**3,071**	**3,899**	**4,707**	**5,957**	**4.6**
China.	551	923	956	1,521	2,045	2,588	3,450	5.7
India	257	379	397	626	788	937	1,154	4.7
South Korea	95	181	197	234	269	299	337	2.4
Other Asia	357	524	552	690	796	883	1,016	2.7
Middle East.	**272**	**394**	**412**	**502**	**601**	**688**	**816**	**3.0**
Turkey	51	85	94	163	242	296	423	6.7
Other Middle East	221	309	318	339	359	392	393	0.9
Africa	**285**	**334**	**350**	**464**	**553**	**660**	**811**	**3.7**
Central and South America . . .	**448**	**594**	**624**	**875**	**1,092**	**1,272**	**1,619**	**4.2**
Brazil.	229	304	323	451	587	724	959	4.9
Other Central/South America . .	219	291	301	424	505	548	660	3.5
Total Developing	**2,265**	**3,328**	**3,489**	**4,911**	**6,145**	**7,328**	**9,203**	**4.3**
Total World	**10,524**	**12,059**	**12,260**	**14,713**	**16,826**	**18,828**	**21,574**	**2.5**
Annex I								
Industrialized	6,246	7,082	7,132	8,050	8,706	9,318	9,880	1.4
EE/FSU	1,577	1,263	1,249	1,304	1,449	1,578	1,783	1.6
Total Annex I	**7,822**	**8,345**	**8,381**	**9,354**	**10,155**	**10,895**	**11,663**	**1.4**

[a]Includes the 50 States and the District of Columbia. U.S. Territories are included in Australasia.

Notes: EE/FSU = Eastern Europe/Former Soviet Union. Electricity consumption equals generation plus imports minus exports minus distribution losses.

Sources: **History:** Energy Information Administration (EIA), *International Energy Annual 1997*, DOE/EIA-0219(97) (Washington, DC, April 1999). **Projections:** EIA, *Annual Energy Outlook 2000*, DOE/EIA-0383(2000) (Washington, DC, December 1999), Table A8; and World Energy Projection System (2000).

Table A10. World Carbon Emissions by Region, Reference Case, 1990-2020
(Million Metric Tons)

Region/Country	History			Projections				Average Annual Percent Change, 1997-2020
	1990	1996	1997	2005	2010	2015	2020	
Industrialized Countries								
North America	**1,553**	**1,688**	**1,716**	**1,959**	**2,090**	**2,227**	**2,344**	**1.4**
United States[a]	1,345	1,461	1,480	1,683	1,787	1,893	1,979	1.3
Canada.	127	139	142	155	160	169	177	1.0
Mexico	81	89	94	121	143	164	188	3.1
Western Europe	**934**	**910**	**918**	**979**	**1,016**	**1,056**	**1,094**	**0.8**
United Kingdom	166	153	156	167	175	183	189	0.8
France	103	103	102	111	114	120	127	1.0
Germany	267	234	234	248	257	267	274	0.7
Italy.	113	116	116	123	128	131	134	0.6
Netherlands	60	62	64	68	69	71	72	0.5
Other Western Europe	224	242	246	262	272	284	297	0.8
Industrialized Asia.	**364**	**402**	**405**	**438**	**457**	**475**	**490**	**0.8**
Japan	274	304	297	318	331	344	354	0.8
Australasia	90	98	108	120	126	131	137	1.0
Total Industrialized	**2,850**	**3,000**	**3,039**	**3,377**	**3,563**	**3,758**	**3,928**	**1.1**
EE/FSU								
Former Soviet Union	1,034	661	646	668	728	781	875	1.3
Eastern Europe	303	236	231	258	264	269	276	0.8
Total EE/FSU	**1,337**	**897**	**878**	**927**	**992**	**1,050**	**1,151**	**1.2**
Developing Countries								
Developing Asia	**1,067**	**1,487**	**1,522**	**2,071**	**2,479**	**2,812**	**3,380**	**3.5**
China.	620	801	822	1,186	1,457	1,685	2,091	4.1
India	153	234	236	328	385	425	487	3.2
South Korea	61	109	116	136	157	174	187	2.1
Other Asia	232	343	348	421	480	527	615	2.5
Middle East.	**229**	**284**	**297**	**362**	**422**	**473**	**552**	**2.7**
Turkey	35	42	45	53	61	69	76	2.4
Other Middle East	194	242	252	309	361	405	476	2.8
Africa	**180**	**208**	**214**	**260**	**292**	**329**	**380**	**2.5**
Central and South America . . .	**174**	**220**	**225**	**310**	**399**	**479**	**617**	**4.5**
Brazil.	57	73	76	104	136	165	213	4.6
Other Central/South America . .	117	147	149	206	263	314	404	4.4
Total Developing	**1,649**	**2,200**	**2,258**	**3,004**	**3,591**	**4,093**	**4,930**	**3.5**
Total World	**5,836**	**6,096**	**6,175**	**7,308**	**8,146**	**8,901**	**10,009**	**2.1**
Annex I								
Industrialized	2,769	2,912	2,945	3,256	3,420	3,594	3,740	1.0
EE/FSU	1,135	771	751	784	835	880	962	1.1
Total Annex I	**3,904**	**3,683**	**3,697**	**4,040**	**4,255**	**4,474**	**4,702**	**1.1**

[a]Includes the 50 States and the District of Columbia. U.S. Territories are included in Australasia.

Notes: EE/FSU = Eastern Europe/Former Soviet Union. The U.S. numbers include carbon emissions attributable to renewable energy sources.

Sources: **History:** Energy Information Administration (EIA), *International Energy Annual 1997*, DOE/EIA-0219(97) (Washington, DC, April 1999). **Projections:** EIA, *Annual Energy Outlook 2000*, DOE/EIA-0383(2000) (Washington, DC, December 1999), Table A19; and World Energy Projection System (2000).

Table A11. World Carbon Emissions from Oil Use by Region, Reference Case, 1990-2020
(Million Metric Tons)

Region/Country	History			Projections				Average Annual Percent Change, 1997-2020
	1990	1996	1997	2005	2010	2015	2020	
Industrialized Countries								
North America	714	749	762	858	927	993	1,058	1.4
United States[a]	591	621	627	700	747	792	833	1.2
Canada.	61	63	65	71	74	76	77	0.7
Mexico	61	66	70	87	105	125	148	3.3
Western Europe	473	500	506	535	546	554	559	0.4
United Kingdom	66	66	64	71	73	75	78	0.8
France	67	69	70	76	78	79	79	0.5
Germany	98	106	106	113	115	116	116	0.4
Italy.	74	76	76	80	82	83	84	0.4
Netherlands	29	28	29	31	32	33	34	0.6
Other Western Europe	140	155	161	165	167	168	169	0.2
Industrialized Asia	218	234	230	239	250	260	267	0.6
Japan	178	191	186	187	194	201	205	0.4
Australasia	40	44	45	52	55	59	62	1.4
Total Industrialized	1,405	1,484	1,499	1,633	1,723	1,807	1,884	1.0
EE/FSU								
Former Soviet Union	332	159	168	174	209	250	299	2.5
Eastern Europe	68	52	53	60	63	66	67	1.0
Total EE/FSU	400	210	221	235	273	316	365	2.2
Developing Countries								
Developing Asia	308	459	480	629	749	867	940	3.0
China.	98	138	148	213	277	345	371	4.1
India	45	61	65	98	116	133	150	3.7
South Korea	38	71	73	89	100	109	116	2.0
Other Asia	127	188	193	229	257	280	303	2.0
Middle East.	153	179	183	207	247	289	343	2.8
Turkey	17	22	22	30	34	40	46	3.1
Other Middle East	136	156	160	178	212	250	297	2.7
Africa	83	93	97	127	157	192	234	3.9
Central and South America . . .	127	158	163	202	243	289	348	3.4
Brazil.	46	62	65	82	104	129	160	4.0
Other Central/South America . .	81	96	98	120	139	160	187	2.9
Total Developing	671	889	923	1,165	1,395	1,637	1,864	3.1
Total World	2,476	2,583	2,643	3,032	3,391	3,760	4,114	1.9
Annex I								
Industrialized	1,343	1,418	1,429	1,546	1,618	1,682	1,736	0.8
EE/FSU	325	171	179	191	221	254	293	2.2
Total Annex I	1,669	1,589	1,608	1,736	1,838	1,936	2,029	1.0

[a]Includes the 50 States and the District of Columbia. U.S. Territories are included in Australasia.
Notes: EE/FSU = Eastern Europe/Former Soviet Union.
Sources: **History:** Energy Information Administration (EIA), *International Energy Annual 1997*, DOE/EIA-0219(97) (Washington, DC, April 1999). **Projections:** EIA, *Annual Energy Outlook 2000*, DOE/EIA-0383(2000) (Washington, DC, December 1999), Table A19; and World Energy Projection System (2000).

Table A12. World Carbon Emissions from Natural Gas Use by Region, Reference Case, 1990-2020
(Million Metric Tons)

Region/Country	History			Projections				Average Annual Percent Change, 1997-2020
	1990	1996	1997	2005	2010	2015	2020	
Industrialized Countries								
North America	323	381	382	430	480	529	559	1.7
United States[a]	273	319	319	352	396	439	464	1.6
Canada.	35	44	44	51	54	58	63	1.6
Mexico	15	18	19	28	30	32	33	2.4
Western Europe	143	194	193	251	286	327	373	2.9
United Kingdom	32	48	49	62	70	78	89	2.6
France	16	20	19	26	30	35	43	3.5
Germany	32	46	44	55	65	76	85	2.9
Italy.	25	29	30	33	36	38	41	1.4
Netherlands	20	24	23	26	28	30	31	1.4
Other Western Europe	18	27	29	49	57	70	84	4.8
Industrialized Asia.	41	51	51	63	69	74	81	2.0
Japan	29	38	37	46	50	53	59	2.0
Australasia	12	14	14	17	19	21	23	2.1
Total Industrialized	507	626	626	745	835	931	1,014	2.1
EE/FSU								
Former Soviet Union	369	304	287	326	368	398	467	2.1
Eastern Europe	45	40	36	58	81	101	127	5.6
Total EE/FSU	414	344	323	383	449	499	594	2.7
Developing Countries								
Developing Asia	45	80	87	164	230	296	415	7.0
China.	8	11	12	40	65	97	142	11.2
India	7	12	14	29	44	56	79	7.9
South Korea	2	7	8	9	14	18	24	4.6
Other Asia	29	50	52	85	108	126	171	5.3
Middle East.	56	84	90	129	147	155	180	3.0
Turkey	2	4	5	5	7	9	10	3.0
Other Middle East	54	80	85	124	140	146	170	3.0
Africa	22	28	29	36	38	39	44	1.8
Central and South America . . .	32	44	47	91	136	169	246	7.5
Brazil.	2	3	3	14	23	26	41	11.6
Other Central/South America . .	30	41	43	77	113	143	204	7.0
Total Developing	155	236	253	420	551	659	885	5.6
Total World	1,077	1,206	1,202	1,548	1,836	2,089	2,492	3.2
Annex I								
Industrialized	492	609	607	717	805	899	981	2.1
EE/FSU	357	294	273	320	373	416	498	2.6
Total Annex I	849	903	880	1,037	1,179	1,315	1,479	2.3

[a]Includes the 50 States and the District of Columbia. U.S. Territories are included in Australasia.

Notes: EE/FSU = Eastern Europe/Former Soviet Union.

Sources: **History:** Energy Information Administration (EIA), *International Energy Annual 1997*, DOE/EIA-0219(97) (Washington, DC, April 1999). **Projections:** EIA, *Annual Energy Outlook 2000*, DOE/EIA-0383(2000) (Washington, DC, December 1999), Table A19; and World Energy Projection System (2000).

Table A13. World Carbon Emissions from Coal Use by Region, Reference Case, 1990-2020
(Million Metric Tons)

Region/Country	History			Projections				Average Annual Percent Change, 1997-2020
	1990	1996	1997	2005	2010	2015	2020	
Industrialized Countries								
North America	517	558	572	670	681	702	724	1.0
United States[a]	481	521	533	631	641	660	680	1.1
Canada.	31	32	34	33	32	35	37	0.4
Mexico	4	5	5	6	7	7	7	1.4
Western Europe	317	215	219	193	184	175	162	-1.3
United Kingdom	68	38	43	34	32	29	23	-2.7
France	20	14	13	9	6	6	5	-3.7
Germany	137	82	85	80	78	76	73	-0.7
Italy.	15	11	10	10	10	10	9	-0.4
Netherlands	11	10	12	11	10	9	7	-2.4
Other Western Europe	66	60	56	48	48	46	44	-1.0
Industrialized Asia	105	116	123	136	138	141	143	0.6
Japan	66	75	74	85	87	89	91	0.9
Australasia	39	41	49	51	51	52	52	0.2
Total Industrialized	939	890	914	999	1,003	1,018	1,028	0.5
EE/FSU								
Former Soviet Union	333	198	192	168	151	133	110	-2.4
Eastern Europe	189	145	142	140	120	102	82	-2.4
Total EE/FSU	523	343	334	309	270	235	192	-2.4
Developing Countries								
Developing Asia	713	949	954	1,278	1,499	1,648	2,026	3.3
China.	514	652	661	932	1,115	1,244	1,578	3.9
India	101	161	157	200	225	237	259	2.2
South Korea	21	31	35	38	43	47	47	1.4
Other Asia	76	104	102	107	115	121	141	1.4
Middle East.	20	21	24	25	28	29	29	0.9
Turkey	16	15	17	18	20	20	21	0.9
Other Middle East	4	6	7	8	8	9	9	1.2
Africa	74	86	88	97	98	99	102	0.7
Central and South America . . .	15	18	16	18	20	21	24	1.8
Brazil.	9	8	8	8	9	9	11	1.4
Other Central/South America . .	6	10	8	10	11	11	12	2.1
Total Developing	822	1,076	1,082	1,419	1,645	1,796	2,181	3.1
Total World	2,283	2,308	2,330	2,727	2,918	3,050	3,401	1.7
Annex I								
Industrialized	934	885	909	992	996	1,011	1,021	0.5
EE/FSU	453	306	299	274	241	209	171	-2.4
Total Annex I	1,387	1,191	1,208	1,266	1,236	1,220	1,192	-0.1

[a]Includes the 50 States and the District of Columbia. U.S. Territories are included in Australasia.

Notes: EE/FSU = Eastern Europe/Former Soviet Union.

Sources: **History:** Energy Information Administration (EIA), *International Energy Annual 1997*, DOE/EIA-0219(97) (Washington, DC, April 1999). **Projections:** EIA, *Annual Energy Outlook 2000*, DOE/EIA-0383(2000) (Washington, DC, December 1999), Table A19; and World Energy Projection System (2000).

Table A14. World Nuclear Generating Capacity by Region and Country, Reference Case, 1997-2020
(Megawatts)

Region/Country	History		Projections			
	1997	1998	2005	2010	2015	2020
Industrialized Countries						
North America						
United States	99,046	97,133	93,401	84,137	67,352	56,967
Canada	11,994	10,298	12,358	12,358	12,358	10,056
Mexico	1,308	1,308	1,308	1,308	1,308	1,308
Industrialized Asia						
Japan	43,850	43,691	44,487	53,493	49,296	43,816
Western Europe						
Belgium	5,712	5,712	5,712	5,712	4,358	3,966
Finland	2,560	2,656	2,656	2,656	2,656	1,328
France	62,853	61,653	62,870	62,870	62,870	60,005
Germany	22,282	22,282	21,942	20,135	19,364	14,294
Netherlands	449	449	449	449	0	0
Spain	7,415	7,350	7,197	6,751	6,751	2,912
Sweden	10,040	10,040	8,840	7,790	6,085	6,085
Switzerland	3,079	3,079	3,194	3,194	3,194	2,829
United Kingdom	12,968	12,968	12,968	11,882	11,647	10,558
Total Industrialized	**283,556**	**278,619**	**277,382**	**272,735**	**247,239**	**214,124**
EE/FSU						
Eastern Europe						
Bulgaria	3,538	3,538	3,538	2,722	1,906	1,906
Czech Republic	1,648	1,648	3,472	3,472	3,472	3,060
Hungary	1,729	1,729	1,729	1,729	1,729	1,729
Romania	650	650	1,300	1,300	1,300	1,300
Slovak Republic	1,632	2,020	2,408	1,592	1,592	1,592
Slovenia	632	632	632	632	632	0
Former Soviet Union						
Armenia	376	376	752	752	376	376
Kazakhstan	70	70	0	0	0	0
Lithuania	2,370	2,370	2,370	1,185	0	0
Russia	19,843	19,843	22,668	19,804	17,018	11,925
Ukraine	13,765	13,765	13,090	13,090	13,090	8,550
Total EE/FSU	**46,253**	**46,641**	**51,959**	**46,278**	**41,115**	**30,438**

See notes at end of table.

Table A14. World Nuclear Generating Capacity by Region and Country, Reference Case, 1997-2020 (Continued)
(Megawatts)

Region/Country	History		Projections			
	1997	1998	2005	2010	2015	2020
Developing Countries						
Developing Asia						
China	2,167	2,167	6,587	10,457	12,927	17,597
India	1,695	1,695	2,653	5,463	6,913	8,726
Korea, North	0	0	0	950	950	950
Korea, South.	9,770	11,380	14,890	16,790	16,234	16,234
Pakistan	125	125	300	600	600	600
Taiwan	4,884	4,884	7,384	7,384	7,384	6,176
Central and South America						
Argentina	935	935	1,292	1,292	1,292	692
Brazil.	626	626	1,855	1,855	3,100	2,474
Middle East						
Iran.	0	0	0	2,146	2,146	2,146
Turkey	0	0	0	0	1,300	1,300
Africa						
South Africa	1,842	1,842	1,842	1,842	1,842	1,842
Total Developing	**22,044**	**23,654**	**36,803**	**48,779**	**54,688**	**58,737**
Total World	**351,853**	**348,914**	**366,144**	**367,792**	**343,042**	**303,299**

Sources: **History:** International Atomic Energy Agency, *Nuclear Power Reactors in the World 1998* (Vienna, Austria, April 1999). **Projections:** Energy Information Administration, Office of Coal, Nuclear, Electric and Alternate Fuels, based on detailed assessments of country-specific nuclear power plants.

Table A15. World Total Energy Consumption in Oil-Equivalent Units by Region, Reference Case, 1990-2020
(Million Tons Oil Equivalent)

Region/Country	History			Projections				Average Annual Percent Change, 1997-2020
	1990	1996	1997	2005	2010	2015	2020	
Industrialized Countries								
North America	**2,517**	**2,821**	**2,835**	**3,199**	**3,403**	**3,588**	**3,741**	**1.2**
United States[a]	2,116	2,366	2,373	2,653	2,804	2,940	3,048	1.1
Canada.	274	316	316	358	379	397	410	1.1
Mexico	126	139	145	188	220	251	284	3.0
Western Europe	**1,510**	**1,594**	**1,614**	**1,753**	**1,829**	**1,901**	**1,975**	**0.9**
United Kingdom	235	245	250	274	288	304	318	1.1
France	235	265	262	285	297	308	318	0.8
Germany	372	357	358	385	401	418	434	0.8
Italy.	168	183	183	196	206	213	221	0.8
Netherlands	83	93	94	104	107	112	116	0.9
Other Western Europe	418	452	467	511	529	548	569	0.9
Industrialized Asia	**579**	**678**	**684**	**736**	**783**	**810**	**834**	**0.9**
Japan	456	541	536	570	607	625	641	0.8
Australasia	123	137	147	167	176	185	193	1.2
Total Industrialized	**4,606**	**5,093**	**5,132**	**5,689**	**6,015**	**6,300**	**6,550**	**1.1**
EE/FSU								
Former Soviet Union	1,538	1,053	1,029	1,086	1,193	1,286	1,445	1.5
Eastern Europe	386	320	314	363	394	421	462	1.7
Total EE/FSU	**1,924**	**1,373**	**1,343**	**1,449**	**1,587**	**1,707**	**1,907**	**1.5**
Developing Countries								
Developing Asia	**1,296**	**1,851**	**1,897**	**2,646**	**3,185**	**3,636**	**4,350**	**3.7**
China.	680	903	925	1,387	1,715	1,996	2,452	4.3
India	196	292	297	429	514	582	687	3.7
South Korea	93	178	188	234	269	299	337	2.6
Other Asia	327	479	487	596	686	759	874	2.6
Middle East.	**330**	**432**	**451**	**566**	**661**	**739**	**864**	**2.9**
Turkey	50	66	69	84	98	109	129	2.8
Other Middle East	280	366	382	482	564	630	735	2.9
Africa	**235**	**279**	**288**	**355**	**398**	**448**	**518**	**2.6**
Central and South America . . .	**346**	**448**	**461**	**611**	**759**	**889**	**1,125**	**4.0**
Brazil.	136	174	182	224	273	316	392	3.4
Other Central/South America . .	210	274	280	387	486	574	734	4.3
Total Developing	**2,207**	**3,009**	**3,097**	**4,178**	**5,002**	**5,712**	**6,858**	**3.5**
Total World	**8,737**	**9,475**	**9,572**	**11,315**	**12,605**	**13,719**	**15,314**	**2.1**
Annex I								
Industrialized	4,479	4,954	4,987	5,501	5,795	6,048	6,266	1.0
EE/FSU	1,635	1,179	1,147	1,223	1,333	1,433	1,596	1.4
Total Annex I	**6,114**	**6,134**	**6,134**	**6,724**	**7,129**	**7,481**	**7,862**	**1.1**

[a]Includes the 50 States and the District of Columbia. U.S. Territories are included in Australasia.
Notes: EE/FSU = Eastern Europe/Former Soviet Union.
Sources: **History:** Energy Information Administration (EIA), *International Energy Annual 1997*, DOE/EIA-0219(97) (Washington, DC, April 1999). **Projections:** EIA, *Annual Energy Outlook 2000*, DOE/EIA-0383(2000) (Washington, DC, December 1999), Table A1; and World Energy Projection System (2000).

Table A16. World Population by Region, Reference Case, 1990-2020
(Millions)

Region/Country	History			Projections				Annual Average Percent Change, 1997-2020
	1990	1996	1997	2005	2010	2015	2020	
Industrialized Countries								
North America	**365**	**392**	**393**	**425**	**445**	**465**	**485**	**0.9**
United States[a]	254	269	268	287	298	311	323	0.8
Canada.	28	30	30	33	34	35	37	0.8
Mexico	83	93	94	106	113	119	125	1.2
Western Europe	**377**	**385**	**387**	**390**	**389**	**387**	**385**	**0.0**
United Kingdom	58	58	59	59	59	60	60	0.1
France	57	58	58	60	61	61	62	0.2
Germany	79	82	82	82	82	82	81	-0.1
Italy.	57	57	57	57	56	54	53	-0.4
Netherlands	15	16	16	16	16	16	16	0.1
Other Western Europe	112	114	114	116	115	115	114	0.0
Industrialized Asia.	**148**	**152**	**152**	**156**	**157**	**156**	**158**	**0.2**
Japan	124	126	126	127	127	126	127	0.0
Australasia	24	26	26	28	29	30	32	0.8
Total Industrialized	**890**	**929**	**932**	**970**	**991**	**1,009**	**1,028**	**0.4**
EE/FSU								
Former Soviet Union	290	292	292	292	294	295	295	0.1
Eastern Europe	122	121	121	121	121	120	119	-0.1
Total EE/FSU	**412**	**413**	**413**	**413**	**414**	**415**	**414**	**0.0**
Developing Countries								
Developing Asia	**2,800**	**3,075**	**3,165**	**3,464**	**3,657**	**3,842**	**4,015**	**1.0**
China.	1,155	1,232	1,244	1,326	1,373	1,418	1,454	0.7
India	851	950	966	1,087	1,152	1,212	1,272	1.2
South Korea	43	45	46	49	50	51	52	0.6
Other Asia	752	847	909	1,001	1,082	1,161	1,236	1.3
Middle East.	**196**	**224**	**229**	**268**	**295**	**323**	**350**	**1.9**
Turkey	56	62	63	72	76	80	84	1.2
Other Middle East	140	162	166	197	219	243	266	2.1
Africa	**615**	**714**	**731**	**876**	**973**	**1,078**	**1,187**	**2.1**
Central and South America . . .	**354**	**391**	**398**	**447**	**478**	**508**	**536**	**1.3**
Brazil.	148	162	164	181	191	201	210	1.1
Other Central/South America . .	206	230	234	267	287	307	326	1.5
Total Developing	**3,965**	**4,405**	**4,523**	**5,055**	**5,403**	**5,750**	**6,088**	**1.3**
Total World	**5,266**	**5,747**	**5,868**	**6,439**	**6,809**	**7,174**	**7,530**	**1.1**
Annex I								
Industrialized	807	836	838	864	878	890	903	0.3
EE/FSU	311	309	308	303	300	297	292	-0.2
Total Annex I	**1,117**	**1,145**	**1,146**	**1,167**	**1,178**	**1,187**	**1,195**	**0.2**

[a]Includes the 50 States and the District of Columbia. U.S. Territories are included in Australasia.
Notes: EE/FSU = Eastern Europe/Former Soviet Union. Totals may not equal sum of components due to independent rounding.
Sources: **United States:** Energy Information Administration, *Annual Energy Outlook 2000*, DOE/EIA-0383(2000) (Washington, DC, December 1999), Table A20. **Other Countries:** United Nations, *The Sex and Age Distribution of the World Populations: The 1996 Revision* (New York, NY, 1997).

High Economic Growth Case Projections:

- World Energy Consumption
- Gross Domestic Product
- Carbon Emissions
- Nuclear Power Capacity

Table B1. World Total Energy Consumption by Region, High Economic Growth Case, 1990-2020
(Quadrillion Btu)

Region/Country	History			Projections				Average Annual Percent Change, 1997-2020
	1990	1996	1997	2005	2010	2015	2020	
Industrialized Countries								
North America	99.9	112.0	112.5	131.0	141.5	151.2	160.5	1.6
United States[a]	84.0	93.9	94.2	108.2	115.9	122.8	129.4	1.4
Canada	10.9	12.6	12.5	14.8	16.0	17.2	18.1	1.6
Mexico	5.0	5.5	5.8	7.9	9.5	11.2	13.0	3.6
Western Europe	59.9	63.3	64.0	72.0	76.5	80.9	85.6	1.3
United Kingdom	9.3	9.7	9.9	11.3	12.2	13.1	13.9	1.5
France	9.3	10.5	10.4	11.7	12.4	13.1	13.8	1.2
Germany	14.8	14.2	14.2	15.8	16.8	17.8	18.9	1.2
Italy	6.7	7.3	7.3	8.0	8.6	9.0	9.5	1.2
Netherlands	3.3	3.7	3.7	4.3	4.5	4.7	5.0	1.2
Other Western Europe	16.6	17.9	18.5	20.9	22.0	23.2	24.5	1.2
Industrialized Asia	23.0	26.9	27.1	30.2	33.2	34.9	36.5	1.3
Japan	18.1	21.4	21.3	23.3	25.7	26.9	27.9	1.2
Australasia	4.9	5.4	5.8	6.9	7.5	8.0	8.6	1.7
Total Industrialized	182.8	202.1	203.7	233.2	251.2	267.0	282.6	1.4
EE/FSU								
Former Soviet Union	61.0	41.8	40.8	46.4	51.9	56.5	64.4	2.0
Eastern Europe	15.3	12.7	12.5	16.0	18.4	20.5	24.1	2.9
Total EE/FSU	76.4	54.5	53.3	62.4	70.2	77.0	88.5	2.2
Developing Countries								
Developing Asia	51.4	73.5	75.3	114.3	144.0	172.2	215.4	4.7
China	27.0	35.8	36.7	59.8	77.6	94.7	122.0	5.4
India	7.8	11.6	11.8	18.7	23.5	27.8	34.3	4.7
South Korea	3.7	7.1	7.5	10.7	12.7	14.7	17.1	3.7
Other Asia	13.0	19.0	19.3	25.1	30.2	35.0	42.0	3.4
Middle East	13.1	17.1	17.9	25.6	31.7	37.3	45.7	4.2
Turkey	2.0	2.6	2.7	3.7	4.5	5.3	6.5	3.9
Other Middle East	11.1	14.5	15.2	21.9	27.2	32.0	39.1	4.2
Africa	9.3	11.1	11.4	15.4	18.0	21.4	25.9	3.6
Central and South America . . .	13.7	17.8	18.3	28.4	37.9	47.6	64.6	5.6
Brazil	5.4	6.9	7.2	14.1	18.3	22.6	29.8	6.4
Other Central/South America . .	8.3	10.9	11.1	14.3	19.5	25.1	34.8	5.1
Total Developing	87.6	119.4	122.9	183.7	231.5	278.4	351.5	4.7
Total World	346.7	376.0	379.9	479.3	553.0	622.5	722.6	2.8

[a]Includes the 50 States and the District of Columbia. U.S. Territories are included in Australasia.

Notes: EE/FSU = Eastern Europe/Former Soviet Union. Energy totals include net imports of coal coke and electricity generated from biomass in the United States. Totals may not equal sum of components due to independent rounding. The electricity portion of the national fuel consumption values consists of generation for domestic use plus an adjustment for electricity trade based on a fuel's share of total generation in the exporting country.

Sources: **History:** Energy Information Administration (EIA), *International Energy Annual 1997*, DOE/EIA-0219(97) (Washington, DC, April 1999). **Projections:** EIA, *Annual Energy Outlook 2000*, DOE/EIA-0383(2000) (Washington, DC, December 1999), Table B1; and World Energy Projection System (2000).

Table B2. World Total Energy Consumption by Region and Fuel, High Economic Growth Case, 1990-2020
(Quadrillion Btu)

Region/Country	History			Projections				Average Annual Percent Change, 1997-2020
	1990	1996	1997	2005	2010	2015	2020	
Industrialized Countries								
North America								
Oil	40.4	43.0	43.8	51.9	57.0	61.9	67.3	1.9
Natural Gas	22.7	26.8	26.9	31.2	35.4	39.2	41.1	1.9
Coal	20.5	22.5	23.0	26.8	27.5	28.9	31.0	1.3
Nuclear.	7.0	8.2	7.7	8.1	7.7	6.4	5.5	-1.4
Other	9.2	11.4	11.2	12.9	14.0	14.8	15.6	1.5
Total	99.9	112.0	112.5	131.0	141.5	151.2	160.5	1.6
Western Europe								
Oil	25.8	28.2	28.5	31.2	32.4	33.5	34.4	0.8
Natural Gas	10.0	13.5	13.4	18.1	20.9	24.4	28.3	3.3
Coal	12.4	8.4	8.7	7.9	7.7	7.4	7.0	-0.9
Nuclear.	7.4	8.7	8.8	9.2	9.1	8.5	8.0	-0.4
Other	4.4	4.5	4.7	5.7	6.4	7.1	7.9	2.3
Total	59.9	63.3	64.0	72.0	76.5	80.9	85.6	1.3
Industrialized Asia								
Oil	12.5	14.3	14.1	15.1	16.2	17.1	17.8	1.0
Natural Gas	2.9	3.6	3.5	4.6	5.1	5.6	6.2	2.5
Coal	4.2	4.7	5.0	5.7	6.0	6.2	6.4	1.1
Nuclear.	2.0	2.9	3.1	3.2	4.0	4.0	4.0	1.1
Other	1.4	1.4	1.4	1.7	1.8	1.9	2.0	1.6
Total	23.0	26.9	27.1	30.2	33.2	34.9	36.5	1.3
Total Industrialized								
Oil	78.7	85.5	86.4	98.2	105.6	112.5	119.6	1.4
Natural Gas	35.6	43.9	43.8	53.8	61.4	69.2	75.7	2.4
Coal	37.2	35.6	36.6	40.4	41.2	42.6	44.3	0.8
Nuclear.	16.3	19.9	19.6	20.5	20.7	19.0	17.5	-0.5
Other.	15.0	17.3	17.2	20.3	22.2	23.8	25.5	1.7
Total	182.8	202.1	203.7	233.2	251.2	267.0	282.6	1.4
EE/FSU								
Oil	21.0	11.4	11.9	13.8	16.4	19.3	22.8	2.9
Natural Gas	28.8	23.9	22.4	28.8	34.6	39.2	48.0	3.4
Coal	20.8	13.6	13.3	13.4	12.1	10.9	9.2	-1.6
Nuclear.	2.9	2.8	2.8	3.2	3.3	3.5	3.1	0.5
Other.	2.8	2.9	2.8	3.2	3.7	4.1	5.4	2.8
Total	76.4	54.5	53.3	62.4	70.2	77.0	88.5	2.2
Developing Countries								
Developing Asia								
Oil	16.0	25.1	26.3	37.6	46.8	56.6	64.1	3.9
Natural Gas	3.2	5.5	6.0	12.3	18.1	24.4	35.8	8.0
Coal	28.1	37.5	37.7	55.0	67.6	78.0	100.3	4.3
Nuclear.	0.9	1.3	1.3	2.3	3.0	3.6	4.0	4.9
Other.	3.2	4.0	3.9	7.1	8.5	9.5	11.3	4.7
Total	51.4	73.5	75.3	114.3	144.0	172.2	215.4	4.7

See notes at end of table.

Table B2. World Total Energy Consumption by Region and Fuel, High Economic Growth Case, 1990-2020 (Continued)
(Quadrillion Btu)

Region/Country	History			Projections				Average Annual Percent Change, 1997-2020
	1990	1996	1997	2005	2010	2015	2020	
Developing Countries (Continued)								
Middle East								
Oil	8.1	9.9	10.1	13.0	16.4	20.3	25.2	4.1
Natural Gas	3.9	5.8	6.3	10.2	12.3	13.7	16.7	4.3
Coal	0.8	0.9	0.9	1.2	1.4	1.5	1.6	2.2
Nuclear.	0.0	0.0	0.0	0.0	0.1	0.2	0.2	--
Other	0.4	0.6	0.6	1.2	1.4	1.6	2.0	5.6
Total	13.1	17.1	17.9	25.6	31.7	37.3	45.7	4.2
Africa								
Oil	4.2	4.9	5.1	7.3	9.4	12.1	15.5	5.0
Natural Gas	1.5	1.9	2.0	2.8	3.0	3.2	3.8	2.8
Coal	3.0	3.5	3.5	4.3	4.5	4.8	5.2	1.7
Nuclear.	0.1	0.1	0.1	0.1	0.1	0.1	0.1	0.5
Other	0.6	0.6	0.6	0.9	1.0	1.1	1.3	3.1
Total	9.3	11.1	11.4	15.4	18.0	21.4	25.9	3.6
Central and South America								
Oil	7.0	8.7	8.9	12.9	16.7	21.4	27.6	5.0
Natural Gas	2.2	3.1	3.2	7.4	11.9	15.8	24.7	9.2
Coal	0.6	0.7	0.6	0.9	1.0	1.1	1.4	3.4
Nuclear.	0.1	0.1	0.1	0.2	0.2	0.3	0.2	2.4
Other	3.9	5.2	5.4	7.0	8.0	9.0	10.7	3.0
Total	13.7	17.8	18.3	28.4	37.9	47.6	64.6	5.6
Total Developing Countries								
Oil	35.2	48.5	50.4	70.8	89.3	110.4	132.3	4.3
Natural Gas	10.8	16.4	17.6	32.7	45.3	57.2	81.0	6.9
Coal	32.4	42.6	42.9	61.3	74.5	85.3	108.4	4.1
Nuclear.	1.1	1.5	1.6	2.6	3.5	4.2	4.5	4.7
Other	8.1	10.4	10.5	16.2	18.9	21.3	25.3	3.9
Total	87.6	119.4	122.9	183.7	231.5	278.4	351.5	4.7
Total World								
Oil	134.9	145.4	148.7	182.8	211.3	242.2	274.7	2.7
Natural Gas	75.1	84.1	83.9	115.3	141.4	165.6	204.6	4.0
Coal	90.5	91.8	92.8	115.1	127.8	138.7	161.9	2.5
Nuclear.	20.4	24.2	24.0	26.4	27.6	26.7	25.2	0.2
Other	25.9	30.5	30.6	39.8	44.8	49.2	56.2	2.7
Total	346.7	376.0	379.9	479.3	553.0	622.5	722.6	2.8

Notes: EE/FSU = Eastern Europe/Former Soviet Union. Energy totals include net imports of coal coke and electricity generated from biomass in the United States. Totals may not equal sum of components due to independent rounding. The electricity portion of the national fuel consumption values consists of generation for domestic use plus an adjustment for electricity trade based on a fuel's share of total generation in the exporting country.

Sources: **History:** Energy Information Administration (EIA), *International Energy Annual 1997*, DOE/EIA-0219(97) (Washington, DC, April 1999). **Projections:** EIA, *Annual Energy Outlook 2000*, DOE/EIA-0383(2000) (Washington, DC, December 1999), Table B1; and World Energy Projection System (2000).

Table B3. World Gross Domestic Product (GDP) by Region, High Economic Growth Case, 1990-2020
(Billion 1997 Dollars)

Region/Country	History			Projections				Average Annual Percent Change, 1997-2020
	1990	1996	1997	2005	2010	2015	2020	
Industrialized Countries								
North America	**7,726**	**8,784**	**9,145**	**12,393**	**14,573**	**17,080**	**19,725**	**3.4**
United States[a]	6,846	7,804	8,111	10,918	12,726	14,814	16,996	3.3
Canada.	547	603	631	858	1,023	1,190	1,337	3.3
Mexico	333	377	403	617	824	1,076	1,393	5.5
Western Europe	**7,569**	**8,284**	**8,570**	**11,181**	**13,157**	**15,401**	**17,935**	**3.3**
United Kingdom	1,143	1,268	1,311	1,705	2,020	2,427	2,875	3.5
France	1,269	1,361	1,412	1,856	2,188	2,515	2,872	3.1
Germany	1,839	2,045	2,122	2,715	3,173	3,693	4,289	3.1
Italy.	1,060	1,128	1,159	1,484	1,747	2,044	2,379	3.2
Netherlands	306	351	363	480	565	665	781	3.4
Other Western Europe	1,951	2,131	2,202	2,941	3,464	4,057	4,739	3.4
Industrialized Asia.	**4,048**	**4,595**	**4,669**	**5,330**	**6,180**	**7,164**	**8,166**	**2.5**
Japan	3,667	4,142	4,199	4,692	5,432	6,290	7,155	2.3
Australasia	381	453	469	639	748	875	1,011	3.4
Total Industrialized	**19,343**	**21,663**	**22,384**	**28,904**	**33,910**	**39,646**	**45,826**	**3.2**
EE/FSU								
Former Soviet Union	1,049	595	600	814	1,092	1,466	1,968	5.3
Eastern Europe	358	347	368	640	909	1,293	1,813	7.2
Total EE/FSU	**1,407**	**942**	**968**	**1,454**	**2,002**	**2,759**	**3,781**	**6.1**
Developing Countries								
Developing Asia	**1,735**	**2,749**	**2,951**	**4,877**	**6,891**	**9,032**	**12,738**	**6.6**
China.	440	851	926	1,787	2,629	3,541	5,191	7.8
India	266	367	382	673	952	1,247	1,752	6.9
South Korea	273	419	476	740	1,047	1,373	1,911	6.2
Other Asia	755	1,111	1,166	1,677	2,264	2,870	3,884	5.4
Middle East.	**356**	**441**	**454**	**640**	**835**	**1,052**	**1,433**	**5.1**
Turkey	140	176	189	270	359	460	643	5.5
Other Middle East	216	266	265	371	476	592	789	4.9
Africa	**386**	**426**	**434**	**646**	**853**	**1,072**	**1,451**	**5.4**
Central and South America . . .	**1,127**	**1,409**	**1,481**	**2,160**	**2,887**	**3,649**	**4,993**	**5.4**
Brazil.	660	778	803	1,159	1,560	1,982	2,724	5.5
Other Central/South America . .	467	631	678	1,001	1,327	1,667	2,269	3.9
Total Developing	**3,604**	**5,026**	**5,320**	**8,324**	**11,466**	**14,805**	**20,615**	**6.1**
Total World	**24,353**	**27,631**	**28,672**	**38,682**	**47,378**	**57,209**	**70,221**	**4.0**

[a]Includes the 50 States and the District of Columbia. U.S. Territories are included in Australasia.
Notes: EE/FSU = Eastern Europe/Former Soviet Union. Totals may not equal sum of components due to independent rounding.
Sources: **History:** The WEFA Group, *World Economic Outlook: 20-Year Extension* (Eddystone, PA, April 1997). **Projections:** Standard & Poor's DRI, *World Economic Outlook*, Vol. 1 (Lexington, MA, 3rd Quarter 1999); Energy Information Administration (EIA), *Annual Energy Outlook 2000*, DOE/EIA-0383(2000) (Washington, DC, December 1999), Table B20; and EIA, World Energy Projection System (2000).

Table B4. World Oil Consumption by Region, High Economic Growth Case, 1990-2020
(Million Barrels per Day)

Region/Country	History			Projections				Average Annual Percent Change, 1997-2020
	1990	1996	1997	2005	2010	2015	2020	
Industrialized Countries								
North America	20.4	21.9	22.3	26.4	29.0	31.5	34.3	1.9
United States[a]	17.0	18.3	18.6	21.9	23.7	25.4	27.2	1.7
Canada.	1.7	1.8	1.9	2.1	2.3	2.4	2.4	1.2
Mexico	1.7	1.8	1.9	2.5	3.1	3.8	4.6	4.0
Western Europe	12.5	13.7	13.8	15.1	15.7	16.2	16.7	0.8
United Kingdom	1.8	1.8	1.8	2.1	2.2	2.3	2.4	1.2
France	1.8	1.9	2.0	2.2	2.3	2.4	2.4	0.9
Germany	2.7	2.9	2.9	3.2	3.3	3.4	3.5	0.8
Italy.	1.9	2.1	2.0	2.2	2.3	2.4	2.5	0.8
Netherlands	0.7	0.8	0.8	0.9	0.9	1.0	1.0	1.0
Other Western Europe	3.6	4.1	4.3	4.6	4.7	4.8	4.9	0.6
Industrialized Asia.	6.2	7.1	6.9	7.4	8.0	8.5	8.8	1.0
Japan	5.1	5.9	5.7	6.0	6.4	6.7	6.9	0.8
Australasia	1.0	1.2	1.2	1.5	1.6	1.8	1.9	1.9
Total Industrialized	39.0	42.6	43.1	49.0	52.8	56.2	59.8	1.4
EE/FSU								
Former Soviet Union	8.4	4.0	4.3	4.8	5.8	7.0	8.5	3.1
Eastern Europe	1.6	1.4	1.4	1.8	2.0	2.1	2.3	2.2
Total EE/FSU	10.0	5.4	5.7	6.6	7.8	9.2	10.9	2.9
Developing Countries								
Developing Asia	7.6	12.0	12.6	18.0	22.5	27.2	30.8	4.0
China.	2.3	3.5	3.8	5.9	8.0	10.5	11.9	5.1
India	1.2	1.7	1.8	3.0	3.7	4.4	5.2	4.7
South Korea	1.0	2.2	2.3	3.1	3.7	4.2	4.6	3.1
Other Asia	3.1	4.6	4.8	6.0	7.1	8.1	9.2	2.9
Middle East.	3.9	4.7	4.8	6.2	7.8	9.7	12.0	4.1
Turkey	0.5	0.6	0.6	0.9	1.1	1.4	1.6	4.2
Other Middle East	3.4	4.1	4.2	5.3	6.7	8.3	10.3	4.0
Africa	2.1	2.4	2.5	3.5	4.5	5.8	7.5	5.0
Central and South America . . .	3.4	4.2	4.4	6.3	8.2	10.5	13.5	5.0
Brazil.	1.3	1.7	1.8	3.6	4.8	6.4	8.5	7.0
Other Central/South America . .	2.1	2.5	2.6	2.7	3.4	4.1	5.0	3.0
Total Developing	17.0	23.3	24.2	34.1	43.0	53.2	63.7	4.3
Total World	66.0	71.3	73.0	89.6	103.5	118.5	134.3	2.7

[a]Includes the 50 States and the District of Columbia. U.S. Territories are included in Australasia.

Notes: EE/FSU = Eastern Europe/Former Soviet Union. Energy totals include net imports of coal coke and electricity generated from biomass in the United States. Totals may not equal sum of components due to independent rounding. The electricity portion of the national fuel consumption values consists of generation for domestic use plus an adjustment for electricity trade based on a fuel's share of total generation in the exporting country.

Sources: **History:** Energy Information Administration (EIA), *International Energy Annual 1997*, DOE/EIA-0219(97) (Washington, DC, April 1999). **Projections:** EIA, *Annual Energy Outlook 2000*, DOE/EIA-0383(2000) (Washington, DC, December 1999), Table B21; and World Energy Projection System (2000).

Table B5. World Natural Gas Consumption by Region, High Economic Growth Case, 1990-2020
(Trillion Cubic Feet)

Region/Country	History			Projections				Average Annual Percent Change, 1997-2020
	1990	1996	1997	2005	2010	2015	2020	
Industrialized Countries								
North America	**22.0**	**26.1**	**26.1**	**30.2**	**34.3**	**38.0**	**39.9**	**1.9**
United States[a]	18.7	22.0	22.0	24.8	28.3	31.4	32.7	1.7
Canada	2.4	3.0	3.0	3.6	3.9	4.3	4.8	2.1
Mexico	0.9	1.1	1.2	1.9	2.1	2.3	2.4	3.0
Western Europe	**10.1**	**13.7**	**13.5**	**18.1**	**21.0**	**24.4**	**28.3**	**3.3**
United Kingdom	2.1	3.2	3.2	4.3	4.9	5.6	6.5	3.1
France	1.0	1.3	1.3	1.8	2.1	2.5	3.1	3.9
Germany	2.7	3.6	3.4	4.4	5.3	6.3	7.2	3.3
Italy	1.7	2.0	2.0	2.3	2.6	2.8	3.1	1.8
Netherlands	1.5	1.9	1.8	2.1	2.3	2.5	2.7	1.8
Other Western Europe	1.2	1.8	1.8	3.2	3.8	4.8	5.8	5.2
Industrialized Asia	**2.6**	**3.3**	**3.2**	**4.2**	**4.7**	**5.1**	**5.7**	**2.5**
Japan	1.9	2.4	2.3	3.0	3.4	3.6	4.1	2.4
Australasia	0.8	0.9	0.9	1.2	1.3	1.5	1.6	2.6
Total Industrialized	**34.8**	**43.0**	**42.9**	**52.5**	**60.0**	**67.6**	**73.9**	**2.4**
EE/FSU								
Former Soviet Union	25.0	20.8	19.7	24.1	27.6	30.2	35.9	2.7
Eastern Europe	3.1	2.9	2.6	4.6	6.9	9.0	12.1	6.9
Total EE/FSU	**28.1**	**23.7**	**22.3**	**28.7**	**34.6**	**39.2**	**48.0**	**3.4**
Developing Countries								
Developing Asia	**3.0**	**5.2**	**5.7**	**11.4**	**16.7**	**22.3**	**32.6**	**7.9**
China	0.5	0.7	0.7	2.7	4.5	7.0	10.7	12.3
India	0.4	0.7	0.8	1.9	3.0	4.1	6.0	9.0
South Korea	0.1	0.4	0.5	1.0	1.4	1.8	3.0	7.9
Other Asia	1.9	3.4	3.6	5.8	7.8	9.5	12.9	5.8
Middle East	**3.7**	**5.6**	**6.0**	**9.7**	**11.8**	**13.1**	**15.9**	**4.3**
Turkey	0.1	0.3	0.3	0.5	0.7	0.8	1.3	2.8
Other Middle East	3.6	5.3	5.7	9.2	11.1	12.2	14.7	4.2
Africa	**1.4**	**1.8**	**1.8**	**2.5**	**2.7**	**3.0**	**3.5**	**2.8**
Central and South America . . .	**2.0**	**2.8**	**2.9**	**6.7**	**10.7**	**14.3**	**22.3**	**9.2**
Brazil	0.1	0.2	0.2	1.3	2.3	2.8	4.7	14.8
Other Central/South America . .	1.9	2.6	2.7	5.4	8.4	11.5	17.6	8.4
Total Developing	**10.1**	**15.3**	**16.4**	**30.3**	**41.9**	**52.7**	**74.4**	**6.8**
Total World	**73.0**	**82.1**	**81.6**	**111.6**	**136.5**	**159.5**	**196.3**	**3.9**

[a] Includes the 50 States and the District of Columbia. U.S. Territories are included in Australasia.

Notes: EE/FSU = Eastern Europe/Former Soviet Union. Energy totals include net imports of coal coke and electricity generated from biomass in the United States. Totals may not equal sum of components due to independent rounding. The electricity portion of the national fuel consumption values consists of generation for domestic use plus an adjustment for electricity trade based on a fuel's share of total generation in the exporting country.

Sources: **History:** Energy Information Administration (EIA), *International Energy Annual 1997*, DOE/EIA-0219(97) (Washington, DC, April 1999). **Projections:** EIA, *Annual Energy Outlook 2000*, DOE/EIA-0383(2000) (Washington, DC, December 1999), Table B13; and World Energy Projection System (2000).

Table B6. World Coal Consumption by Region, High Economic Growth Case, 1990-2020
(Million Short Tons)

Region/Country	History			Projections				Average Annual Percent Change, 1997-2020
	1990	1996	1997	2005	2010	2015	2020	
Industrialized Countries								
North America	**959**	**1,076**	**1,102**	**1,271**	**1,308**	**1,377**	**1,488**	**1.3**
United States[a]	896	1,006	1,028	1,193	1,226	1,288	1,393	1.3
Canada.	55	59	62	62	63	70	76	0.9
Mexico	9	12	12	16	18	19	19	2.1
Western Europe	**896**	**579**	**583**	**539**	**528**	**514**	**491**	**-0.7**
United Kingdom	119	70	78	65	63	58	46	-2.2
France	36	25	23	17	11	12	10	-3.3
Germany	528	279	277	270	268	265	261	-0.3
Italy.	25	19	17	18	18	18	17	-0.1
Netherlands	15	15	20	19	17	15	12	-2.1
Other Western Europe	173	170	168	150	151	147	144	-0.7
Industrialized Asia.	**233**	**265**	**281**	**317**	**332**	**345**	**355**	**1.0**
Japan	125	145	143	170	179	187	193	1.3
Australasia	108	119	138	148	153	158	163	0.7
Total Industrialized	**2,088**	**1,920**	**1,966**	**2,128**	**2,168**	**2,236**	**2,334**	**0.7**
EE/FSU								
Former Soviet Union	848	471	453	429	391	349	292	-1.9
Eastern Europe	524	438	424	464	418	374	320	-1.2
Total EE/FSU	**1,373**	**909**	**877**	**893**	**809**	**723**	**613**	**-1.5**
Developing Countries								
Developing Asia	**1,552**	**2,117**	**2,126**	**3,120**	**3,843**	**4,435**	**5,722**	**4.4**
China.	1,124	1,514	1,532	2,349	2,946	3,447	4,585	4.9
India	242	352	342	480	566	622	710	3.2
South Korea	42	59	65	82	97	108	114	2.5
Other Asia	145	191	188	209	234	258	314	2.3
Middle East.	**66**	**77**	**81**	**100**	**117**	**126**	**134**	**2.2**
Turkey	60	67	70	80	95	101	108	1.9
Other Middle East	6	10	11	19	23	25	26	3.6
Africa	**152**	**179**	**184**	**222**	**233**	**248**	**269**	**1.7**
Central and South America . . .	**27**	**39**	**35**	**47**	**54**	**61**	**75**	**3.4**
Brazil.	17	21	21	34	39	43	55	4.3
Other Central/South America . .	10	19	14	12	15	17	20	1.5
Total Developing	**1,797**	**2,412**	**2,426**	**3,488**	**4,247**	**4,869**	**6,200**	**4.2**
Total World	**5,258**	**5,240**	**5,269**	**6,509**	**7,223**	**7,829**	**9,147**	**2.4**

[a]Includes the 50 States and the District of Columbia. U.S. Territories are included in Australasia.

Notes: EE/FSU = Eastern Europe/Former Soviet Union. Totals may not equal sum of components due to independent rounding. The electricity portion of the national fuel consumption values consists of generation for domestic use plus an adjustment for electricity trade based on a fuel's share of total generation in the exporting country. To convert short tons to metric tons, divide each number in the table by 1.102.

Sources: **History:** Energy Information Administration (EIA), *International Energy Annual 1997*, DOE/EIA-0219(97) (Washington, DC, April 1999). **Projections:** EIA, *Annual Energy Outlook 2000*, DOE/EIA-0383(2000) (Washington, DC, December 1999), Table B16; and World Energy Projection System (2000).

Table B7. World Nuclear Energy Consumption by Region, High Economic Growth Case, 1990-2020
(Billion Kilowatthours)

Region/Country	History			Projections				Average Annual Percent Change, 1997-2020
	1990	1996	1997	2005	2010	2015	2020	
Industrialized Countries								
North America	649	770	717	758	712	597	511	-1.5
United States[a]	577	675	629	674	627	510	440	-1.5
Canada.	69	88	78	75	76	78	62	-1.0
Mexico	3	7	10	9	9	9	9	-0.3
Western Europe	703	826	835	873	865	812	757	-0.4
United Kingdom	59	86	89	85	85	83	77	-0.6
France	298	377	374	414	430	441	432	0.6
Germany	145	152	162	157	139	114	116	-1.5
Italy.	0	0	0	0	0	0	0	0.0
Netherlands	3	4	2	3	3	0	0	-100.0
Other Western Europe	198	207	207	214	208	175	133	-1.9
Industrialized Asia	192	287	306	315	393	393	393	1.1
Japan	192	287	306	315	393	393	393	1.1
Australasia	0	0	0	0	0	0	0	0.0
Total Industrialized	1,544	1,883	1,859	1,946	1,970	1,803	1,661	-0.5
EE/FSU								
Former Soviet Union	201	194	193	209	221	236	204	0.3
Eastern Europe	54	60	63	80	82	83	80	1.0
Total EE/FSU	256	254	256	290	303	319	284	0.5
Developing Countries								
Developing Asia	88	128	130	228	296	354	385	4.8
China.	0	14	11	42	77	102	133	11.3
India	6	7	10	15	25	39	54	7.4
South Korea	50	70	73	115	128	143	136	2.7
Other Asia	32	37	35	56	65	69	62	2.5
Middle East	0	0	0	0	12	21	22	0.0
Turkey	0	0	0	0	0	8	8	0.0
Other Middle East	0	0	0	0	12	13	14	0.0
Africa	8	12	13	12	13	14	14	0.5
Central and South America . . .	9	9	10	18	22	23	18	2.4
Brazil.	2	2	3	13	15	16	17	7.7
Other Central/South America . .	7	7	7	5	7	8	1	-6.9
Total Developing	105	149	153	258	343	412	440	4.7
Total World	1,905	2,286	2,268	2,494	2,615	2,534	2,385	0.2

[a]Includes the 50 States and the District of Columbia. U.S. Territories are included in Australasia.

Notes: EE/FSU = Eastern Europe/Former Soviet Union. Totals may not equal sum of components due to independent rounding. The electricity portion of the national fuel consumption values consists of generation for domestic use plus an adjustment for electricity trade based on a fuel's share of total generation in the exporting country.

Sources: **History:** Energy Information Administration (EIA), *International Energy Annual 1997*, DOE/EIA-0219(97) (Washington, DC, April 1999). **Projections:** EIA, *Annual Energy Outlook 2000*, DOE/EIA-0383(2000) (Washington, DC, December 1999), Table B16; and World Energy Projection System (2000).

Table B8. World Consumption of Hydroelectricity and Other Renewable Energy by Region, High Economic Growth Case, 1990-2020
(Quadrillion Btu)

Region/Country	History			Projections				Average Annual Percent Change, 1997-2020
	1990	1996	1997	2005	2010	2015	2020	
Industrialized Countries								
North America	**9.2**	**11.4**	**11.2**	**12.9**	**14.0**	**14.8**	**15.6**	**1.5**
United States[a]	5.8	7.3	7.2	7.7	8.0	8.4	8.8	0.9
Canada.	3.1	3.7	3.6	4.8	5.4	5.7	6.1	2.3
Mexico	0.3	0.4	0.4	0.5	0.6	0.7	0.7	3.0
Western Europe	**4.4**	**4.5**	**4.7**	**5.7**	**6.4**	**7.1**	**7.9**	**2.3**
United Kingdom	0.1	0.0	0.0	0.2	0.2	0.2	0.3	8.7
France	0.6	0.7	0.6	0.6	0.8	0.8	0.8	1.2
Germany	0.2	0.2	0.2	0.4	0.6	0.8	0.9	6.1
Italy.	0.4	0.5	0.5	0.6	0.7	0.8	0.9	2.3
Netherlands	0.0	0.0	0.0	0.1	0.1	0.2	0.2	16.0
Other Western Europe	3.2	3.1	3.2	3.8	4.0	4.3	4.7	1.7
Industrialized Asia	**1.4**	**1.4**	**1.4**	**1.7**	**1.8**	**1.9**	**2.0**	**1.6**
Japan	1.0	0.9	1.0	1.2	1.2	1.3	1.3	1.4
Australasia	0.4	0.5	0.4	0.6	0.6	0.7	0.7	2.0
Total Industrialized	**15.0**	**17.3**	**17.2**	**20.3**	**22.2**	**23.8**	**25.5**	**1.7**
EE/FSU								
Former Soviet Union	2.4	2.3	2.2	2.5	2.7	2.7	3.0	1.2
Eastern Europe	0.4	0.6	0.6	0.7	1.0	1.4	2.4	6.1
Total EE/FSU	**2.8**	**2.9**	**2.8**	**3.2**	**3.7**	**4.1**	**5.4**	**2.8**
Developing Countries								
Developing Asia	**3.2**	**4.0**	**3.9**	**7.1**	**8.5**	**9.5**	**11.3**	**4.7**
China.	1.3	1.9	1.8	4.1	4.8	5.2	5.6	5.0
India	0.7	0.7	0.7	1.3	1.6	2.1	3.0	6.6
South Korea	0.1	0.1	0.1	0.1	0.2	0.2	0.4	8.7
Other Asia	1.1	1.3	1.3	1.5	1.8	2.1	2.3	2.4
Middle East.	**0.4**	**0.6**	**0.6**	**1.2**	**1.4**	**1.6**	**2.0**	**5.6**
Turkey	0.2	0.4	0.4	0.5	0.6	0.6	0.7	2.5
Other Middle East	0.1	0.2	0.2	0.7	0.9	1.0	1.3	9.3
Africa	**0.6**	**0.6**	**0.6**	**0.9**	**1.0**	**1.1**	**1.3**	**3.1**
Central and South America . . .	**3.9**	**5.2**	**5.4**	**7.0**	**8.0**	**9.0**	**10.7**	**3.0**
Brazil.	2.2	2.8	3.0	4.8	5.2	5.6	6.2	3.2
Other Central/South America . .	1.7	2.4	2.4	2.2	2.8	3.4	4.6	2.8
Total Developing	**8.1**	**10.4**	**10.5**	**16.2**	**18.9**	**21.3**	**25.3**	**3.9**
Total World	**25.9**	**30.5**	**30.6**	**39.8**	**44.8**	**49.2**	**56.2**	**2.7**

[a]Includes the 50 States and the District of Columbia. U.S. Territories are included in Australasia.

Notes: EE/FSU = Eastern Europe/Former Soviet Union. Totals may not equal sum of components due to independent rounding. The electricity portion of the national fuel consumption values consists of generation for domestic use plus an adjustment for electricity trade based on a fuel's share of total generation in the exporting country. U.S. totals include net electricity imports, methanol, and liquid hydrogen.

Sources: **History:** Energy Information Administration (EIA), *International Energy Annual 1997*, DOE/EIA-0219(97) (Washington, DC, April 1999). **Projections:** EIA, *Annual Energy Outlook 2000*, DOE/EIA-0383(2000) (Washington, DC, December 1999), Table B1; and World Energy Projection System (2000).

Table B9. World Net Electricity Consumption by Region, High Economic Growth Case, 1990-2020
(Billion Kilowatthours)

Region/Country	History			Projections				Average Annual Percent Change, 1997-2020
	1990	1996	1997	2005	2010	2015	2020	
Industrialized Countries								
North America	**3,359**	**3,869**	**3,908**	**4,493**	**4,928**	**5,368**	**5,792**	**1.7**
United States[a]	2,817	3,247	3,279	3,733	4,051	4,364	4,653	1.5
Canada.	435	479	475	547	600	656	705	1.7
Mexico	107	144	154	214	277	348	434	4.6
Western Europe	**2,064**	**2,266**	**2,262**	**2,663**	**2,970**	**3,297**	**3,649**	**2.1**
United Kingdom	286	317	310	355	395	438	483	2.0
France	324	382	376	447	500	554	606	2.1
Germany	485	479	477	533	598	670	744	1.9
Italy.	222	249	257	315	364	410	464	2.6
Netherlands	71	85	88	108	120	135	149	2.3
Other Western Europe	675	755	754	905	993	1,090	1,203	2.0
Industrialized Asia	**930**	**1,090**	**1,117**	**1,366**	**1,543**	**1,667**	**1,786**	**2.1**
Japan	750	884	905	1,054	1,187	1,264	1,337	1.7
Australasia	180	207	212	312	356	403	448	3.3
Total Industrialized	**6,353**	**7,226**	**7,287**	**8,522**	**9,442**	**10,332**	**11,226**	**1.9**
EE/FSU								
Former Soviet Union	1,488	1,106	1,081	1,210	1,351	1,473	1,678	1.9
Eastern Europe	418	399	403	474	572	666	817	3.1
Total EE/FSU	**1,906**	**1,505**	**1,484**	**1,684**	**1,923**	**2,139**	**2,495**	**2.3**
Developing Countries								
Developing Asia	**1,260**	**2,006**	**2,103**	**3,341**	**4,443**	**5,618**	**7,437**	**5.6**
China.	551	923	956	1,653	2,332	3,096	4,324	6.8
India	257	379	397	686	906	1,127	1,450	5.8
South Korea	95	181	197	269	321	371	431	3.5
Other Asia	357	524	552	732	884	1,024	1,232	3.5
Middle East.	**272**	**394**	**412**	**572**	**726**	**874**	**1,087**	**4.3**
Turkey	51	85	94	181	282	362	540	7.9
Other Middle East	221	309	318	392	443	512	547	2.4
Africa.	**285**	**334**	**350**	**508**	**630**	**795**	**1,022**	**4.8**
Central and South America . . .	**448**	**594**	**624**	**1,025**	**1,373**	**1,717**	**2,341**	**5.9**
Brazil.	229	304	323	715	992	1,305	1,838	7.9
Other Central/South America . .	219	291	301	310	381	413	503	2.3
Total Developing	**2,265**	**3,328**	**3,489**	**5,447**	**7,172**	**9,004**	**11,887**	**5.5**
Total World	**10,524**	**12,059**	**12,260**	**15,652**	**18,536**	**21,475**	**25,609**	**3.3**

[a]Includes the 50 States and the District of Columbia. U.S. Territories are included in Australasia.

Notes: EE/FSU = Eastern Europe/Former Soviet Union. Electricity consumption equals generation plus imports minus exports minus distribution losses.

Sources: **History:** Energy Information Administration (EIA), *International Energy Annual 1997*, DOE/EIA-0219(97) (Washington, DC, April 1999). **Projections:** EIA, *Annual Energy Outlook 2000*, DOE/EIA-0383(2000) (Washington, DC, December 1999), Table B8; and World Energy Projection System (2000).

Table B10. World Carbon Emissions by Region, High Economic Growth Case, 1990-2020
(Million Metric Tons)

Region/Country	History			Projections				Average Annual Percent Change, 1997-2020
	1990	1996	1997	2005	2010	2015	2020	
Industrialized Countries								
North America	**1,553**	**1,689**	**1,716**	**2,022**	**2,190**	**2,366**	**2,541**	**1.7**
United States[a]	1,346	1,461	1,480	1,732	1,863	1,997	2,126	1.6
Canada.	127	139	142	161	171	185	197	1.4
Mexico	81	89	94	128	156	185	217	3.7
Western Europe	**934**	**910**	**918**	**1,014**	**1,071**	**1,133**	**1,195**	**1.2**
United Kingdom	166	153	156	174	187	198	209	1.3
France	103	103	102	114	120	128	139	1.4
Germany	267	234	234	257	271	287	300	1.1
Italy.	113	116	116	127	135	141	146	1.0
Netherlands	60	62	64	71	73	76	79	0.9
Other Western Europe	224	242	246	271	285	302	322	1.2
Industrialized Asia.	**364**	**402**	**405**	**454**	**488**	**516**	**541**	**1.3**
Japan	274	304	297	328	353	372	389	1.5
Australasia	90	98	108	125	135	144	153	1.5
Total Industrialized	**2,851**	**3,000**	**3,039**	**3,489**	**3,750**	**4,015**	**4,277**	**1.5**
EE/FSU								
Former Soviet Union	1,034	661	646	721	797	865	983	1.8
Eastern Europe	303	236	231	286	310	331	363	2.0
Total EE/FSU	**1,337**	**897**	**878**	**1,007**	**1,108**	**1,196**	**1,346**	**1.9**
Developing Countries								
Developing Asia	**1,067**	**1,487**	**1,522**	**2,254**	**2,825**	**3,357**	**4,220**	**4.5**
China.	620	801	822	1,289	1,661	2,016	2,621	5.2
India	153	234	236	359	443	511	612	4.2
South Korea	61	109	116	157	187	215	238	3.2
Other Asia	232	343	348	449	535	614	748	3.4
Middle East.	**229**	**284**	**297**	**412**	**509**	**601**	**735**	**4.0**
Turkey	35	42	45	58	71	83	97	3.4
Other Middle East	194	242	252	354	438	518	638	4.1
Africa	**180**	**208**	**214**	**285**	**333**	**396**	**479**	**3.6**
Central and South America . . .	**174**	**220**	**225**	**364**	**501**	**646**	**893**	**6.2**
Brazil.	57	73	76	165	229	296	408	7.6
Other Central/South America . .	117	147	149	198	272	350	485	5.3
Total Developing	**1,649**	**2,200**	**2,258**	**3,316**	**4,168**	**5,000**	**6,327**	**4.6**
Total World	**5,836**	**6,097**	**6,175**	**7,812**	**9,026**	**10,210**	**11,950**	**2.9**

[a]Includes the 50 States and the District of Columbia. U.S. Territories are included in Australasia.

Notes: EE/FSU = Eastern Europe/Former Soviet Union. The U.S. numbers include carbon emissions attributable to renewable energy sources.

Sources: **History:** Energy Information Administration (EIA), *International Energy Annual 1997*, DOE/EIA-0219(97) (Washington, DC, April 1999). **Projections:** EIA, *Annual Energy Outlook 2000*, DOE/EIA-0383(2000) (Washington, DC, December 1999), Table B19; and World Energy Projection System (2000).

Table B11. World Carbon Emissions from Oil Use by Region, High Economic Growth Case, 1990-2020
(Million Metric Tons)

Region/Country	History			Projections				Average Annual Percent Change, 1997-2020
	1990	1996	1997	2005	2010	2015	2020	
Industrialized Countries								
North America	714	750	762	891	980	1,064	1,158	1.8
United States[a]	591	621	627	725	785	841	901	1.6
Canada.	61	63	65	74	79	83	85	1.2
Mexico	61	66	70	92	115	141	172	4.0
Western Europe	473	500	506	554	576	594	610	0.8
United Kingdom	66	66	64	74	77	82	86	1.2
France	67	69	70	78	82	84	86	0.9
Germany	98	106	106	117	121	124	127	0.8
Italy.	74	76	76	83	86	89	91	0.8
Netherlands	29	28	29	32	34	35	37	1.0
Other Western Europe	140	155	161	171	175	179	183	0.6
Industrialized Asia	218	234	230	248	267	282	294	1.1
Japan	178	191	186	194	207	218	225	0.8
Australasia	40	44	45	54	59	64	69	1.9
Total Industrialized	1,405	1,484	1,499	1,693	1,822	1,940	2,063	1.4
EE/FSU								
Former Soviet Union	332	159	168	188	229	277	335	3.1
Eastern Europe	68	52	53	67	74	81	88	2.2
Total EE/FSU	400	210	221	255	304	357	423	2.9
Developing Countries								
Developing Asia	308	459	480	686	854	1,034	1,170	3.9
China.	98	138	148	232	315	413	465	5.1
India	45	61	65	108	133	159	188	4.7
South Korea	38	71	73	103	120	136	148	3.1
Other Asia	127	188	193	244	286	326	368	2.8
Middle East.	153	179	183	236	298	368	456	4.1
Turkey	17	22	22	33	40	48	58	4.2
Other Middle East	136	156	160	203	258	319	398	4.0
Africa	83	93	97	139	179	231	295	5.0
Central and South America . . .	127	158	163	236	305	390	503	5.0
Brazil.	46	62	65	130	175	232	307	7.0
Other Central/South America . .	81	96	98	106	130	158	196	3.1
Total Developing	671	889	923	1,297	1,636	2,023	2,424	4.3
Total World	2,476	2,583	2,643	3,245	3,761	4,320	4,910	2.7

[a]Includes the 50 States and the District of Columbia. U.S. Territories are included in Australasia.
Notes: EE/FSU = Eastern Europe/Former Soviet Union.
Sources: **History:** Energy Information Administration (EIA), *International Energy Annual 1997*, DOE/EIA-0219(97) (Washington, DC, April 1999). **Projections:** EIA, *Annual Energy Outlook 2000*, DOE/EIA-0383(2000) (Washington, DC, December 1999), Table B19; and World Energy Projection System (2000).

Table B12. World Carbon Emissions from Natural Gas Use by Region, High Economic Growth Case, 1990-2020
(Million Metric Tons)

Region/Country	History			Projections				Average Annual Percent Change, 1997-2020
	1990	1996	1997	2005	2010	2015	2020	
Industrialized Countries								
North America	323	381	382	446	506	561	589	1.9
United States[a]	273	319	319	364	416	462	481	1.8
Canada.	35	44	44	53	57	64	70	2.1
Mexico	15	18	19	29	33	36	38	3.0
Western Europe	143	194	193	260	302	351	408	3.1
United Kingdom	32	48	49	65	75	85	98	3.1
France	16	20	19	26	32	37	47	3.9
Germany	32	46	44	57	68	81	93	3.3
Italy.	25	29	30	34	38	41	45	1.8
Netherlands	20	24	23	27	29	32	34	1.8
Other Western Europe	18	27	29	51	60	74	91	5.2
Industrialized Asia.	41	51	51	66	74	81	90	2.5
Japan	29	38	37	47	53	58	64	2.4
Australasia	12	14	14	18	21	23	25	2.6
Total Industrialized	507	626	626	772	882	993	1,087	2.4
EE/FSU								
Former Soviet Union	369	304	287	351	403	441	524	2.7
Eastern Europe	45	40	36	64	96	125	167	6.9
Total EE/FSU	414	344	323	415	499	565	691	3.4
Developing Countries								
Developing Asia	45	80	87	177	261	352	515	8.0
China.	8	11	12	44	74	116	178	12.3
India	7	12	14	32	51	67	99	9.0
South Korea	2	7	8	10	16	22	29	5.6
Other Asia	29	50	52	91	121	147	209	6.2
Middle East.	56	84	90	147	177	197	240	4.3
Turkey	2	4	5	5	8	11	12	3.9
Other Middle East	54	80	85	141	170	187	228	4.4
Africa	22	28	29	40	43	47	55	2.8
Central and South America . . .	32	44	47	106	171	228	356	9.2
Brazil.	2	3	3	22	39	48	80	14.8
Other Central/South America . .	30	41	43	85	132	180	276	8.4
Total Developing	155	236	253	470	652	823	1,166	6.9
Total World	1,077	1,206	1,202	1,657	2,033	2,382	2,943	4.0

[a]Includes the 50 States and the District of Columbia. U.S. Territories are included in Australasia.
Notes: EE/FSU = Eastern Europe/Former Soviet Union.
Sources: **History:** Energy Information Administration (EIA), *International Energy Annual 1997*, DOE/EIA-0219(97) (Washington, DC, April 1999). **Projections:** EIA, *Annual Energy Outlook 2000*, DOE/EIA-0383(2000) (Washington, DC, December 1999), Table B19; and World Energy Projection System (2000).

Table B13. World Carbon Emissions from Coal Use by Region, High Economic Growth Case, 1990-2020
(Million Metric Tons)

Region/Country	History			Projections				Average Annual Percent Change, 1997-2020
	1990	1996	1997	2005	2010	2015	2020	
Industrialized Countries								
North America	517	558	572	683	702	738	790	1.4
United States[a]	481	521	533	642	660	691	740	1.4
Canada.	31	32	34	34	34	38	42	0.9
Mexico	4	5	5	7	8	8	8	2.1
Western Europe	317	215	219	200	194	188	177	-0.9
United Kingdom	68	38	43	36	34	32	25	-2.2
France	20	14	13	10	6	7	6	-3.3
Germany	137	82	85	83	82	81	80	-0.3
Italy.	15	11	10	11	11	10	10	-0.1
Netherlands	11	10	12	12	10	9	8	-2.1
Other Western Europe	66	60	56	50	50	49	48	-0.7
Industrialized Asia.	105	116	123	140	147	153	158	1.1
Japan	66	75	74	88	93	97	100	1.3
Australasia	39	41	49	53	55	56	58	0.7
Total Industrialized	939	890	914	1,024	1,043	1,079	1,125	0.9
EE/FSU								
Former Soviet Union	333	198	192	181	165	147	124	-1.9
Eastern Europe	189	145	142	156	140	126	108	-1.2
Total EE/FSU	523	343	334	337	306	273	231	-1.6
Developing Countries								
Developing Asia	713	949	954	1,391	1,710	1,971	2,535	4.3
China.	514	652	661	1,014	1,271	1,488	1,978	4.9
India	101	161	157	220	259	285	325	3.2
South Korea	21	31	35	44	52	58	61	2.5
Other Asia	76	104	102	114	128	141	171	2.3
Middle East.	20	21	24	29	34	37	39	2.2
Turkey	16	15	17	20	23	25	26	1.9
Other Middle East	4	6	7	9	11	12	13	2.8
Africa	74	86	88	107	112	119	129	1.7
Central and South America . . .	15	18	16	21	25	28	34	3.4
Brazil.	9	8	8	13	15	17	21	4.3
Other Central/South America . .	6	10	8	8	9	11	13	2.2
Total Developing	822	1,076	1,082	1,548	1,880	2,154	2,738	4.1
Total World	2,283	2,308	2,330	2,909	3,229	3,506	4,093	2.5

[a]Includes the 50 States and the District of Columbia. U.S. Territories are included in Australasia.

Notes: EE/FSU = Eastern Europe/Former Soviet Union.

Sources: **History:** Energy Information Administration (EIA), *International Energy Annual 1997*, DOE/EIA-0219(97) (Washington, DC, April 1999). **Projections:** EIA, *Annual Energy Outlook 2000*, DOE/EIA-0383(2000) (Washington, DC, December 1999), Table B19; and World Energy Projection System (2000).

Table B14. World Nuclear Generating Capacity by Region and Country, High Nuclear Case, 1997-2020
(Megawatts)

Region/Country	History		Projections			
	1997	1998	2005	2010	2015	2020
Industrialized Countries						
North America						
United States	99,046	97,133	95,106	90,248	79,735	71,117
Canada	11,994	10,298	12,358	14,054	14,054	14,054
Mexico	1,308	1,308	1,308	1,308	1,308	1,308
Industrialized Asia						
Japan	43,850	43,691	44,487	54,154	55,101	61,043
Western Europe						
Belgium	5,712	5,712	5,712	5,712	5,712	5,712
Finland.	2,560	2,656	2,656	3,956	3,956	3,468
France.	62,853	61,653	63,103	62,870	62,870	67,220
Germany.	22,282	22,282	22,282	21,942	20,135	19,364
Netherlands	449	449	449	449	449	449
Spain.	7,415	7,350	7,350	6,751	6,751	6,751
Sweden	10,040	10,040	9,440	9,440	8,390	7,790
Switzerland	3,079	3,079	3,194	3,194	3,194	3,194
United Kingdom	12,968	12,968	12,968	12,568	12,087	11,852
Total Industrialized	**283,556**	**278,619**	**280,413**	**286,646**	**273,742**	**273,322**
EE/FSU						
Eastern Europe						
Bulgaria	3,538	3,538	3,538	3,130	2,722	2,722
Czech Republic	1,648	1,648	3,472	3,472	3,472	3,472
Hungary	1,729	1,729	1,729	1,729	1,729	1,729
Romania.	650	650	1,300	1,300	1,300	1,300
Slovak Republic	1,632	2,020	2,408	2,000	1,592	1,592
Slovenia	632	632	632	632	632	632
Former Soviet Union						
Armenia	376	376	752	752	752	752
Belarus.	0	0	0	0	1,000	1,000
Kazakhstan	70	70	0	0	600	1,200
Lithuania.	2,370	2,370	2,370	2,370	1,185	0
Russia	19,843	19,843	22,668	23,268	21,929	20,607
Ukraine	13,765	13,765	13,090	13,090	13,690	14,290
Total EE/FSU	**46,253**	**46,641**	**51,959**	**51,743**	**50,603**	**49,296**

See notes at end of table.

Table B14. World Nuclear Generating Capacity by Region and Country, High Nuclear Case, 1997-2020 (Continued)

(Megawatts)

Region/Country	History		Projections			
	1997	1998	2005	2010	2015	2020
Developing Countries						
Developing Asia						
China	2,167	2,167	6,587	10,457	12,927	20,667
India	1,695	1,695	3,103	7,463	8,913	10,813
Indonesia	0	0	0	0	0	1,800
Korea, North	0	0	0	1,900	1,900	1,900
Korea, South	9,770	11,380	15,840	16,790	17,740	21,934
Pakistan	125	125	425	725	900	900
Philippines	0	0	0	0	0	900
Taiwan	4,884	4,884	7,384	7,384	9,884	9,884
Thailand	0	0	0	0	0	1,000
Vietnam	0	0	0	0	1,000	1,000
Central and South America						
Argentina	935	935	1,627	1,627	1,292	1,292
Brazil	626	626	1,855	3,100	3,100	3,100
Cuba	0	0	0	408	816	816
Middle East						
Iran	0	0	1,073	2,146	2,146	2,146
Egypt	0	0	0	0	600	600
Turkey	0	0	0	1,300	2,600	2,600
Africa						
South Africa	1,842	1,842	1,842	1,842	1,842	1,842
Total Developing	**22,044**	**23,654**	**39,736**	**55,142**	**65,660**	**83,194**
Total World	**351,853**	**348,914**	**372,108**	**393,531**	**390,005**	**405,812**

Sources: **History:** International Atomic Energy Agency, *Nuclear Power Reactors in the World 1998* (Vienna, Austria, April 1999). **Projections:** Energy Information Administration, Office of Coal, Nuclear, Electric and Alternate Fuels, based on detailed assessments of country-specific nuclear power plants.

Table B15. World Total Energy Consumption in Oil-Equivalent Units by Region, High Economic Growth Case, 1990-2020

(Million Tons Oil Equivalent)

Region/Country	History			Projections				Average Annual Percent Change, 1997-2020
	1990	1996	1997	2005	2010	2015	2020	
Industrialized Countries								
North America	2,517	2,821	2,835	3,300	3,567	3,811	4,045	1.6
United States[a]	2,116	2,366	2,373	2,728	2,922	3,095	3,260	1.4
Canada.	274	316	316	374	404	433	457	1.6
Mexico	126	139	145	199	240	282	328	3.6
Western Europe	1,510	1,594	1,614	1,815	1,928	2,039	2,157	1.3
United Kingdom	235	245	250	285	307	329	351	1.5
France	235	265	262	294	313	330	348	1.2
Germany	372	357	358	399	423	449	475	1.2
Italy.	168	183	183	202	217	228	241	1.2
Netherlands	83	93	94	108	113	119	126	1.2
Other Western Europe	418	452	467	528	555	584	617	1.2
Industrialized Asia.	579	678	684	762	836	879	920	1.3
Japan	456	541	536	588	648	677	704	1.2
Australasia	123	137	147	174	188	202	216	1.7
Total Industrialized	4,606	5,093	5,132	5,877	6,331	6,729	7,122	1.4
EE/FSU								
Former Soviet Union	1,538	1,053	1,029	1,170	1,307	1,425	1,623	2.0
Eastern Europe	386	320	314	402	463	517	607	2.9
Total EE/FSU	1,924	1,373	1,343	1,573	1,769	1,941	2,230	2.2
Developing Countries								
Developing Asia	1,296	1,851	1,897	2,880	3,629	4,339	5,427	4.7
China.	680	903	925	1,508	1,955	2,387	3,073	5.4
India	196	292	297	471	591	700	864	4.7
South Korea	93	178	188	269	321	370	431	3.7
Other Asia	327	479	487	633	761	881	1,059	3.4
Middle East.	330	432	451	645	798	939	1,150	4.2
Turkey	50	66	69	93	114	134	165	3.9
Other Middle East	280	366	382	552	684	805	985	4.2
Africa	235	279	288	389	453	539	653	3.6
Central and South America . . .	346	448	461	716	954	1,200	1,627	5.6
Brazil.	136	174	182	355	461	568	751	6.4
Other Central/South America . .	210	274	280	361	493	632	877	5.1
Total Developing	2,207	3,009	3,097	4,629	5,834	7,017	8,858	4.7
Total World	8,737	9,475	9,572	12,079	13,934	15,687	18,210	2.8

[a]Includes the 50 States and the District of Columbia. U.S. Territories are included in Australasia.

Notes: EE/FSU = Eastern Europe/Former Soviet Union.

Sources: **History:** Energy Information Administration (EIA), *International Energy Annual 1997*, DOE/EIA-0219(97) (Washington, DC, April 1999). **Projections:** EIA, *Annual Energy Outlook 2000*, DOE/EIA-0383(2000) (Washington, DC, December 1999), Table B1; and World Energy Projection System (2000).

Appendix C

Low Economic Growth Case Projections:
- World Energy Consumption
- Gross Domestic Product
- Carbon Emissions
- Nuclear Power Capacity

Table C1. World Total Energy Consumption by Region, Low Economic Growth Case, 1990-2020
(Quadrillion Btu)

Region/Country	History			Projections				Average Annual Percent Change, 1997-2020
	1990	1996	1997	2005	2010	2015	2020	
Industrialized Countries								
North America	99.9	112.0	112.5	123.2	129.1	133.7	136.8	0.9
United States[a]	84.0	93.9	94.2	103.2	107.7	111.1	113.3	0.8
Canada.	10.9	12.6	12.5	13.3	13.8	14.1	14.2	0.5
Mexico	5.0	5.5	5.8	6.7	7.7	8.5	9.3	2.1
Western Europe	59.9	63.3	64.0	66.4	68.0	69.5	70.9	0.4
United Kingdom	9.3	9.7	9.9	10.4	10.7	11.0	11.3	0.6
France	9.3	10.5	10.4	10.8	11.1	11.2	11.4	0.4
Germany	14.8	14.2	14.2	14.6	14.9	15.2	15.6	0.4
Italy.	6.7	7.3	7.3	7.5	7.7	7.8	8.0	0.4
Netherlands	3.3	3.7	3.7	3.9	4.0	4.1	4.1	0.4
Other Western Europe	16.6	17.9	18.5	19.3	19.7	20.0	20.4	0.4
Industrialized Asia.	23.0	26.9	27.1	27.7	28.6	29.1	29.4	0.4
Japan	18.1	21.4	21.3	21.6	22.3	22.6	22.8	0.3
Australasia	4.9	5.4	5.8	6.2	6.3	6.5	6.7	0.6
Total Industrialized	182.8	202.1	203.7	217.4	225.7	232.3	237.1	0.7
EE/FSU								
Former Soviet Union	61.0	41.8	40.8	42.4	45.3	47.8	52.0	1.1
Eastern Europe	15.3	12.7	12.5	13.5	14.2	14.9	15.8	1.0
Total EE/FSU	76.4	54.5	53.3	55.9	59.5	62.7	67.8	1.1
Developing Countries								
Developing Asia	51.4	73.5	75.3	89.1	99.8	105.9	117.9	2.0
China.	27.0	35.8	36.7	44.1	49.6	52.3	58.5	2.0
India	7.8	11.6	11.8	14.4	16.4	17.8	20.1	2.3
South Korea	3.7	7.1	7.5	8.3	9.2	9.8	10.7	1.6
Other Asia	13.0	19.0	19.3	22.4	24.6	26.0	28.7	1.7
Middle East.	13.1	17.1	17.9	19.8	21.8	23.2	25.8	1.6
Turkey	2.0	2.6	2.7	3.0	3.3	3.6	4.0	1.7
Other Middle East	11.1	14.5	15.2	16.8	18.5	19.6	21.8	1.6
Africa	9.3	11.1	11.4	12.7	13.7	14.6	16.1	1.5
Central and South America . . .	13.7	17.8	18.3	21.4	24.7	26.9	31.8	2.4
Brazil.	5.4	6.9	7.2	7.1	8.2	8.9	10.3	1.6
Other Central/South America . .	8.3	10.9	11.1	14.2	16.5	18.1	21.5	2.9
Total Developing	87.6	119.4	122.9	143.0	160.0	170.6	191.6	1.9
Total World	346.7	376.0	379.9	416.2	445.3	465.6	496.6	1.2

[a]Includes the 50 States and the District of Columbia. U.S. Territories are included in Australasia.

Notes: EE/FSU = Eastern Europe/Former Soviet Union. Energy totals include net imports of coal coke and electricity generated from biomass in the United States. Totals may not equal sum of components due to independent rounding. The electricity portion of the national fuel consumption values consists of generation for domestic use plus an adjustment for electricity trade based on a fuel's share of total generation in the exporting country.

Sources: **History:** Energy Information Administration (EIA), *International Energy Annual 1997*, DOE/EIA-0219(97) (Washington, DC, April 1999). **Projections:** EIA, *Annual Energy Outlook 2000*, DOE/EIA-0383(2000) (Washington, DC, December 1999), Table B1; and World Energy Projection System (2000).

Table C2. World Total Energy Consumption by Region and Fuel, Low Economic Growth Case, 1990-2020
(Quadrillion Btu)

Region/Country	History			Projections				Average Annual Percent Change, 1997-2020
	1990	1996	1997	2005	2010	2015	2020	
Industrialized Countries								
North America								
Oil	40.4	43.0	43.8	48.1	50.9	53.3	55.4	1.0
Natural Gas	22.7	26.8	26.9	29.0	31.9	34.6	36.0	1.3
Coal	20.5	22.5	23.0	25.9	26.1	26.5	26.9	0.7
Nuclear.	7.0	8.2	7.7	8.0	7.5	6.2	5.2	-1.7
Other	9.2	11.4	11.2	12.2	12.7	13.1	13.4	0.8
Total	99.9	112.0	112.5	123.2	129.1	133.7	136.8	0.9
Western Europe								
Oil	25.8	28.2	28.5	28.8	28.8	28.7	28.5	0.0
Natural Gas	10.0	13.5	13.4	16.6	18.6	20.9	23.4	2.5
Coal	12.4	8.4	8.7	7.3	6.8	6.4	5.8	-1.7
Nuclear.	7.4	8.7	8.8	8.5	8.1	7.3	6.6	-1.2
Other	4.4	4.5	4.7	5.2	5.7	6.1	6.6	1.5
Total	59.9	63.3	64.0	66.4	68.0	69.4	70.8	0.4
Industrialized Asia								
Oil	12.5	14.3	14.1	13.8	14.0	14.3	14.4	0.1
Natural Gas	2.9	3.6	3.5	4.2	4.4	4.7	5.0	1.5
Coal	4.2	4.7	5.0	5.2	5.1	5.1	5.1	0.1
Nuclear.	2.0	2.9	3.1	3.0	3.5	3.4	3.3	0.2
Other	1.4	1.4	1.4	1.6	1.6	1.6	1.6	0.6
Total	23.0	26.9	27.1	27.7	28.6	29.1	29.4	0.4
Total Industrialized								
Oil	78.7	85.5	86.4	90.7	93.7	96.3	98.3	0.6
Natural Gas	35.6	43.9	43.8	49.8	54.9	60.2	64.4	1.7
Coal	37.2	35.6	36.6	38.3	38.0	38.0	37.8	0.1
Nuclear.	16.3	19.9	19.6	19.5	19.1	17.0	15.0	-1.1
Other	15.0	17.3	17.2	19.0	20.0	20.8	21.6	1.0
Total	182.8	202.1	203.7	217.4	225.7	232.3	237.1	0.7
EE/FSU								
Oil	21.0	11.4	11.9	12.3	13.9	15.8	17.7	1.7
Natural Gas	28.8	23.9	22.4	26.0	29.6	32.2	37.0	2.2
Coal	20.8	13.6	13.3	11.8	10.1	8.6	6.8	-2.9
Nuclear.	2.9	2.8	2.8	2.9	2.8	2.9	2.4	-0.7
Other	2.8	2.9	2.8	2.9	3.1	3.3	4.0	1.5
Total	76.4	54.5	53.3	55.9	59.5	62.7	67.8	1.1
Developing Countries								
Developing Asia								
Oil	16.0	25.1	26.3	30.1	33.6	36.4	37.1	1.5
Natural Gas	3.2	5.5	6.0	10.1	13.3	16.0	21.0	5.6
Coal	28.1	37.5	37.7	41.6	44.9	45.4	51.2	1.3
Nuclear.	0.9	1.3	1.3	1.8	2.2	2.3	2.3	2.4
Other	3.2	4.0	3.9	5.5	5.8	5.9	6.2	2.1
Total	51.4	73.5	75.3	89.1	99.8	105.9	117.9	2.0

See notes at end of table.

Table C2. World Total Energy Consumption by Region and Fuel, Low Economic Growth Case, 1990-2020 (Continued)

(Quadrillion Btu)

Region/Country	History			Projections				Average Annual Percent Change, 1997-2020
	1990	1996	1997	2005	2010	2015	2020	
Developing Countries (Continued)								
Middle East								
Oil	8.1	9.9	10.1	10.1	11.3	12.6	14.3	1.5
Natural Gas	3.9	5.8	6.3	7.9	8.5	8.5	9.4	1.8
Coal	0.8	0.9	0.9	0.9	0.9	0.9	0.9	-0.3
Nuclear.	0.0	0.0	0.0	0.0	0.1	0.1	0.1	--
Other	0.4	0.6	0.6	0.9	1.0	1.0	1.1	3.0
Total	13.1	17.1	17.9	19.8	21.8	23.2	25.8	1.6
Africa								
Oil	4.2	4.9	5.1	6.0	7.1	8.3	9.7	2.8
Natural Gas	1.5	1.9	2.0	2.3	2.3	2.2	2.4	0.7
Coal	3.0	3.5	3.5	3.5	3.4	3.3	3.2	-0.4
Nuclear.	0.1	0.1	0.1	0.1	0.1	0.1	0.1	-1.5
Other	0.6	0.6	0.6	0.8	0.8	0.7	0.8	1.0
Total	9.3	11.1	11.4	12.7	13.7	14.6	16.1	1.5
Central and South America								
Oil	7.0	8.7	8.9	9.7	10.9	12.1	13.6	1.8
Natural Gas	2.2	3.1	3.2	5.6	7.8	9.0	12.2	5.9
Coal	0.6	0.7	0.6	0.6	0.6	0.6	0.7	0.3
Nuclear.	0.1	0.1	0.1	0.2	0.2	0.1	0.1	-0.7
Other	3.9	5.2	5.4	5.3	5.2	5.1	5.3	-0.1
Total	13.7	17.8	18.3	21.4	24.7	26.9	31.8	2.4
Total Developing Countries								
Oil	35.2	48.5	50.4	55.9	63.0	69.3	74.6	1.7
Natural Gas	10.8	16.4	17.6	25.8	31.8	35.7	45.0	4.2
Coal	32.4	42.6	42.9	46.7	49.9	50.2	55.9	1.2
Nuclear.	1.1	1.5	1.6	2.1	2.5	2.7	2.6	2.2
Other	8.1	10.4	10.5	12.5	12.8	12.7	13.5	1.1
Total	87.6	119.4	122.9	143.0	160.0	170.6	191.6	1.9
Total World								
Oil	134.9	145.4	148.7	158.9	170.6	181.4	190.6	1.1
Natural Gas	75.1	84.1	83.9	101.7	116.3	128.0	146.4	2.5
Coal	90.5	91.8	92.8	96.3	98.0	98.7	100.5	0.3
Nuclear.	20.4	24.2	24.0	24.4	24.4	22.5	20.0	-0.8
Other	25.9	30.5	30.6	34.4	35.8	36.8	39.0	1.1
Total	346.7	376.0	379.9	416.2	445.3	465.6	496.6	1.2

Notes: EE/FSU = Eastern Europe/Former Soviet Union. Energy totals include net imports of coal coke and electricity generated from biomass in the United States. Totals may not equal sum of components due to independent rounding. The electricity portion of the national fuel consumption values consists of generation for domestic use plus an adjustment for electricity trade based on a fuel's share of total generation in the exporting country.

Sources: **History:** Energy Information Administration (EIA), *International Energy Annual 1997*, DOE/EIA-0219(97) (Washington, DC, April 1999). **Projections:** EIA, *Annual Energy Outlook 2000*, DOE/EIA-0383(2000) (Washington, DC, December 1999), Table B1; and World Energy Projection System (2000).

Table C3. World Gross Domestic Product (GDP) by Region, Low Economic Growth Case, 1990-2020
(Billion 1997 Dollars)

Region/Country	History			Projections				Average Annual Percent Change, 1997-2020
	1990	1996	1997	2005	2010	2015	2020	
Industrialized Countries								
North America	**7,726**	**8,784**	**9,145**	**10,608**	**11,312**	**12,023**	**12,589**	**1.4**
United States[a]	6,846	7,804	8,111	9,344	9,875	10,423	10,839	1.3
Canada.	547	603	631	735	794	838	853	1.3
Mexico	333	377	403	529	643	762	897	3.5
Western Europe	**7,569**	**8,284**	**8,570**	**9,563**	**10,206**	**10,833**	**11,438**	**1.3**
United Kingdom	1,143	1,268	1,311	1,458	1,567	1,708	1,835	1.5
France	1,269	1,361	1,412	1,588	1,697	1,769	1,831	1.1
Germany	1,839	2,045	2,122	2,322	2,460	2,596	2,734	1.1
Italy.	1,060	1,128	1,159	1,269	1,355	1,437	1,516	1.2
Netherlands	306	351	363	410	438	468	499	1.4
Other Western Europe	1,951	2,131	2,202	2,516	2,688	2,855	3,024	1.4
Industrialized Asia.	**4,048**	**4,595**	**4,669**	**4,547**	**4,780**	**5,023**	**5,189**	**0.5**
Japan	3,667	4,142	4,199	4,000	4,199	4,408	4,544	0.3
Australasia	381	453	469	547	581	616	645	1.4
Total Industrialized	**19,343**	**21,663**	**22,384**	**24,718**	**26,298**	**27,880**	**29,216**	**1.2**
EE/FSU								
Former Soviet Union	1,049	595	600	598	715	862	1,043	2.4
Eastern Europe	358	347	368	454	521	597	676	2.7
Total EE/FSU	**1,407**	**942**	**968**	**1,052**	**1,236**	**1,459**	**1,719**	**2.5**
Developing Countries								
Developing Asia	**1,735**	**2,749**	**2,951**	**3,726**	**4,448**	**4,913**	**5,844**	**3.0**
China.	440	851	926	1,273	1,514	1,643	1,946	3.3
India	266	367	382	537	659	747	910	3.9
South Korea	273	419	476	587	721	819	989	3.2
Other Asia	755	1,111	1,166	1,329	1,554	1,704	1,999	2.4
Middle East.	**356**	**441**	**454**	**507**	**572**	**623**	**736**	**2.1**
Turkey	140	176	189	214	246	273	331	2.5
Other Middle East	216	266	265	293	326	350	405	1.9
Africa	**386**	**426**	**434**	**513**	**586**	**637**	**746**	**2.4**
Central and South America . . .	**1,127**	**1,409**	**1,481**	**1,712**	**1,982**	**2,167**	**2,570**	**2.4**
Brazil.	660	778	803	918	1,071	1,177	1,402	2.5
Other Central/South America . .	467	631	678	794	911	990	1,168	2.4
Total Developing	**3,604**	**5,026**	**5,320**	**6,459**	**7,589**	**8,340**	**9,897**	**2.7**
Total World	**24,353**	**27,631**	**28,672**	**32,229**	**35,123**	**37,679**	**40,832**	**1.5**

[a]Includes the 50 States and the District of Columbia. U.S. Territories are included in Australasia.

Notes: EE/FSU = Eastern Europe/Former Soviet Union. Totals may not equal sum of components due to independent rounding.

Sources: **History:** The WEFA Group, *World Economic Outlook: 20-Year Extension* (Eddystone, PA, April 1997). **Projections:** Standard & Poor's DRI, *World Economic Outlook*, Vol. 1 (Lexington, MA, 3rd Quarter 1999); Energy Information Administration (EIA), *Annual Energy Outlook 2000*, DOE/EIA-0383(2000) (Washington, DC, December 1999), Table B20; and EIA, World Energy Projection System (2000).

Table C4. World Oil Consumption by Region, Low Economic Growth Case, 1990-2020
(Million Barrels per Day)

Region/Country	History			Projections				Average Annual Percent Change, 1997-2020
	1990	1996	1997	2005	2010	2015	2020	
Industrialized Countries								
North America	**20.4**	**21.9**	**22.3**	**24.5**	**26.0**	**27.2**	**28.2**	**1.0**
United States[a]	17.0	18.3	18.6	20.5	21.5	22.4	23.0	0.9
Canada.	1.7	1.8	1.9	1.9	1.9	2.0	1.9	0.1
Mexico	1.7	1.8	1.9	2.1	2.5	2.8	3.3	2.5
Western Europe	**12.5**	**13.7**	**13.8**	**14.0**	**14.0**	**13.9**	**13.8**	**0.0**
United Kingdom	1.8	1.8	1.8	1.9	1.9	1.9	1.9	0.3
France	1.8	1.9	2.0	2.0	2.1	2.0	2.0	0.1
Germany	2.7	2.9	2.9	3.0	2.9	2.9	2.9	0.0
Italy.	1.9	2.1	2.0	2.1	2.1	2.1	2.1	0.0
Netherlands	0.7	0.8	0.8	0.8	0.8	0.8	0.8	0.2
Other Western Europe	3.6	4.1	4.3	4.2	4.2	4.1	4.1	-0.2
Industrialized Asia.	**6.2**	**7.1**	**6.9**	**6.8**	**6.9**	**7.1**	**7.1**	**0.1**
Japan	5.1	5.9	5.7	5.5	5.5	5.6	5.6	0.0
Australasia	1.0	1.2	1.2	1.3	1.4	1.4	1.5	0.8
Total Industrialized	**39.0**	**42.6**	**43.1**	**45.3**	**46.8**	**48.2**	**49.2**	**0.6**
EE/FSU								
Former Soviet Union	8.4	4.0	4.3	4.3	5.1	5.9	6.9	2.1
Eastern Europe	1.6	1.4	1.4	1.5	1.5	1.6	1.5	0.3
Total EE/FSU	**10.0**	**5.4**	**5.7**	**5.9**	**6.6**	**7.5**	**8.4**	**1.7**
Developing Countries								
Developing Asia	**7.6**	**12.0**	**12.6**	**14.4**	**16.1**	**17.4**	**17.8**	**1.5**
China.	2.3	3.5	3.8	4.4	5.1	5.8	5.7	1.8
India	1.2	1.7	1.8	2.3	2.6	2.8	3.0	2.3
South Korea	1.0	2.2	2.3	2.4	2.6	2.8	2.8	1.0
Other Asia	3.1	4.6	4.8	5.3	5.8	6.0	6.3	1.2
Middle East.	**3.9**	**4.7**	**4.8**	**4.8**	**5.4**	**6.0**	**6.8**	**1.5**
Turkey	0.5	0.6	0.6	0.8	0.8	0.9	1.0	2.0
Other Middle East	3.4	4.1	4.2	4.1	4.6	5.1	5.8	1.4
Africa	**2.1**	**2.4**	**2.5**	**2.9**	**3.4**	**4.0**	**4.6**	**2.8**
Central and South America . . .	**3.4**	**4.2**	**4.4**	**4.8**	**5.3**	**5.9**	**6.6**	**1.8**
Brazil.	1.3	1.7	1.8	1.8	2.1	2.5	2.9	2.2
Other Central/South America . .	2.1	2.5	2.6	2.9	3.2	3.4	3.7	1.6
Total Developing	**17.0**	**23.3**	**24.2**	**26.9**	**30.3**	**33.3**	**35.9**	**1.7**
Total World	**66.0**	**71.3**	**73.0**	**78.0**	**83.7**	**89.0**	**93.4**	**1.1**

[a]Includes the 50 States and the District of Columbia. U.S. Territories are included in Australasia.

Notes: EE/FSU = Eastern Europe/Former Soviet Union. Energy totals include net imports of coal coke and electricity generated from biomass in the United States. Totals may not equal sum of components due to independent rounding. The electricity portion of the national fuel consumption values consists of generation for domestic use plus an adjustment for electricity trade based on a fuel's share of total generation in the exporting country.

Sources: **History:** Energy Information Administration (EIA), *International Energy Annual 1997*, DOE/EIA-0219(97) (Washington, DC, April 1999). **Projections:** EIA, *Annual Energy Outlook 2000*, DOE/EIA-0383(2000) (Washington, DC, December 1999), Table B21; and World Energy Projection System (2000).

Table C5. World Natural Gas Consumption by Region, Low Economic Growth Case, 1990-2020
(Trillion Cubic Feet)

Region/Country	History			Projections				Average Annual Percent Change, 1997-2020
	1990	1996	1997	2005	2010	2015	2020	
Industrialized Countries								
North America	**22.0**	**26.1**	**26.1**	**28.2**	**30.9**	**33.6**	**34.9**	**1.3**
United States[a]	18.7	22.0	22.0	23.3	25.9	28.4	29.5	1.3
Canada.	2.4	3.0	3.0	3.2	3.3	3.5	3.8	1.0
Mexico	0.9	1.1	1.2	1.6	1.7	1.7	1.7	1.5
Western Europe	**10.1**	**13.7**	**13.5**	**16.7**	**18.6**	**20.9**	**23.4**	**2.4**
United Kingdom	2.1	3.2	3.2	3.9	4.3	4.7	5.3	2.2
France	1.0	1.3	1.3	1.6	1.9	2.1	2.6	3.1
Germany	2.7	3.6	3.4	4.1	4.7	5.4	6.0	2.5
Italy.	1.7	2.0	2.0	2.2	2.3	2.4	2.6	1.0
Netherlands	1.5	1.9	1.8	1.9	2.0	2.1	2.2	1.0
Other Western Europe	1.2	1.8	1.8	3.0	3.4	4.1	4.9	4.3
Industrialized Asia	**2.6**	**3.3**	**3.2**	**3.8**	**4.1**	**4.3**	**4.6**	**1.5**
Japan	1.9	2.4	2.3	2.8	2.9	3.1	3.3	1.5
Australasia	0.8	0.9	0.9	1.1	1.1	1.2	1.3	1.5
Total Industrialized	**34.8**	**43.0**	**42.9**	**48.7**	**53.7**	**58.8**	**63.0**	**1.7**
EE/FSU								
Former Soviet Union	25.0	20.8	19.7	22.0	24.2	25.6	29.0	1.7
Eastern Europe	3.1	2.9	2.6	3.9	5.4	6.5	7.9	4.9
Total EE/FSU	**28.1**	**23.7**	**22.3**	**25.9**	**29.5**	**32.1**	**36.9**	**2.2**
Developing Countries								
Developing Asia	**3.0**	**5.2**	**5.7**	**9.4**	**12.3**	**14.7**	**19.3**	**5.5**
China.	0.5	0.7	0.7	2.0	2.8	3.8	5.1	8.7
India	0.4	0.7	0.8	1.5	2.1	2.6	3.5	6.5
South Korea	0.1	0.4	0.5	0.7	1.0	1.2	1.9	5.7
Other Asia	1.9	3.4	3.6	5.2	6.3	7.1	8.8	4.0
Middle East.	**3.7**	**5.6**	**6.0**	**7.6**	**8.1**	**8.1**	**9.0**	**1.8**
Turkey	0.1	0.3	0.3	0.4	0.5	0.6	0.8	0.7
Other Middle East	3.6	5.3	5.7	7.1	7.6	7.6	8.2	5.9
Africa	**1.4**	**1.8**	**1.8**	**2.1**	**2.1**	**2.0**	**2.2**	**0.7**
Central and South America . . .	**2.0**	**2.8**	**2.9**	**5.0**	**7.0**	**8.1**	**11.0**	**5.9**
Brazil.	0.1	0.2	0.2	0.7	1.0	1.1	1.6	9.7
Other Central/South America . .	1.9	2.6	2.7	4.4	6.0	7.0	9.4	5.5
Total Developing	**10.1**	**15.3**	**16.4**	**24.0**	**29.5**	**33.0**	**41.5**	**4.1**
Total World	**73.0**	**82.1**	**81.6**	**98.6**	**112.7**	**123.9**	**141.4**	**2.4**

[a]Includes the 50 States and the District of Columbia. U.S. Territories are included in Australasia.

Notes: EE/FSU = Eastern Europe/Former Soviet Union. Energy totals include net imports of coal coke and electricity generated from biomass in the United States. Totals may not equal sum of components due to independent rounding. The electricity portion of the national fuel consumption values consists of generation for domestic use plus an adjustment for electricity trade based on a fuel's share of total generation in the exporting country.

Sources: **History:** Energy Information Administration (EIA), *International Energy Annual 1997*, DOE/EIA-0219(97) (Washington, DC, April 1999). **Projections:** EIA, *Annual Energy Outlook 2000*, DOE/EIA-0383(2000) (Washington, DC, December 1999), Table B13; and World Energy Projection System (2000).

Table C6. World Coal Consumption by Region, Low Economic Growth Case, 1990-2020
(Million Short Tons)

Region/Country	History			Projections				Average Annual Percent Change, 1997-2020
	1990	1996	1997	2005	2010	2015	2020	
Industrialized Countries								
North America	959	1,076	1,102	1,231	1,247	1,265	1,293	0.7
United States[a]	896	1,006	1,028	1,162	1,178	1,193	1,219	0.7
Canada.	55	59	62	56	54	57	60	-0.1
Mexico	9	12	12	14	15	15	14	0.6
Western Europe	896	579	583	497	470	441	406	**-1.6**
United Kingdom	119	70	78	60	55	49	38	-3.1
France	36	25	23	16	10	10	9	-4.1
Germany	528	279	277	249	239	227	215	-1.1
Italy.	25	19	17	17	16	15	14	-0.8
Netherlands	15	15	20	17	15	13	10	-2.8
Other Western Europe	173	170	168	138	135	127	121	-1.4
Industrialized Asia.	233	265	281	288	285	286	283	**0.0**
Japan	125	145	143	157	155	158	157	0.4
Australasia	108	119	138	132	130	128	126	-0.4
Total Industrialized	**2,088**	**1,920**	**1,966**	**2,016**	**2,001**	**1,992**	**1,983**	**0.0**
EE/FSU								
Former Soviet Union	848	471	453	392	341	295	236	-2.8
Eastern Europe	524	438	424	392	324	271	209	-3.0
Total EE/FSU	**1,373**	**909**	**877**	**784**	**666**	**567**	**446**	**-2.9**
Developing Countries								
Developing Asia	1,552	2,117	2,126	2,352	2,538	2,566	2,899	**1.4**
China.	1,124	1,514	1,532	1,733	1,882	1,904	2,198	1.6
India	242	352	342	370	396	398	416	0.8
South Korea	42	59	65	63	70	72	71	0.4
Other Asia	145	191	188	186	190	192	214	0.6
Middle East.	66	77	81	77	81	78	76	**-0.3**
Turkey	60	67	70	66	70	68	67	-0.2
Other Middle East	6	10	11	11	11	11	9	-0.8
Africa	152	179	184	183	177	169	167	**-0.4**
Central and South America . . .	27	39	35	35	35	34	37	**0.3**
Brazil.	17	21	21	17	17	17	19	-0.4
Other Central/South America . .	10	19	14	18	18	17	18	1.1
Total Developing	**1,797**	**2,412**	**2,426**	**2,647**	**2,831**	**2,847**	**3,179**	**1.2**
Total World	**5,258**	**5,240**	**5,269**	**5,447**	**5,498**	**5,405**	**5,607**	**0.3**

[a]Includes the 50 States and the District of Columbia. U.S. Territories are included in Australasia.

Notes: EE/FSU = Eastern Europe/Former Soviet Union. Totals may not equal sum of components due to independent rounding. The electricity portion of the national fuel consumption values consists of generation for domestic use plus an adjustment for electricity trade based on a fuel's share of total generation in the exporting country. To convert short tons to metric tons, divide each number in the table by 1.102.

Sources: **History:** Energy Information Administration (EIA), *International Energy Annual 1997*, DOE/EIA-0219(97) (Washington, DC, April 1999). **Projections:** EIA, *Annual Energy Outlook 2000*, DOE/EIA-0383(2000) (Washington, DC, December 1999), Table B16; and World Energy Projection System (2000).

Table C7. World Nuclear Energy Consumption by Region, Low Economic Growth Case, 1990-2020
(Billion Kilowatthours)

Region/Country	History			Projections				Average Annual Percent Change, 1997-2020
	1990	1996	1997	2005	2010	2015	2020	
Industrialized Countries								
North America	649	770	717	749	700	582	483	-1.7
United States[a]	577	675	629	674	627	511	428	-1.7
Canada.	69	88	78	67	65	64	49	-2.0
Mexico	3	7	10	7	7	7	7	-1.7
Western Europe	703	826	835	805	769	697	625	-1.2
United Kingdom	59	86	89	78	74	70	63	-1.5
France	298	377	374	383	383	378	356	-0.2
Germany	145	152	162	145	123	97	95	-2.3
Italy.	0	0	0	0	0	0	0	0.0
Netherlands	3	4	2	3	3	0	0	-100.0
Other Western Europe	198	207	207	197	185	152	111	-2.7
Industrialized Asia	192	287	306	291	340	331	320	0.2
Japan	192	287	306	291	340	331	320	0.2
Australasia	0	0	0	0	0	0	0	0.0
Total Industrialized	1,544	1,883	1,859	1,845	1,809	1,609	1,429	-1.1
EE/FSU								
Former Soviet Union	201	194	193	191	193	199	165	-0.7
Eastern Europe	54	60	63	68	63	60	52	-0.8
Total EE/FSU	256	254	256	259	257	260	217	-0.7
Developing Countries								
Developing Asia	88	128	130	181	213	228	223	2.4
China.	0	14	11	31	49	56	64	7.8
India	6	7	10	12	18	25	32	4.9
South Korea	50	70	73	89	92	95	85	0.6
Other Asia	32	37	35	50	53	52	42	0.8
Middle East.	0	0	0	0	8	13	13	0.0
Turkey	0	0	0	0	0	5	5	0.0
Other Middle East	0	0	0	0	8	8	8	0.0
Africa	8	12	13	10	10	9	9	-1.5
Central and South America . . .	9	9	10	13	14	13	9	-0.7
Brazil.	2	2	3	7	7	6	6	2.8
Other Central/South America . .	7	7	7	7	8	7	3	-3.7
Total Developing	105	149	153	205	245	264	253	2.2
Total World	1,905	2,286	2,268	2,309	2,310	2,133	1,899	-0.8

[a]Includes the 50 States and the District of Columbia. U.S. Territories are included in Australasia.

Notes: EE/FSU = Eastern Europe/Former Soviet Union. Totals may not equal sum of components due to independent rounding. The electricity portion of the national fuel consumption values consists of generation for domestic use plus an adjustment for electricity trade based on a fuel's share of total generation in the exporting country.

Sources: **History:** Energy Information Administration (EIA), *International Energy Annual 1997*, DOE/EIA-0219(97) (Washington, DC, April 1999). **Projections:** EIA, *Annual Energy Outlook 2000*, DOE/EIA-0383(2000) (Washington, DC, December 1999), Table B16; and World Energy Projection System (2000).

Table C8. World Consumption of Hydroelectricity and Other Renewable Energy by Region, Low Economic Growth Case, 1990-2020

(Quadrillion Btu)

Region/Country	History			Projections				Average Annual Percent Change, 1997-2020
	1990	1996	1997	2005	2010	2015	2020	
Industrialized Countries								
North America	**9.2**	**11.4**	**11.2**	**12.2**	**12.7**	**13.1**	**13.4**	**0.8**
United States[a]	5.8	7.3	7.2	7.5	7.6	7.8	8.1	0.6
Canada.	3.1	3.7	3.6	4.3	4.6	4.7	4.8	1.2
Mexico	0.3	0.4	0.4	0.4	0.5	0.5	0.5	1.5
Western Europe	**4.4**	**4.5**	**4.7**	**5.2**	**5.7**	**6.1**	**6.6**	**1.5**
United Kingdom	0.1	0.0	0.0	0.1	0.2	0.2	0.3	2.8
France	0.6	0.7	0.6	0.6	0.7	0.7	0.7	0.4
Germany	0.2	0.2	0.2	0.4	0.5	0.7	0.8	5.3
Italy.	0.4	0.5	0.5	0.6	0.6	0.7	0.7	1.5
Netherlands	0.0	0.0	0.0	0.1	0.1	0.1	0.2	15.0
Other Western Europe	3.2	3.1	3.2	3.5	3.6	3.7	3.9	0.9
Industrialized Asia.	**1.4**	**1.4**	**1.4**	**1.6**	**1.6**	**1.6**	**1.6**	**0.6**
Japan	1.0	0.9	1.0	1.1	1.1	1.1	1.1	0.5
Australasia	0.4	0.5	0.4	0.5	0.5	0.5	0.5	0.9
Total Industrialized	**15.0**	**17.3**	**17.2**	**19.0**	**20.0**	**20.8**	**21.6**	**1.0**
EE/FSU								
Former Soviet Union	2.4	2.3	2.2	2.3	2.3	2.3	2.4	0.3
Eastern Europe	0.4	0.6	0.6	0.6	0.8	1.0	1.6	4.2
Total EE/FSU	**2.8**	**2.9**	**2.8**	**2.9**	**3.1**	**3.3**	**4.0**	**1.5**
Developing Countries								
Developing Asia	**3.2**	**4.0**	**3.9**	**5.5**	**5.8**	**5.9**	**6.2**	**2.1**
China.	1.3	1.9	1.8	3.0	3.1	2.9	2.7	1.7
India	0.7	0.7	0.7	1.0	1.1	1.3	1.7	4.2
South Korea	0.1	0.1	0.1	0.1	0.1	0.1	0.2	6.5
Other Asia	1.1	1.3	1.3	1.4	1.5	1.5	1.4	0.8
Middle East.	**0.4**	**0.6**	**0.6**	**0.9**	**1.0**	**1.0**	**1.1**	**3.0**
Turkey	0.2	0.4	0.4	0.4	0.4	0.4	0.4	0.4
Other Middle East	0.1	0.2	0.2	0.5	0.6	0.6	0.7	6.4
Africa	**0.6**	**0.6**	**0.6**	**0.8**	**0.8**	**0.7**	**0.8**	**1.0**
Central and South America . . .	**3.9**	**5.2**	**5.4**	**5.3**	**5.2**	**5.1**	**5.3**	**-0.1**
Brazil.	2.2	2.8	3.0	2.4	2.3	2.2	2.1	-1.4
Other Central/South America . .	1.7	2.4	2.4	2.9	2.9	2.9	3.1	1.1
Total Developing	**8.1**	**10.4**	**10.5**	**12.5**	**12.8**	**12.7**	**13.5**	**1.1**
Total World	**25.9**	**30.5**	**30.6**	**34.4**	**35.9**	**36.8**	**39.0**	**1.1**

[a]Includes the 50 States and the District of Columbia. U.S. Territories are included in Australasia.

Notes: EE/FSU = Eastern Europe/Former Soviet Union. Totals may not equal sum of components due to independent rounding. The electricity portion of the national fuel consumption values consists of generation for domestic use plus an adjustment for electricity trade based on a fuel's share of total generation in the exporting country. U.S. totals include net electricity imports, methanol, and liquid hydrogen.

Sources: **History:** Energy Information Administration (EIA), *International Energy Annual 1997*, DOE/EIA-0219(97) (Washington, DC, April 1999). **Projections:** EIA, *Annual Energy Outlook 2000*, DOE/EIA-0383(2000) (Washington, DC, December 1999), Table B1; and World Energy Projection System (2000).

Table C9. World Net Electricity Consumption by Region, Low Economic Growth Case, 1990-2020
(Billion Kilowatthours)

Region/Country	History			Projections				Average Annual Percent Change, 1997-2020
	1990	1996	1997	2005	2010	2015	2020	
Industrialized Countries								
North America	**3,359**	**3,869**	**3,908**	**4,259**	**4,542**	**4,781**	**4,950**	**1.0**
United States[a]	2,817	3,247	3,279	3,586	3,805	3,979	4,087	1.0
Canada.	435	479	475	491	515	538	553	0.7
Mexico	107	144	154	183	222	264	310	3.1
Western Europe	**2,064**	**2,266**	**2,262**	**2,456**	**2,642**	**2,833**	**3,024**	**1.3**
United Kingdom	286	317	310	325	347	371	393	1.0
France	324	382	376	413	445	475	500	1.3
Germany	485	479	477	492	531	573	613	1.1
Italy.	222	249	257	294	325	356	388	1.8
Netherlands	71	85	88	99	107	115	124	1.5
Other Western Europe	675	755	754	833	887	943	1,006	1.3
Industrialized Asia	**930**	**1,090**	**1,117**	**1,252**	**1,331**	**1,390**	**1,440**	**1.1**
Japan	750	884	905	974	1,028	1,062	1,091	0.8
Australasia	180	207	212	278	303	328	348	2.2
Total Industrialized	**6,353**	**7,226**	**7,287**	**7,967**	**8,515**	**9,004**	**9,414**	**1.1**
EE/FSU								
Former Soviet Union	1,488	1,106	1,081	1,104	1,180	1,246	1,356	1.0
Eastern Europe	418	399	403	401	444	483	534	1.2
Total EE/FSU	**1,906**	**1,505**	**1,484**	**1,504**	**1,624**	**1,729**	**1,890**	**1.1**
Developing Countries								
Developing Asia	**1,260**	**2,006**	**2,103**	**2,608**	**3,075**	**3,440**	**4,031**	**2.9**
China.	551	923	956	1,220	1,489	1,710	2,073	3.4
India	257	379	397	528	635	721	849	3.4
South Korea	95	181	197	208	231	247	269	1.4
Other Asia	357	524	552	652	720	763	841	1.8
Middle East.	**272**	**394**	**412**	**443**	**500**	**544**	**615**	**1.8**
Turkey	51	85	94	148	208	243	331	5.6
Other Middle East	285	334	350	419	480	542	636	2.6
Africa	**448**	**594**	**624**	**771**	**896**	**971**	**1,152**	**2.7**
Central and South America . . .	**229**	**304**	**323**	**362**	**442**	**512**	**636**	**3.0**
Brazil.	219	291	301	409	454	460	516	2.4
Other Central/South America . .	2,265	3,328	3,489	4,241	4,951	5,497	6,435	2.7
Total Developing	**2,265**	**3,328**	**3,489**	**4,241**	**4,951**	**5,497**	**6,435**	**2.7**
Total World	**10,524**	**12,059**	**12,260**	**13,712**	**15,089**	**16,231**	**17,739**	**1.6**

[a]Includes the 50 States and the District of Columbia. U.S. Territories are included in Australasia.

Notes: EE/FSU = Eastern Europe/Former Soviet Union. Electricity consumption equals generation plus imports minus exports minus distribution losses.

Sources: **History:** Energy Information Administration (EIA), *International Energy Annual 1997*, DOE/EIA-0219(97) (Washington, DC, April 1999). **Projections:** EIA, *Annual Energy Outlook 2000*, DOE/EIA-0383(2000) (Washington, DC, December 1999), Table B8; and World Energy Projection System (2000).

Table C10. World Carbon Emissions by Region, Low Economic Growth Case, 1990-2020
(Million Metric Tons)

Region/Country	History			Projections				Average Annual Percent Change, 1997-2020
	1990	1996	1997	2005	2010	2015	2020	
Industrialized Countries								
North America	1,553	1,689	1,716	1,903	2,000	2,093	2,161	1.0
United States[a]	1,346	1,461	1,480	1,649	1,728	1,801	1,851	1.0
Canada.	127	139	142	145	147	152	155	0.4
Mexico	81	89	94	110	125	140	155	2.2
Western Europe	934	910	918	938	957	978	997	0.4
United Kingdom	166	153	156	159	164	168	170	0.4
France	103	103	102	109	112	116	122	0.8
Germany	267	234	234	237	241	246	248	0.2
Italy.	113	116	116	119	121	122	122	0.2
Netherlands	60	62	64	65	65	65	65	0.1
Other Western Europe	224	242	246	249	254	261	269	0.4
Industrialized Asia.	364	402	405	415	421	430	436	0.3
Japan	274	304	297	303	306	313	317	0.4
Australasia	90	98	108	112	114	117	119	0.4
Total Industrialized	2,851	3,000	3,039	3,257	3,377	3,501	3,594	0.7
EE/FSU								
Former Soviet Union	1,034	661	646	657	697	732	794	0.9
Eastern Europe	303	236	231	242	241	240	237	0.1
Total EE/FSU	1,337	897	878	900	937	972	1,031	0.7
Developing Countries								
Developing Asia	1,067	1,487	1,522	1,746	1,939	2,041	2,274	1.8
China.	620	801	822	951	1,061	1,113	1,256	1.9
India	153	234	236	277	310	327	358	1.8
South Korea	61	109	116	125	139	148	161	1.4
Other Asia	232	343	348	394	470	452	499	1.6
Middle East.	229	284	297	319	351	374	416	1.5
Turkey	35	42	45	49	54	57	64	1.6
Other Middle East	194	242	252	270	297	317	353	1.5
Africa	180	208	214	235	253	270	298	1.4
Central and South America . . .	174	220	225	274	327	366	440	2.9
Brazil.	57	73	76	84	102	116	141	2.7
Other Central/South America . .	117	147	149	190	225	249	298	3.1
Total Developing	1,649	2,200	2,258	2,575	2,871	3,051	3,428	1.8
Total World	5,836	6,097	6,175	6,731	7,185	7,523	8,053	1.2

[a]Includes the 50 States and the District of Columbia. U.S. Territories are included in Australasia.
Notes: EE/FSU = Eastern Europe/Former Soviet Union. The U.S. numbers include carbon emissions attributable to renewable energy sources.
Sources: **History:** Energy Information Administration (EIA), *International Energy Annual 1997*, DOE/EIA-0219(97) (Washington, DC, April 1999). **Projections:** EIA, *Annual Energy Outlook 2000*, DOE/EIA-0383(2000) (Washington, DC, December 1999), Table B19; and World Energy Projection System (2000).

Table C11. World Carbon Emissions from Oil Use by Region, Low Economic Growth Case, 1990-2020
(Million Metric Tons)

Region/Country	History			Projections				Average Annual Percent Change, 1997-2020
	1990	1996	1997	2005	2010	2015	2020	
Industrialized Countries								
North America	714	749	762	827	876	919	957	1.0
United States[a]	591	621	627	681	716	745	767	0.9
Canada.	61	63	65	67	68	68	67	0.1
Mexico	61	66	70	79	92	106	123	2.5
Western Europe	473	500	506	514	517	516	514	0.1
United Kingdom	66	66	64	67	68	69	70	0.3
France	67	69	70	76	78	79	79	0.5
Germany	98	106	106	108	107	106	105	0.0
Italy.	74	76	76	77	77	77	76	0.0
Netherlands	29	28	29	29	30	30	31	0.2
Other Western Europe	140	155	161	157	157	155	153	-0.2
Industrialized Asia	218	234	230	227	230	235	237	0.1
Japan	178	191	186	179	180	183	183	0.0
Australasia	40	44	45	48	50	52	54	0.8
Total Industrialized	1,405	1,484	1,499	1,568	1,623	1,671	1,708	0.6
EE/FSU								
Former Soviet Union	332	159	168	171	200	234	271	2.1
Eastern Europe	68	52	53	57	58	58	57	0.3
Total EE/FSU	400	210	221	228	258	293	328	1.7
Developing Countries								
Developing Asia	308	459	480	549	613	664	678	1.5
China.	98	138	148	171	201	228	223	1.8
India	45	61	65	83	93	102	110	2.3
South Korea	38	71	73	79	86	90	92	1.0
Other Asia	127	188	193	216	233	243	253	1.2
Middle East.	153	179	183	183	205	229	258	1.5
Turkey	17	22	22	27	29	32	36	2.0
Other Middle East	136	156	160	156	176	196	223	1.4
Africa	83	93	97	114	136	157	184	2.8
Central and South America . . .	127	158	163	178	199	221	248	1.8
Brazil.	46	62	65	66	78	91	106	2.2
Other Central/South America . .	81	96	98	112	121	130	141	1.6
Total Developing	671	889	923	1,024	1,153	1,270	1,368	1.7
Total World	2,476	2,583	2,643	2,820	3,035	3,234	3,404	1.1

[a]Includes the 50 States and the District of Columbia. U.S. Territories are included in Australasia.
Notes: EE/FSU = Eastern Europe/Former Soviet Union.
Sources: **History:** Energy Information Administration (EIA), *International Energy Annual 1997*, DOE/EIA-0219(97) (Washington, DC, April 1999). **Projections:** EIA, *Annual Energy Outlook 2000*, DOE/EIA-0383(2000) (Washington, DC, December 1999), Table B19; and World Energy Projection System (2000).

Table C12. World Carbon Emissions from Natural Gas Use by Region, Low Economic Growth Case, 1990-2020

(Million Metric Tons)

Region/Country	History			Projections				Average Annual Percent Change, 1997-2020
	1990	1996	1997	2005	2010	2015	2020	
Industrialized Countries								
North America	323	381	382	416	457	496	516	1.3
United States[a]	273	319	319	343	381	417	434	1.3
Canada.	35	44	44	48	49	52	55	1.0
Mexico	15	18	19	25	26	27	27	1.5
Western Europe	143	194	193	240	268	301	337	2.2
United Kingdom	32	48	49	59	66	72	80	2.2
France	16	20	19	24	28	32	39	3.1
Germany	32	46	44	53	61	70	77	2.5
Italy.	25	29	30	32	34	36	38	1.0
Netherlands	20	24	23	25	26	28	28	1.0
Other Western Europe	18	27	29	47	53	64	76	4.3
Industrialized Asia	41	51	51	60	64	67	72	1.5
Japan	29	38	37	44	46	49	53	1.5
Australasia	12	14	14	16	18	19	20	1.5
Total Industrialized	507	626	626	715	788	864	925	1.7
EE/FSU								
Former Soviet Union	369	304	287	321	352	373	423	1.7
Eastern Europe	45	40	36	54	74	90	109	4.9
Total EE/FSU	414	344	323	374	426	463	533	2.2
Developing Countries								
Developing Asia	45	80	87	145	192	230	303	5.6
China.	8	11	12	32	47	64	85	8.7
India	7	12	14	24	35	43	58	6.5
South Korea	2	7	8	12	16	19	31	5.7
Other Asia	29	50	52	77	93	104	129	4.0
Middle East	56	84	90	114	122	123	136	1.8
Turkey	2	4	5	6	8	8	12	3.6
Other Middle East	54	80	85	107	115	114	124	1.6
Africa	22	28	29	33	33	32	34	0.7
Central and South America . . .	32	44	47	80	112	129	175	5.9
Brazil.	2	3	3	11	17	19	28	9.7
Other Central/South America . .	30	41	43	69	94	110	148	5.5
Total Developing	155	236	253	372	458	514	648	4.2
Total World	1,077	1,206	1,202	1,462	1,673	1,841	2,106	2.5

[a]Includes the 50 States and the District of Columbia. U.S. Territories are included in Australasia.

Notes: EE/FSU = Eastern Europe/Former Soviet Union.

Sources: **History:** Energy Information Administration (EIA), *International Energy Annual 1997*, DOE/EIA-0219(97) (Washington, DC, April 1999). **Projections:** EIA, *Annual Energy Outlook 2000*, DOE/EIA-0383(2000) (Washington, DC, December 1999), Table B19; and World Energy Projection System (2000).

Table C13. World Carbon Emissions from Coal Use by Region, Low Economic Growth Case, 1990-2020
(Million Metric Tons)

Region/Country	History			Projections				Average Annual Percent Change, 1997-2020
	1990	1996	1997	2005	2010	2015	2020	
Industrialized Countries								
North America	517	558	572	659	665	675	686	0.8
United States[a]	481	521	533	623	629	638	647	0.8
Canada.	31	32	34	31	30	31	33	-0.1
Mexico	4	5	5	6	6	6	6	0.6
Western Europe	317	215	219	184	172	161	146	-1.7
United Kingdom	68	38	43	33	30	27	21	-3.1
France	20	14	13	9	6	6	5	-4.1
Germany	137	82	85	76	73	70	66	-1.1
Italy.	15	11	10	10	10	9	8	-0.8
Netherlands	11	10	12	11	9	8	6	-2.8
Other Western Europe	66	60	56	46	45	42	40	-1.4
Industrialized Asia.	105	116	123	128	127	127	126	0.1
Japan	66	75	74	81	80	81	81	0.4
Australasia	39	41	49	47	46	46	45	-0.4
Total Industrialized	939	890	914	972	964	964	958	0.2
EE/FSU								
Former Soviet Union	333	198	192	166	144	125	100	-2.8
Eastern Europe	189	145	142	132	109	91	70	-3.0
Total EE/FSU	523	343	334	297	253	216	170	-2.9
Developing Countries								
Developing Asia	713	949	954	1,052	1,135	1,147	1,298	1.3
China.	514	652	661	748	812	821	948	1.6
India	101	161	157	169	181	182	190	0.8
South Korea	21	31	35	34	37	38	38	0.4
Other Asia	76	104	102	101	104	105	117	0.6
Middle East.	20	21	24	22	24	23	22	-0.3
Turkey	16	15	17	16	17	17	16	-0.2
Other Middle East	4	6	7	6	6	6	6	-0.5
Africa	74	86	88	88	85	81	80	-0.4
Central and South America . . .	15	18	16	16	16	16	17	0.3
Brazil.	9	8	8	7	7	7	7	-0.4
Other Central/South America . .	6	10	8	9	9	9	9	0.8
Total Developing	822	1,076	1,082	1,178	1,259	1,266	1,412	1.2
Total World	2,283	2,308	2,330	2,447	2,476	2,446	2,541	0.4

[a]Includes the 50 States and the District of Columbia. U.S. Territories are included in Australasia.
Notes: EE/FSU = Eastern Europe/Former Soviet Union.
Sources: **History:** Energy Information Administration (EIA), *International Energy Annual 1997*, DOE/EIA-0219(97) (Washington, DC, April 1999). **Projections:** EIA, *Annual Energy Outlook 2000*, DOE/EIA-0383(2000) (Washington, DC, December 1999), Table B19; and World Energy Projection System (2000).

Table C14. World Nuclear Generating Capacity by Region and Country, Low Nuclear Case, 1997-2020
(Megawatts)

Region/Country	History		Projections			
	1997	1998	2005	2010	2015	2020
Industrialized Countries						
North America						
United States	99,046	97,133	85,371	72,487	53,508	43,667
Canada	11,994	10,298	10,298	10,298	6,276	2,643
Mexico	1,308	1,308	1,308	1,308	1,308	654
Industrialized Asia						
Japan	43,850	43,691	43,826	39,629	30,199	28,065
Western Europe						
Belgium	5,712	5,712	5,712	3,966	3,966	2,000
Finland	2,560	2,656	2,656	2,168	840	0
France	62,853	61,653	62,870	62,870	55,655	51,130
Germany	22,282	22,282	18,145	16,120	13,075	5,250
Netherlands	449	449	0	0	0	0
Spain	7,415	7,350	6,751	6,751	4,854	966
Sweden	10,040	10,040	7,790	4,202	0	0
Switzerland	3,079	3,079	3,194	2,829	2,115	2,115
United Kingdom	12,968	12,968	12,322	11,412	8,148	6,603
Total Industrialized	**283,556**	**278,619**	**260,243**	**234,040**	**179,944**	**143,093**
EE/FSU						
Eastern Europe						
Bulgaria	3,538	3,538	2,722	1,906	1,906	1,906
Czech Republic	1,648	1,648	3,472	3,472	3,472	1,824
Hungary	1,729	1,729	1,729	1,729	866	0
Romania	650	650	650	1,300	1,300	1,300
Slovak Republic	1,632	2,020	1,592	1,592	776	776
Slovenia	632	632	632	632	0	0
Former Soviet Union						
Armenia	376	376	376	0	0	0
Kazakhstan	70	70	0	0	0	0
Lithuania	2,370	2,370	1,185	0	0	0
Russia	19,843	19,843	19,204	17,357	10,375	4,725
Ukraine	13,765	13,765	11,910	12,727	8,550	3,800
Total EE/FSU	**46,253**	**46,641**	**43,472**	**40,715**	**27,245**	**14,331**

See notes at end of table.

Table C14. World Nuclear Generating Capacity by Region and Country, Low Nuclear Case, 1997-2020 (Continued)

(Megawatts)

Region/Country	History		Projections			
	1997	1998	2005	2010	2015	2020
Developing Countries						
Developing Asia						
China	2,167	2,167	6,587	8,587	8,587	8,587
India	1,695	1,695	2,653	4,826	4,516	4,516
Korea, South.	9,770	11,380	14,890	14,334	15,000	13,210
Pakistan	125	125	300	300	300	300
Taiwan.	4,884	4,884	6,134	7,384	7,384	6,176
Central and South America						
Argentina	935	935	600	1,292	1,292	692
Brazil.	626	626	1,855	1,855	1,229	1,229
Middle East						
Iran.	0	0	0	1,073	1,073	1,073
Africa						
South Africa	1,842	1,842	1,842	1,842	0	0
Total Developing	**22,044**	**23,654**	**34,861**	**41,493**	**39,381**	**35,783**
Total World	**351,853**	**348,914**	**338,576**	**316,248**	**246,570**	**193,207**

Sources: **History:** International Atomic Energy Agency, *Nuclear Power Reactors in the World 1998* (Vienna, Austria, April 1999). **Projections:** Energy Information Administration, Office of Coal, Nuclear, Electric and Alternate Fuels, based on detailed assessments of country-specific nuclear power plants.

Table C15. World Total Energy Consumption in Oil-Equivalent Units by Region, Low Economic Growth Case, 1990-2020
(Million Tons Oil Equivalent)

Region/Country	History			Projections				Average Annual Percent Change, 1997-2020
	1990	1996	1997	2005	2010	2015	2020	
Industrialized Countries								
North America	2,517	2,821	2,835	3,106	3,254	3,370	3,448	0.9
United States[a]	2,116	2,366	2,373	2,601	2,714	2,800	2,855	0.8
Canada.	274	316	316	335	347	356	359	0.5
Mexico	126	139	145	170	193	214	235	2.1
Western Europe	1,510	1,594	1,614	1,673	1,714	1,750	1,786	0.4
United Kingdom	235	245	250	261	269	278	286	0.6
France	235	265	262	272	279	283	287	0.4
Germany	372	357	358	368	376	384	392	0.4
Italy.	168	183	183	189	194	198	201	0.4
Netherlands	83	93	94	98	100	102	104	0.4
Other Western Europe	418	452	467	486	495	505	515	0.4
Industrialized Asia	579	678	684	699	721	733	742	0.4
Japan	456	541	536	543	561	569	574	0.3
Australasia	123	137	147	155	160	164	168	0.6
Total Industrialized	4,606	5,093	5,132	5,477	5,688	5,853	5,976	0.7
EE/FSU								
Former Soviet Union	1,538	1,053	1,029	1,068	1,142	1,205	1,311	1.1
Eastern Europe	386	320	314	340	359	374	397	1.0
Total EE/FSU	1,924	1,373	1,343	1,408	1,500	1,580	1,708	1.1
Developing Countries								
Developing Asia	1,296	1,851	1,897	2,246	2,514	2,669	2,970	2.0
China.	680	903	925	1,112	1,249	1,318	1,473	2.0
India	196	292	297	362	414	448	506	2.3
South Korea	93	178	188	208	231	247	269	1.6
Other Asia	327	479	487	563	620	656	723	1.7
Middle East.	330	432	451	499	550	584	651	1.6
Turkey	50	66	69	76	84	90	101	1.7
Other Middle East	280	366	382	424	466	495	550	1.6
Africa	235	279	288	320	345	367	406	1.5
Central and South America . . .	346	448	461	538	622	679	801	2.4
Brazil.	136	174	182	179	205	223	260	1.6
Other Central/South America . .	210	274	280	359	417	456	541	2.9
Total Developing	2,207	3,009	3,097	3,604	4,032	4,300	4,829	1.9
Total World	8,737	9,475	9,572	10,489	11,220	11,733	12,513	1.2

[a]Includes the 50 States and the District of Columbia. U.S. Territories are included in Australasia.
Notes: EE/FSU = Eastern Europe/Former Soviet Union.
Sources: **History:** Energy Information Administration (EIA), *International Energy Annual 1997*, DOE/EIA-0219(97) (Washington, DC, April 1999). **Projections:** EIA, *Annual Energy Outlook 2000*, DOE/EIA-0383(2000) (Washington, DC, December 1999), Table B1; and World Energy Projection System (2000).

Projections of Oil Production Capacity and Oil Production in Five Cases:

- Reference
- High World Oil Price
- Low World Oil Price
- High Non-OPEC Supply
- Low Non-OPEC Supply

Table D1. World Oil Production Capacity by Region and Country, Reference Case, 1990-2020
(Million Barrels per Day)

Region/Country	History (Estimates)		Projections			
	1990	1998	2005	2010	2015	2020
OPEC						
Persian Gulf						
Iran	3.2	3.9	4.3	4.5	4.9	5.5
Iraq	2.2	2.8	3.2	4.2	5.0	6.2
Kuwait	1.7	2.6	3.1	3.8	4.5	5.2
Qatar.	0.5	0.6	0.6	0.6	0.6	0.7
Saudi Arabia.	8.6	11.4	13.6	14.7	17.7	22.1
United Arab Emitates	2.5	2.7	3.2	3.6	4.2	5.1
Total Persian Gulf.	**18.7**	**24.0**	**28.0**	**31.4**	**36.9**	**44.8**
Other OPEC						
Algeria.	1.3	1.4	2.0	2.2	2.1	2.0
Indonesia	1.5	1.7	1.5	1.5	1.4	1.3
Libya.	1.5	1.5	1.6	1.7	1.6	1.5
Nigeria.	1.8	2.2	2.9	3.2	3.3	3.1
Venezuela[a]	2.4	3.4	4.6	5.1	5.5	6.0
Total Other OPEC.	**8.5**	**10.2**	**12.6**	**13.7**	**13.9**	**14.0**
Total OPEC	**27.2**	**34.2**	**40.6**	**45.1**	**50.8**	**58.8**
Non-OPEC						
Industrialized						
United States	9.7	9.3	8.5	8.6	8.9	9.1
Canada	2.0	2.7	3.0	3.2	3.4	3.4
Mexico.	3.0	3.5	3.7	4.0	3.9	3.9
Australia	0.7	0.8	0.8	0.8	0.8	0.7
North Sea	4.2	6.3	7.1	6.8	6.4	5.8
Other.	0.5	0.7	0.8	0.8	0.7	0.7
Total Industrialized	**20.1**	**23.3**	**23.9**	**24.2**	**24.1**	**23.6**
Eurasia						
China	2.8	3.2	3.3	3.5	3.6	3.6
Former Soviet Union.	11.4	7.2	7.6	10.1	12.1	13.1
Eastern Europe	0.3	0.3	0.3	0.4	0.4	0.5
Total Eurasia	**14.5**	**10.7**	**11.2**	**14.0**	**16.1**	**17.2**
Other Non-OPEC						
Central and South America	2.4	3.6	4.2	4.4	4.8	5.0
Middle East	1.4	1.9	2.2	2.3	2.1	2.0
Africa	2.2	2.8	3.1	3.7	4.5	5.5
Asia	1.7	2.2	2.5	2.9	3.2	3.3
Total Other Non-OPEC	**7.7**	**10.5**	**12.0**	**13.3**	**14.6**	**15.8**
Total Non-OPEC.	**42.3**	**44.5**	**47.1**	**51.5**	**54.8**	**56.6**
Total World	**69.5**	**78.7**	**87.7**	**96.6**	**105.6**	**115.4**

[a]These estimates were made prior to recent statements by Venezuela's current regime indicating a potential unwillingness to support the outside investment activity required for the substantial capacity expansion shown in this table.
Note: OPEC = Organization of Petroleum Exporting Countries.
Sources: **History:** Energy Information Administration (EIA), Energy Markets and Contingency Information Division. **Projections:** EIA, Office of Integrated Analysis and Forecasting, World Energy Projection System and "DESTINY" International Energy Forecast Software (Dallas, TX: Petroconsultants, Fourth Quarter 1999).

Table D2. World Oil Production Capacity by Region and Country, High Oil Price Case, 1990-2020
(Million Barrels per Day)

Region/Country	History (Estimates)		Projections			
	1990	1998	2005	2010	2015	2020
OPEC						
Persian Gulf						
Iran	3.2	3.9	4.1	4.2	4.5	4.9
Iraq	2.2	2.8	3.0	3.5	4.1	5.2
Kuwait	1.7	2.6	2.9	3.4	3.8	4.3
Qatar.	0.5	0.6	0.6	0.6	0.6	0.7
Saudi Arabia.	8.6	11.4	12.4	13.7	15.8	19.9
United Arab Emitates	2.5	2.7	3.0	3.1	3.3	3.7
Total Persian Gulf.	**18.7**	**24.0**	**26.0**	**28.5**	**32.1**	**38.7**
Other OPEC						
Algeria	1.3	1.4	2.0	2.0	2.1	2.0
Indonesia	1.5	1.7	1.5	1.5	1.4	1.3
Libya	1.5	1.5	1.6	1.6	1.6	1.5
Nigeria.	1.8	2.2	2.7	2.8	3.1	3.0
Venezuela[a]	2.4	3.4	4.5	5.0	5.4	5.8
Total Other OPEC.	**8.5**	**10.2**	**12.3**	**12.9**	**13.6**	**14.0**
Total OPEC	**27.2**	**34.2**	**38.3**	**41.4**	**45.7**	**52.7**
Non-OPEC						
Industrialized						
United States	9.7	9.3	8.7	9.1	9.6	10.0
Canada	2.0	2.7	3.1	3.3	3.4	3.5
Mexico.	3.0	3.5	3.9	4.1	4.0	3.9
Australia	0.7	0.8	0.8	0.8	0.9	0.8
North Sea	4.2	6.3	7.1	7.0	6.5	5.9
Other.	0.5	0.7	0.7	0.8	0.7	0.7
Total Industrialized	**20.1**	**23.3**	**24.3**	**25.1**	**25.1**	**24.8**
Eurasia						
China	2.8	3.2	3.3	3.6	3.7	3.7
Former Soviet Union.	11.4	7.2	7.6	10.3	12.4	13.3
Eastern Europe	0.3	0.3	0.3	0.4	0.4	0.5
Total Eurasia	**14.5**	**10.7**	**11.2**	**14.3**	**16.5**	**17.5**
Other Non-OPEC						
Central and South America	2.4	3.6	4.2	4.5	4.9	5.1
Middle East	1.4	1.9	2.2	2.2	2.2	2.1
Africa	2.2	2.8	3.2	3.5	4.5	5.6
Asia	1.7	2.2	2.6	3.0	3.3	3.4
Total Other Non-OPEC	**7.7**	**10.5**	**12.2**	**13.2**	**14.9**	**16.2**
Total Non-OPEC.	**42.3**	**44.5**	**47.7**	**52.6**	**56.5**	**58.5**
Total World	**69.5**	**78.7**	**86.0**	**94.0**	**102.2**	**111.2**

[a]These estimates were made prior to recent statements by Venezuela's current regime indicating a potential unwillingness to support the outside investment activity required for the substantial capacity expansion shown in this table.

Note: OPEC = Organization of Petroleum Exporting Countries.

Sources: **History:** Energy Information Administration (EIA), Energy Markets and Contingency Information Division. **Projections:** EIA, Office of Integrated Analysis and Forecasting, World Energy Projection System and "DESTINY" International Energy Forecast Software (Dallas, TX: Petroconsultants, Fourth Quarter 1999).

Table D3. World Oil Production Capacity by Region and Country, Low Oil Price Case, 1990-2020
(Million Barrels per Day)

Region/Country	History (Estimates)		Projections			
	1990	1998	2005	2010	2015	2020
OPEC						
Persian Gulf						
Iran	3.2	3.9	4.7	5.3	5.6	6.5
Iraq	2.2	2.8	3.5	4.4	5.3	6.9
Kuwait	1.7	2.6	3.3	4.0	4.8	5.5
Qatar.	0.5	0.6	0.6	0.7	0.8	0.7
Saudi Arabia.	8.6	11.4	14.9	17.1	21.6	27.4
United Arab Emitates	2.5	2.7	3.4	4.0	4.6	5.7
Total Persian Gulf.	18.7	24.0	30.4	35.5	42.7	52.7
Other OPEC						
Algeria	1.3	1.4	2.3	2.3	2.2	2.1
Indonesia	1.5	1.7	1.7	1.6	1.5	1.4
Libya	1.5	1.5	1.8	1.9	1.8	1.6
Nigeria	1.8	2.2	3.0	3.5	3.6	3.8
Venezuela[a]	2.4	3.4	5.4	6.3	6.8	6.7
Total Other OPEC.	8.5	10.2	14.2	15.6	15.8	15.6
Total OPEC	27.2	34.2	44.6	51.1	58.5	68.3
Non-OPEC						
Industrialized						
United States	9.7	9.3	8.0	8.1	8.2	8.3
Canada	2.0	2.7	3.0	3.2	3.3	3.3
Mexico.	3.0	3.5	3.7	3.8	3.8	3.7
Australia	0.7	0.8	0.8	0.8	0.8	0.7
North Sea	4.2	6.3	7.1	6.7	6.4	5.8
Other.	0.5	0.7	0.7	0.7	0.6	0.6
Total Industrialized	20.1	23.3	23.3	23.3	23.1	22.4
Eurasia						
China	2.8	3.2	3.3	3.4	3.5	3.5
Former Soviet Union.	11.4	7.2	7.5	10.0	11.7	12.7
Eastern Europe	0.3	0.3	0.3	0.4	0.4	0.4
Total Eurasia	14.5	10.7	11.1	13.8	15.6	16.6
Other Non-OPEC						
Central and South America	2.4	3.6	4.1	4.3	4.6	4.8
Middle East	1.4	1.9	2.2	2.2	2.1	2.0
Africa	2.2	2.8	2.8	3.5	4.4	5.3
Asia	1.7	2.2	2.4	2.8	3.1	3.2
Total Other Non-OPEC	7.7	10.5	11.5	12.8	14.2	15.3
Total Non-OPEC.	42.3	44.5	45.9	49.9	52.9	54.3
Total World	69.5	78.7	90.5	101.0	111.4	122.6

[a]These estimates were made prior to recent statements by Venezuela's current regime indicating a potential unwillingness to support the outside investment activity required for the substantial capacity expansion shown in this table.

Note: OPEC = Organization of Petroleum Exporting Countries.

Sources: **History:** Energy Information Administration (EIA), Energy Markets and Contingency Information Division. **Projections:** EIA, Office of Integrated Analysis and Forecasting, World Energy Projection System and "DESTINY" International Energy Forecast Software (Dallas, TX: Petroconsultants, Fourth Quarter 1999).

Table D4. World Oil Production Capacity by Region and Country, High Non-OPEC Supply Case, 1990-2020
(Million Barrels per Day)

Region/Country	History (Estimates)		Projections			
	1990	1998	2005	2010	2015	2020
OPEC						
Persian Gulf						
Iran	3.2	3.9	4.0	4.3	4.5	4.8
Iraq	2.2	2.8	2.8	3.3	3.9	5.2
Kuwait	1.7	2.6	2.7	3.2	3.6	4.3
Qatar.	0.5	0.6	0.5	0.6	0.6	0.6
Saudi Arabia.	8.6	11.4	11.3	12.5	14.1	19.6
United Arab Emitates	2.5	2.7	2.8	3.1	4.2	4.8
Total Persian Gulf.	**18.7**	**24.0**	**24.1**	**27.0**	**30.9**	**39.3**
Other OPEC						
Algeria.	1.3	1.4	1.6	1.9	2.1	2.0
Indonesia	1.5	1.7	1.5	1.4	1.4	1.3
Libya.	1.5	1.5	1.5	1.6	1.5	1.5
Nigeria.	1.8	2.2	2.5	2.8	3.3	3.1
Venezuela[a]	2.4	3.4	3.8	4.3	5.5	5.9
Total Other OPEC.	**8.5**	**10.2**	**10.9**	**11.2**	**13.3**	**13.8**
Total OPEC	**27.2**	**34.2**	**35.0**	**38.2**	**44.2**	**53.1**
Non-OPEC						
Industrialized						
United States	9.7	9.3	8.7	9.1	9.6	9.9
Canada	2.0	2.7	3.2	3.4	3.6	3.7
Mexico.	3.0	3.5	4.1	4.3	4.5	4.5
Australia.	0.7	0.8	0.8	0.8	0.8	0.8
North Sea	4.2	6.3	7.7	7.5	6.8	6.1
Other.	0.5	0.7	0.8	0.8	0.8	0.8
Total Industrialized	**20.1**	**23.3**	**25.3**	**25.9**	**26.1**	**25.8**
Eurasia						
China	2.8	3.2	3.6	3.9	4.0	3.9
Former Soviet Union.	11.4	7.2	9.1	11.8	13.4	14.4
Eastern Europe	0.3	0.3	0.4	0.5	0.5	0.5
Total Eurasia	**14.5**	**10.7**	**13.1**	**16.2**	**17.9**	**18.8**
Other Non-OPEC						
Central and South America	2.4	3.6	4.6	5.1	5.5	5.7
Middle East	1.4	1.9	2.3	2.3	2.2	2.2
Africa	2.2	2.8	3.8	4.8	5.6	5.9
Asia	1.7	2.2	3.0	3.6	3.8	3.7
Total Other Non-OPEC	**7.7**	**10.5**	**13.7**	**15.8**	**17.1**	**17.5**
Total Non-OPEC.	**42.3**	**44.5**	**52.1**	**57.9**	**61.1**	**62.1**
Total World	**69.5**	**78.7**	**87.1**	**96.1**	**105.3**	**115.2**

[a]These estimates were made prior to recent statements by Venezuela's current regime indicating a potential unwillingness to support the outside investment activity required for the substantial capacity expansion shown in this table.

Note: OPEC = Organization of Petroleum Exporting Countries.

Sources: **History:** Energy Information Administration (EIA), Energy Markets and Contingency Information Division. **Projections:** EIA, Office of Integrated Analysis and Forecasting, World Energy Projection System and "DESTINY" International Energy Forecast Software (Dallas, TX: Petroconsultants, Fourth Quarter 1999).

Table D5. World Oil Production Capacity by Region and Country, Low Non-OPEC Supply Case, 1990-2020
(Million Barrels per Day)

Region/Country	History (Estimates)		Projections			
	1990	1998	2005	2010	2015	2020
OPEC						
Persian Gulf						
Iran	3.2	3.9	4.5	5.3	5.7	6.7
Iraq	2.2	2.8	3.4	4.4	5.5	7.2
Kuwait	1.7	2.6	3.2	4.0	4.9	5.6
Qatar	0.5	0.6	0.6	0.7	0.8	0.8
Saudi Arabia	8.6	11.4	14.5	17.1	21.8	27.6
United Arab Emirates	2.5	2.7	3.4	4.0	4.7	5.8
Total Persian Gulf	18.7	24.0	29.6	35.5	43.4	53.7
Other OPEC						
Algeria	1.3	1.4	2.1	2.3	2.2	2.1
Indonesia	1.5	1.7	1.5	1.6	1.5	1.4
Libya	1.5	1.5	1.6	1.9	1.8	1.6
Nigeria	1.8	2.2	3.1	3.5	3.6	4.0
Venezuela[a]	2.4	3.4	4.9	6.3	6.9	7.1
Total Other OPEC	8.5	10.2	13.2	15.6	16.0	16.2
Total OPEC	27.2	34.2	42.8	51.1	59.4	69.9
Non-OPEC						
Industrialized						
United States	9.7	9.3	8.3	8.2	8.2	8.3
Canada	2.0	2.7	2.8	2.9	3.0	2.8
Mexico	3.0	3.5	3.6	3.7	3.6	3.4
Australia	0.7	0.8	0.8	0.8	0.7	0.6
North Sea	4.2	6.3	6.6	5.9	5.2	4.2
Other	0.5	0.7	0.8	0.8	0.7	0.7
Total Industrialized	20.1	23.3	22.9	22.3	21.4	20.0
Eurasia						
China	2.8	3.2	3.2	3.3	3.3	3.2
Former Soviet Union	11.4	7.2	7.5	8.3	9.1	9.6
Eastern Europe	0.3	0.3	0.3	0.3	0.3	0.3
Total Eurasia	14.5	10.7	11.0	11.9	12.7	13.1
Other Non-OPEC						
Central and South America	2.4	3.6	4.0	4.2	4.3	4.3
Middle East	1.4	1.9	2.1	2.1	2.0	1.9
Africa	2.2	2.8	2.7	2.8	3.5	4.1
Asia	1.7	2.2	2.4	2.6	2.8	2.7
Total Other Non-OPEC	7.7	10.5	11.2	11.7	12.6	13.0
Total Non-OPEC	42.3	44.5	45.1	45.9	46.7	46.1
Total World	69.5	78.7	87.9	97.0	106.1	116.0

[a]These estimates were made prior to recent statements by Venezuela's current regime indicating a potential unwillingness to support the outside investment activity required for the substantial capacity expansion shown in this table.

Note: OPEC = Organization of Petroleum Exporting Countries.

Sources: **History:** Energy Information Administration (EIA), Energy Markets and Contingency Information Division. **Projections:** EIA, Office of Integrated Analysis and Forecasting, World Energy Projection System and "DESTINY" International Energy Forecast Software (Dallas, TX: Petroconsultants, Fourth Quarter 1999).

Table D6. World Oil Production by Region and Country, Reference Case, 1990-2020
  (Million Barrels per Day)

Region/Country	History (Estimates)		Projections			
	1990	1998	2005	2010	2015	2020
OPEC						
Persian Gulf.	16.2	20.5	24.5	28.3	34.3	41.6
Other OPEC	8.3	9.9	12.0	13.4	14.0	14.3
Total OPEC	24.5	30.4	36.5	41.7	48.3	55.9
Non-OPEC						
Industrialized						
United States	9.7	9.3	8.5	8.6	8.9	9.1
Canada	2.0	2.7	3.0	3.2	3.4	3.4
Mexico.	3.0	3.5	3.7	4.0	3.9	3.9
Western Europe	4.6	7.0	7.8	7.5	7.0	6.4
Other.	0.8	0.8	0.9	0.9	0.9	0.8
Total Industrialized	20.1	23.3	23.9	24.2	24.1	23.6
Eurasia						
China	2.8	3.2	3.3	3.5	3.6	3.6
Former Soviet Union.	11.4	7.2	7.6	10.1	12.1	13.1
Eastern Europe	0.3	0.3	0.3	0.4	0.4	0.5
Total Eurasia	14.5	10.7	11.2	14.0	16.1	17.2
Other Non-OPEC						
Central and South America	2.4	3.6	4.2	4.4	4.8	5.0
Pacific Rim.	1.7	2.2	2.5	2.9	3.2	3.3
Other.	3.5	4.7	5.3	6.0	6.6	7.5
Total Other Non-OPEC	7.6	10.5	12.0	13.3	14.6	15.8
Total World	66.7	74.9	83.6	93.2	103.1	112.5
Persian Gulf Production as a Percentage of World Consumption	24.6	27.8	29.2	30.3	33.2	36.9

Note: OPEC = Organization of Petroleum Exporting Countries. Production includes crude oil (including lease condensates), natural gas liquids, other hydrogen hydrocarbons for refinery feedstocks, refinery gains, alcohol, and liquids produced from coal and other sources. Totals may not equal sum of components due to independent rounding.

Sources: **History:** Energy Information Administration (EIA), Energy Markets and Contingency Information Division. **Projections:** EIA, Office of Integrated Analysis and Forecasting, World Energy Projection System and "DESTINY" International Energy Forecast Software (Dallas, TX: Petroconsultants, Fourth Quarter 1999).

Table D7. World Oil Production by Region and Country, High Oil Price Case, 1990-2020
(Million Barrels per Day)

Region/Country	History (Estimates)		Projections			
	1990	1998	2005	2010	2015	2020
OPEC						
Persian Gulf.	16.2	20.5	22.9	25.4	29.7	36.2
Other OPEC	8.3	9.9	11.6	12.9	13.7	13.9
Total **OPEC**	**24.5**	**30.4**	**34.5**	**38.3**	**43.4**	**50.1**
Non-OPEC						
Industrialized						
United States	9.7	9.3	8.7	9.1	9.6	10.0
Canada	2.0	2.7	3.1	3.3	3.4	3.5
Mexico.	3.0	3.5	3.9	4.1	4.0	3.9
Western Europe	4.6	7.0	7.7	7.7	7.1	6.5
Other.	0.8	0.8	0.9	0.9	1.0	0.9
Total **Industrialized**	**20.1**	**23.3**	**24.3**	**25.1**	**25.1**	**24.8**
Eurasia						
China	2.8	3.2	3.3	3.6	3.7	3.7
Former Soviet Union.	11.4	7.2	7.6	10.3	12.4	13.3
Eastern Europe	0.3	0.3	0.3	0.4	0.4	0.5
Total **Eurasia**	**14.5**	**10.7**	**11.2**	**14.3**	**16.5**	**17.5**
Other Non-OPEC						
Central and South America	2.4	3.6	4.2	4.5	4.9	5.1
Pacific Rim.	1.7	2.2	2.6	3.0	3.3	3.4
Other.	3.5	4.7	5.4	5.7	6.7	7.7
Total **Other Non-OPEC**	**7.6**	**10.5**	**12.2**	**13.2**	**14.9**	**16.2**
Total World	**66.7**	**74.9**	**82.2**	**90.9**	**99.9**	**108.6**
Persian Gulf Production as a Percentage of World Consumption	**24.6**	**27.8**	**27.7**	**27.8**	**29.6**	**33.2**

Note: OPEC = Organization of Petroleum Exporting Countries. Production includes crude oil (including lease condensates), natural gas liquids, other hydrogen hydrocarbons for refinery feedstocks, refinery gains, alcohol, and liquids produced from coal and other sources. Totals may not equal sum of components due to independent rounding.

Sources: **History:** Energy Information Administration (EIA), Energy Markets and Contingency Information Division. **Projections:** EIA, Office of Integrated Analysis and Forecasting, World Energy Projection System and "DESTINY" International Energy Forecast Software (Dallas, TX: Petroconsultants, Fourth Quarter 1999).

Table D8. World Oil Production by Region and Country, Low Oil Price Case, 1990-2020
(Million Barrels per Day)

Region/Country	History (Estimates)		Projections			
	1990	1998	2005	2010	2015	2020
OPEC						
Persian Gulf.	16.2	20.5	27.3	33.0	40.8	49.8
Other OPEC	8.3	9.9	12.8	14.3	14.8	15.1
Total OPEC	24.5	30.4	40.1	47.3	55.6	64.9
Non-OPEC						
Industrialized						
United States	9.7	9.3	8.0	8.1	8.2	8.3
Canada	2.0	2.7	3.0	3.2	3.3	3.3
Mexico.	3.0	3.5	3.7	3.8	3.8	3.7
Western Europe	4.6	7.0	7.7	7.3	7.0	6.3
Other.	0.8	0.8	0.9	0.9	0.8	0.8
Total Industrialized	20.1	23.3	23.3	23.3	23.1	22.4
Eurasia						
China	2.8	3.2	3.3	3.4	3.5	3.5
Former Soviet Union.	11.4	7.2	7.5	10.0	11.7	12.7
Eastern Europe	0.3	0.3	0.3	0.4	0.4	0.4
Total Eurasia	14.5	10.7	11.1	13.8	15.6	16.6
Other Non-OPEC						
Central and South America	2.4	3.6	4.1	4.3	4.6	4.8
Pacific Rim.	1.7	2.2	2.4	2.8	3.1	3.2
Other.	3.5	4.7	5.0	5.7	6.3	7.3
Total Other Non-OPEC	7.6	10.5	11.5	12.8	14.0	15.3
Total World	66.7	74.9	86.0	97.2	108.3	119.2
Persian Gulf Production as a Percentage of World Consumption	24.6	27.8	31.6	33.9	37.6	41.7

Note: OPEC = Organization of Petroleum Exporting Countries. Production includes crude oil (including lease condensates), natural gas liquids, other hydrogen hydrocarbons for refinery feedstocks, refinery gains, alcohol, and liquids produced from coal and other sources. Totals may not equal sum of components due to independent rounding.

Sources: **History:** Energy Information Administration (EIA), Energy Markets and Contingency Information Division. **Projections:** EIA, Office of Integrated Analysis and Forecasting, World Energy Projection System and "DESTINY" International Energy Forecast Software (Dallas, TX: Petroconsultants, Fourth Quarter 1999).

Table D9. World Oil Production by Region and Country, High Non-OPEC Supply Case, 1990-2020
(Million Barrels per Day)

Region/Country	History (Estimates)		Projections			
	1990	1998	2005	2010	2015	2020
OPEC						
Persian Gulf.	16.2	20.5	21.7	25.0	29.4	37.3
Other OPEC	8.3	9.9	9.8	10.3	12.6	13.1
Total OPEC	24.5	30.4	31.5	35.3	42.0	50.4
Non-OPEC						
Industrialized						
United States	9.7	9.3	8.7	9.1	9.6	9.9
Canada	2.0	2.7	3.2	3.4	3.6	3.7
Mexico.	3.0	3.5	4.1	4.3	4.5	4.5
Western Europe	4.6	7.0	8.4	8.2	7.5	6.8
Other.	0.8	0.8	0.9	0.9	0.9	0.9
Total Industrialized	20.1	23.3	25.3	25.9	26.1	25.8
Eurasia						
China	2.8	3.2	3.6	3.9	4.0	3.9
Former Soviet Union.	11.4	7.2	9.1	11.8	13.4	14.4
Eastern Europe	0.3	0.3	0.4	0.5	0.5	0.5
Total Eurasia	14.5	10.7	13.1	16.2	17.9	18.8
Other Non-OPEC						
Central and South America	2.4	3.6	4.6	5.1	5.5	5.7
Pacific Rim.	1.7	2.2	3.0	3.6	3.8	3.7
Other.	3.5	4.7	6.1	7.1	7.8	8.1
Total Other Non-OPEC	7.6	10.5	13.7	15.8	17.1	17.5
Total World	66.7	74.9	83.6	93.2	103.1	112.5
Persian Gulf Production as a Percentage of World Consumption	24.6	27.8	29.2	30.3	33.2	36.9

Note: OPEC = Organization of Petroleum Exporting Countries. Production includes crude oil (including lease condensates), natural gas liquids, other hydrogen hydrocarbons for refinery feedstocks, refinery gains, alcohol, and liquids produced from coal and other sources. Totals may not equal sum of components due to independent rounding.

Sources: **History:** Energy Information Administration (EIA), Energy Markets and Contingency Information Division. **Projections:** EIA, Office of Integrated Analysis and Forecasting, World Energy Projection System and "DESTINY" International Energy Forecast Software (Dallas, TX: Petroconsultants, Fourth Quarter 1999).

Table D10. World Oil Production by Region and Country, Low Non-OPEC Supply Case, 1990-2020
(Million Barrels per Day)

Region/Country	History (Estimates)		Projections			
	1990	1998	2005	2010	2015	2020
OPEC						
Persian Gulf.	16.2	20.5	26.6	32.8	41.2	51.0
Other OPEC	8.3	9.9	11.9	14.5	15.2	15.4
Total OPEC	**24.5**	**30.4**	**38.5**	**47.3**	**56.4**	**66.4**
Non-OPEC						
Industrialized						
United States	9.7	9.3	8.3	8.2	8.2	8.3
Canada	2.0	2.7	2.8	2.9	3.0	2.8
Mexico.	3.0	3.5	3.6	3.7	3.6	3.4
Western Europe	4.6	7.0	7.3	6.6	5.8	4.8
Other.	0.8	0.8	0.9	0.9	0.8	0.7
Total Industrialized	**20.1**	**23.3**	**22.9**	**22.3**	**21.4**	**20.0**
Eurasia						
China	2.8	3.2	3.2	3.3	3.3	3.2
Former Soviet Union.	11.4	7.2	7.5	8.3	9.1	9.6
Eastern Europe	0.3	0.3	0.3	0.3	0.3	0.3
Total Eurasia	**14.5**	**10.7**	**11.0**	**11.9**	**12.7**	**13.1**
Other Non-OPEC						
Central and South America	2.4	3.6	4.0	4.2	4.3	4.3
Pacific Rim.	1.7	2.2	2.4	2.6	2.8	2.7
Other.	3.5	4.7	4.8	4.9	5.5	6.0
Total Other Non-OPEC	**7.6**	**10.5**	**11.2**	**11.7**	**12.6**	**13.0**
Total World	**66.7**	**74.9**	**83.6**	**93.2**	**103.1**	**112.5**
Persian Gulf Production as a Percentage of World Consumption	**24.6**	**27.8**	**29.2**	**30.3**	**33.2**	**36.9**

Note: OPEC = Organization of Petroleum Exporting Countries. Production includes crude oil (including lease condensates), natural gas liquids, other hydrogen hydrocarbons for refinery feedstocks, refinery gains, alcohol, and liquids produced from coal and other sources. Totals may not equal sum of components due to independent rounding.

Sources: **History:** Energy Information Administration (EIA), Energy Markets and Contingency Information Division. **Projections:** EIA, Office of Integrated Analysis and Forecasting, World Energy Projection System and "DESTINY" International Energy Forecast Software (Dallas, TX: Petroconsultants, Fourth Quarter 1999).

Projections of Transportation Energy Use in the Reference Case

Table E1. World Total Energy Consumption for Transportation by Region, Reference Case, 1990-2020
(Million Barrels of Oil per Day)

Region/Country	History			Projections				Average Annual Percent Change, 1997-2020
	1990	1996	1997	2005	2010	2015	2020	
Industrialized Countries								
North America	**12.5**	**13.9**	**14.0**	**17.1**	**18.9**	**20.6**	**22.2**	**2.0**
United States[a]	11.0	12.1	12.2	14.9	16.2	17.5	18.6	1.8
Canada.	0.9	1.0	1.0	1.2	1.3	1.3	1.4	1.2
Mexico	0.6	0.7	0.8	1.1	1.4	1.8	2.2	4.7
Western Europe	**6.2**	**6.9**	**7.1**	**7.9**	**8.3**	**8.7**	**9.0**	**1.0**
United Kingdom	1.0	1.0	1.1	1.2	1.3	1.4	1.5	1.5
France	0.9	1.0	1.0	1.2	1.2	1.3	1.3	1.1
Germany	1.2	1.3	1.4	1.5	1.6	1.7	1.7	1.0
Italy.	0.7	0.8	0.8	0.9	0.9	1.0	1.0	0.7
Netherlands	0.4	0.5	0.5	0.5	0.6	0.6	0.6	1.0
Other Western Europe	2.0	2.3	2.3	2.6	2.7	2.8	2.9	1.0
Industrialized Asia.	**2.1**	**2.5**	**2.5**	**2.8**	**2.9**	**3.1**	**3.2**	**0.9**
Japan	1.6	1.9	1.9	2.1	2.1	2.2	2.3	0.7
Australasia	0.5	0.6	0.6	0.7	0.8	0.9	0.9	1.7
Total Industrialized	**20.9**	**23.3**	**23.6**	**27.8**	**30.2**	**32.3**	**34.4**	**1.6**
EE/FSU								
Former Soviet Union	2.7	1.3	1.3	1.7	2.0	2.3	2.7	3.0
Eastern Europe	0.6	0.5	0.6	0.8	0.9	1.0	1.1	2.9
Total EE/FSU	**3.3**	**1.8**	**1.9**	**2.6**	**3.0**	**3.4**	**3.8**	**3.0**
Developing Countries								
Developing Asia	**3.1**	**4.9**	**5.2**	**7.9**	**10.0**	**12.2**	**13.3**	**4.1**
China.	0.8	1.2	1.3	2.4	3.7	5.2	5.6	6.4
India	0.5	0.8	0.9	1.4	1.6	1.8	1.9	3.5
South Korea	0.3	0.7	0.7	1.0	1.1	1.2	1.3	2.7
Other Asia	1.4	2.2	2.3	3.1	3.6	4.0	4.4	2.9
Middle East.	**1.2**	**1.7**	**1.7**	**1.9**	**2.2**	**2.5**	**3.0**	**2.5**
Turkey	0.2	0.3	0.3	0.4	0.4	0.5	0.5	2.5
Other Middle East	1.0	1.4	1.4	1.5	1.8	2.1	2.5	2.5
Africa	**0.9**	**1.0**	**1.0**	**1.4**	**1.7**	**2.1**	**2.6**	**4.1**
Central and South America . . .	**1.6**	**2.0**	**2.1**	**2.8**	**3.6**	**4.4**	**5.5**	**4.2**
Brazil.	0.7	0.9	0.9	1.1	1.4	1.8	2.3	4.3
Other Central/South America . .	0.9	1.2	1.3	1.7	2.1	2.6	3.2	4.1
Total Developing	**6.8**	**9.6**	**10.1**	**14.0**	**17.5**	**21.2**	**24.3**	**3.9**
Total World	**31.0**	**34.7**	**35.6**	**44.4**	**50.6**	**56.9**	**62.5**	**2.5**

[a]Includes the 50 States and the District of Columbia. U.S. Territories are included in Australasia.
Notes: EE/FSU = Eastern Europe/Former Soviet Union. Totals may not equal sum of components due to independent rounding.
Sources: **History:** Derived from Energy Information Administration (EIA), *International Energy Annual 1997*, DOE/EIA-0219(97) (Washington, DC, April 1999). **Projections:** EIA, *Annual Energy Outlook 2000*, DOE/EIA-0383(2000) (Washington, DC, December 1999), Table A2; and World Energy Projection System (2000).

Table E2. World Total Gasoline Consumption for Transportation by Region, Reference Case, 1990-2020
(Million Barrels of Oil per Day)

Region/Country	History			Projections				Average Annual Percent Change, 1997-2020
	1990	1996	1997	2005	2010	2015	2020	
Industrialized Countries								
North America	7.6	8.5	8.7	10.3	11.2	12.0	12.8	1.7
United States[a]	6.7	7.4	7.6	8.9	9.6	10.1	10.6	1.5
Canada.	0.5	0.6	0.6	0.6	0.7	0.7	0.7	1.1
Mexico	0.4	0.5	0.5	0.7	0.9	1.2	1.5	4.6
Western Europe	2.6	2.7	2.8	3.0	3.0	3.0	3.0	0.4
United Kingdom	0.5	0.5	0.5	0.5	0.6	0.6	0.6	0.6
France	0.4	0.3	0.3	0.4	0.4	0.4	0.4	0.3
Germany	0.7	0.6	0.6	0.7	0.7	0.7	0.7	0.4
Italy.	0.3	0.4	0.4	0.4	0.4	0.4	0.4	0.3
Netherlands	0.1	0.1	0.1	0.1	0.1	0.1	0.1	0.7
Other Western Europe	0.7	0.8	0.8	0.9	0.9	0.9	0.9	0.3
Industrialized Asia.	1.0	1.1	1.2	1.2	1.3	1.3	1.3	0.5
Japan	0.7	0.8	0.8	0.9	0.9	0.9	0.9	0.2
Australasia	0.3	0.3	0.3	0.4	0.4	0.4	0.4	1.1
Total Industrialized	11.3	12.3	12.6	14.5	15.5	16.3	17.1	1.3
EE/FSU								
Former Soviet Union	1.1	0.3	0.4	0.6	0.7	0.8	0.8	3.4
Eastern Europe	0.3	0.3	0.3	0.4	0.5	0.5	0.5	2.4
Total EE/FSU	1.3	0.6	0.7	1.0	1.2	1.3	1.3	3.0
Developing Countries								
Developing Asia	1.0	1.6	1.7	2.9	3.9	5.0	5.2	4.9
China.	0.4	0.7	0.7	1.5	2.3	3.3	3.4	6.9
India	0.1	0.1	0.1	0.2	0.2	0.2	0.2	2.5
South Korea	0.1	0.2	0.2	0.2	0.3	0.3	0.3	2.3
Other Asia	0.4	0.6	0.7	1.0	1.1	1.2	1.2	2.5
Middle East	0.6	0.7	0.7	0.8	1.0	1.1	1.3	2.6
Turkey	0.1	0.1	0.1	0.1	0.1	0.2	0.2	1.9
Other Middle East	0.5	0.6	0.6	0.7	0.8	0.9	1.1	2.7
Africa	0.4	0.5	0.5	0.6	0.8	1.0	1.3	4.4
Central and South America . . .	0.6	0.9	0.9	1.2	1.6	2.0	2.4	4.3
Brazil.	0.1	0.2	0.2	0.3	0.4	0.5	0.6	4.4
Other Central/South America . .	0.5	0.6	0.7	0.9	1.2	1.5	1.8	4.3
Total Developing	2.5	3.6	3.8	5.6	7.2	9.0	10.2	4.3
Total World	15.1	16.6	17.1	21.1	23.9	26.6	28.6	2.3

[a]Includes the 50 States and the District of Columbia. U.S. Territories are included in Australasia.
Notes: EE/FSU = Eastern Europe/Former Soviet Union. Totals may not equal sum of components due to independent rounding.
Sources: **History:** Derived from Energy Information Administration (EIA), *International Energy Annual 1997*, DOE/EIA-0219(97) (Washington, DC, April 1999). **Projections:** EIA, *Annual Energy Outlook 2000*, DOE/EIA-0383(2000) (Washington, DC, December 1999), Table A2; and World Energy Projection System (2000).

Table E3. World Total Diesel Fuel Consumption for Transportation by Region, Reference Case, 1990-2020
(Million Barrels of Oil per Day)

Region/Country	History			Projections				Average Annual Percent Change, 1997-2020
	1990	1996	1997	2005	2010	2015	2020	
Industrialized Countries								
North America	**2.4**	**2.7**	**2.7**	**3.3**	**3.7**	**4.0**	**4.3**	**2.0**
United States[a]	2.0	2.3	2.3	2.8	3.1	3.3	3.5	1.8
Canada.	0.2	0.2	0.2	0.3	0.3	0.3	0.3	1.6
Mexico	0.2	0.2	0.2	0.3	0.3	0.4	0.5	3.9
Western Europe	**2.2**	**2.6**	**2.7**	**3.1**	**3.2**	**3.4**	**3.5**	**1.1**
United Kingdom	0.3	0.3	0.3	0.4	0.4	0.4	0.5	1.5
France	0.4	0.5	0.5	0.6	0.6	0.7	0.7	1.3
Germany	0.4	0.5	0.5	0.6	0.6	0.7	0.7	1.4
Italy.	0.3	0.3	0.3	0.4	0.4	0.4	0.4	0.8
Netherlands	0.1	0.1	0.1	0.1	0.2	0.2	0.2	0.8
Other Western Europe	0.7	0.9	0.9	1.0	1.0	1.1	1.1	0.9
Industrialized Asia.	**0.6**	**0.8**	**0.8**	**0.9**	**1.0**	**1.0**	**1.1**	**1.2**
Japan	0.5	0.7	0.7	0.8	0.8	0.8	0.9	1.1
Australasia	0.1	0.1	0.1	0.2	0.2	0.2	0.2	2.0
Total Industrialized	**5.2**	**6.1**	**6.2**	**7.3**	**7.9**	**8.4**	**8.8**	**1.5**
EE/FSU								
Former Soviet Union	0.6	0.3	0.3	0.5	0.5	0.6	0.7	3.0
Eastern Europe	0.2	0.2	0.2	0.3	0.3	0.4	0.4	3.1
Total EE/FSU	**0.8**	**0.5**	**0.5**	**0.8**	**0.9**	**1.0**	**1.1**	**3.0**
Developing Countries								
Developing Asia	**1.2**	**2.1**	**2.2**	**3.2**	**3.9**	**4.6**	**5.0**	**3.7**
China.	0.2	0.3	0.3	0.6	1.0	1.4	1.7	7.4
India	0.4	0.6	0.7	1.1	1.2	1.3	1.4	3.1
South Korea	0.2	0.3	0.3	0.5	0.5	0.6	0.6	2.5
Other Asia	0.5	0.8	0.8	1.1	1.2	1.3	1.4	2.2
Middle East.	**0.3**	**0.5**	**0.5**	**0.7**	**0.8**	**0.9**	**1.1**	**3.2**
Turkey	0.1	0.1	0.1	0.2	0.2	0.2	0.3	2.5
Other Middle East	0.2	0.4	0.4	0.5	0.6	0.7	0.9	3.5
Africa	**0.3**	**0.3**	**0.3**	**0.3**	**0.4**	**0.5**	**0.6**	**3.2**
Central and South America . . .	**0.6**	**0.7**	**0.7**	**0.9**	**1.2**	**1.5**	**1.9**	**4.1**
Brazil.	0.3	0.4	0.4	0.5	0.6	0.8	1.0	3.9
Other Central/South America . .	0.2	0.3	0.3	0.4	0.6	0.7	0.9	4.4
Total Developing	**2.4**	**3.6**	**3.7**	**5.2**	**6.3**	**7.5**	**8.6**	**3.7**
Total World	**8.4**	**10.2**	**10.5**	**13.2**	**15.0**	**16.8**	**18.4**	**2.5**

[a]Includes the 50 States and the District of Columbia. U.S. Territories are included in Australasia.

Notes: EE/FSU = Eastern Europe/Former Soviet Union. Totals may not equal sum of components due to independent rounding.

Sources: **History:** Derived from Energy Information Administration (EIA), *International Energy Annual 1997*, DOE/EIA-0219(97) (Washington, DC, April 1999). **Projections:** EIA, *Annual Energy Outlook 2000*, DOE/EIA-0383(2000) (Washington, DC, December 1999), Table A2; and World Energy Projection System (2000).

Table E4. World Total Jet Fuel Consumption for Transportation by Region, Reference Case, 1990-2020
(Million Barrels of Oil per Day)

Region/Country	History			Projections				Average Annual Percent Change, 1997-2020
	1990	1996	1997	2005	2010	2015	2020	
Industrialized Countries								
North America	**1.7**	**1.8**	**1.8**	**2.3**	**2.8**	**3.1**	**3.5**	**2.9**
United States[a]	1.6	1.6	1.7	2.2	2.5	2.8	3.1	2.8
Canada.	0.1	0.1	0.1	0.1	0.1	0.1	0.1	1.4
Mexico	0.0	0.0	0.1	0.1	0.1	0.2	0.2	6.1
Western Europe	**0.6**	**0.8**	**0.8**	**1.0**	**1.2**	**1.4**	**1.7**	**3.3**
United Kingdom	0.1	0.2	0.2	0.2	0.3	0.3	0.4	3.3
France	0.1	0.1	0.1	0.1	0.1	0.2	0.2	2.9
Germany	0.1	0.1	0.1	0.2	0.2	0.2	0.3	3.4
Italy.	0.0	0.1	0.1	0.1	0.1	0.1	0.1	3.3
Netherlands	0.0	0.1	0.1	0.1	0.1	0.1	0.1	3.3
Other Western Europe	0.2	0.3	0.3	0.4	0.4	0.5	0.6	3.4
Industrialized Asia	**0.2**	**0.3**	**0.3**	**0.3**	**0.4**	**0.5**	**0.5**	**2.5**
Japan	0.1	0.2	0.2	0.2	0.2	0.3	0.3	2.2
Australasia	0.1	0.1	0.1	0.1	0.1	0.2	0.2	3.1
Total Industrialized	**2.5**	**2.8**	**2.9**	**3.7**	**4.4**	**5.0**	**5.7**	**3.0**
EE/FSU								
Former Soviet Union	0.4	0.2	0.2	0.2	0.2	0.3	0.3	1.7
Eastern Europe	0.0	0.0	0.0	0.1	0.1	0.1	0.1	5.7
Total EE/FSU	**0.5**	**0.3**	**0.3**	**0.3**	**0.3**	**0.4**	**0.5**	**2.5**
Developing Countries								
Developing Asia	**0.3**	**0.5**	**0.5**	**0.7**	**1.0**	**1.4**	**2.0**	**6.0**
China.	0.0	0.1	0.1	0.1	0.2	0.3	0.5	8.7
India	0.0	0.0	0.1	0.1	0.1	0.2	0.3	7.2
South Korea	0.0	0.0	0.1	0.1	0.1	0.1	0.2	6.2
Other Asia	0.2	0.3	0.3	0.4	0.6	0.8	1.0	5.0
Middle East.	**0.1**	**0.2**	**0.2**	**0.2**	**0.3**	**0.4**	**0.5**	**5.4**
Turkey	0.0	0.0	0.0	0.0	0.1	0.1	0.1	5.8
Other Middle East	0.1	0.1	0.1	0.2	0.3	0.3	0.4	5.4
Africa	**0.1**	**0.1**	**0.1**	**0.2**	**0.2**	**0.3**	**0.3**	**4.8**
Central and South America . . .	**0.1**	**0.1**	**0.2**	**0.2**	**0.3**	**0.4**	**0.6**	**5.7**
Brazil.	0.0	0.0	0.0	0.1	0.1	0.1	0.2	5.2
Other Central/South America . .	0.1	0.1	0.1	0.2	0.2	0.3	0.4	5.9
Total Developing	**0.6**	**0.9**	**0.9**	**1.4**	**1.9**	**2.5**	**3.4**	**5.7**
Total World	**3.6**	**4.0**	**4.1**	**5.4**	**6.6**	**7.9**	**9.5**	**3.7**

[a]Includes the 50 States and the District of Columbia. U.S. Territories are included in Australasia.
Notes: EE/FSU = Eastern Europe/Former Soviet Union. Totals may not equal sum of components due to independent rounding.
Sources: **History:** Derived from Energy Information Administration (EIA), *International Energy Annual 1997*, DOE/EIA-0219(97) (Washington, DC, April 1999). **Projections:** EIA, *Annual Energy Outlook 2000*, DOE/EIA-0383(2000) (Washington, DC, December 1999), Table A2; and World Energy Projection System (2000).

Table E5. World Total Residual Fuel Consumption for Transportation by Region, Reference Case, 1990-2020
(Million Barrels of Oil per Day)

Region/Country	History			Projections				Average Annual Percent Change, 1997-2020
	1990	1996	1997	2005	2010	2015	2020	
Industrialized Countries								
North America	0.6	0.5	0.4	0.5	0.6	0.6	0.7	2.0
United States[a]	0.5	0.5	0.4	0.5	0.6	0.6	0.7	2.0
Canada.	0.0	0.0	0.0	0.0	0.0	0.0	0.0	0.6
Mexico	0.0	0.0	0.0	0.0	0.0	0.0	0.0	0.7
Western Europe	0.6	0.6	0.6	0.6	0.6	0.6	0.6	0.3
United Kingdom	0.0	0.0	0.0	0.0	0.0	0.0	0.0	0.3
France	0.0	0.0	0.0	0.1	0.1	0.1	0.1	0.3
Germany	0.0	0.0	0.0	0.0	0.0	0.0	0.0	0.3
Italy.	0.0	0.0	0.0	0.0	0.0	0.0	0.0	0.3
Netherlands	0.2	0.2	0.2	0.2	0.2	0.2	0.2	0.3
Other Western Europe	0.2	0.2	0.2	0.2	0.2	0.3	0.3	0.3
Industrialized Asia	0.1	0.2	0.2	0.2	0.2	0.2	0.2	0.5
Japan	0.1	0.2	0.2	0.2	0.2	0.2	0.2	0.5
Australasia	0.0	0.0	0.0	0.0	0.0	0.0	0.0	0.6
Total Industrialized	1.3	1.3	1.2	1.3	1.4	1.4	1.5	1.0
EE/FSU								
Former Soviet Union	0.2	0.1	0.1	0.1	0.1	0.1	0.1	1.1
Eastern Europe	0.0	0.0	0.0	0.0	0.0	0.0	0.0	1.9
Total EE/FSU	0.2	0.1	0.1	0.1	0.1	0.1	0.1	1.3
Developing Countries								
Developing Asia	0.3	0.5	0.5	0.6	0.6	0.7	0.8	1.6
China.	0.0	0.1	0.1	0.1	0.1	0.1	0.1	2.3
India	0.0	0.0	0.0	0.0	0.0	0.0	0.0	2.0
South Korea	0.0	0.1	0.1	0.1	0.2	0.2	0.2	1.6
Other Asia	0.3	0.3	0.3	0.3	0.4	0.4	0.4	1.4
Middle East	0.2	0.3	0.3	0.3	0.4	0.4	0.4	1.4
Turkey	0.0	0.0	0.0	0.0	0.0	0.0	0.0	1.4
Other Middle East	0.2	0.3	0.3	0.3	0.4	0.4	0.4	1.4
Africa	0.1	0.1	0.1	0.1	0.2	0.2	0.2	1.3
Central and South America . . .	0.1	0.1	0.1	0.2	0.2	0.2	0.2	0.8
Brazil.	0.0	0.0	0.0	0.1	0.1	0.1	0.1	0.9
Other Central/South America . .	0.1	0.1	0.1	0.1	0.1	0.1	0.1	0.8
Total Developing	0.8	1.1	1.1	1.2	1.3	1.4	1.5	1.4
Total World	2.3	2.4	2.4	2.6	2.8	2.9	3.1	1.2

[a]Includes the 50 States and the District of Columbia. U.S. Territories are included in Australasia.
Notes: EE/FSU = Eastern Europe/Former Soviet Union. Totals may not equal sum of components due to independent rounding.
Sources: **History:** Derived from Energy Information Administration (EIA), *International Energy Annual 1997*, DOE/EIA-0219(97) (Washington, DC, April 1999). **Projections:** EIA, *Annual Energy Outlook 2000*, DOE/EIA-0383(2000) (Washington, DC, December 1999), Table A2; and World Energy Projection System (2000).

Table E6. World Total Other Fuel Consumption for Transportation by Region, Reference Case, 1990-2020
(Million Barrels of Oil per Day)

Region/Country	History			Projections				Average Annual Percent Change, 1997-2020
	1990	1996	1997	2005	2010	2015	2020	
Industrialized Countries								
North America	**0.5**	**0.6**	**0.6**	**0.6**	**0.7**	**0.8**	**0.8**	**1.6**
United States[a]	0.4	0.4	0.4	0.5	0.5	0.5	0.6	1.5
Canada	0.1	0.1	0.1	0.2	0.2	0.2	0.2	1.5
Mexico	0.0	0.0	0.0	0.0	0.0	0.0	0.0	3.8
Western Europe	**0.3**	**0.3**	**0.3**	**0.3**	**0.3**	**0.3**	**0.3**	**0.5**
United Kingdom	0.0	0.0	0.0	0.0	0.0	0.0	0.0	0.5
France	0.0	0.0	0.0	0.0	0.0	0.0	0.0	0.3
Germany	0.0	0.0	0.0	0.0	0.0	0.0	0.0	0.2
Italy	0.0	0.1	0.1	0.1	0.1	0.1	0.1	0.5
Netherlands	0.0	0.0	0.0	0.0	0.0	0.0	0.0	0.7
Other Western Europe	0.1	0.1	0.1	0.1	0.1	0.1	0.1	0.5
Industrialized Asia	**0.1**	**0.1**	**0.1**	**0.1**	**0.2**	**0.2**	**0.2**	**0.8**
Japan	0.1	0.1	0.1	0.1	0.1	0.1	0.1	0.6
Australasia	0.0	0.0	0.0	0.0	0.0	0.1	0.1	1.3
Total Industrialized	**0.9**	**1.0**	**1.0**	**1.1**	**1.2**	**1.2**	**1.3**	**1.2**
EE/FSU								
Former Soviet Union	0.5	0.4	0.4	0.4	0.4	0.5	0.5	0.9
Eastern Europe	0.0	0.0	0.0	0.0	0.0	0.0	0.0	1.4
Total EE/FSU	**0.5**	**0.4**	**0.4**	**0.5**	**0.5**	**0.5**	**0.5**	**1.0**
Developing Countries								
Developing Asia	**0.4**	**0.3**	**0.3**	**0.3**	**0.4**	**0.4**	**0.4**	**1.2**
China	0.2	0.2	0.2	0.2	0.2	0.2	0.2	0.9
India	0.1	0.0	0.0	0.0	0.0	0.0	0.0	1.0
South Korea	0.0	0.0	0.1	0.1	0.1	0.1	0.1	2.3
Other Asia	0.0	0.1	0.1	0.1	0.1	0.1	0.1	0.9
Middle East	**0.0**	**0.0**	**0.0**	**0.0**	**0.0**	**0.0**	**0.0**	**1.5**
Turkey	0.0	0.0	0.0	0.0	0.0	0.0	0.0	1.4
Other Middle East	0.0	0.0	0.0	0.0	0.0	0.0	0.0	3.5
Africa	**0.0**	**0.0**	**0.0**	**0.0**	**0.0**	**0.0**	**0.0**	**2.9**
Central and South America . . .	**0.1**	**0.2**	**0.2**	**0.2**	**0.3**	**0.3**	**0.4**	**3.9**
Brazil	0.1	0.1	0.1	0.2	0.2	0.3	0.4	3.9
Other Central/South America . .	0.0	0.0	0.0	0.0	0.1	0.1	0.1	3.8
Total Developing	**0.5**	**0.5**	**0.5**	**0.6**	**0.7**	**0.8**	**0.9**	**2.4**
Total World	**1.9**	**1.9**	**1.9**	**2.1**	**2.3**	**2.5**	**2.7**	**1.5**

[a]Includes the 50 States and the District of Columbia. U.S. Territories are included in Australasia.
Notes: EE/FSU = Eastern Europe/Former Soviet Union. Totals may not equal sum of components due to independent rounding.
Sources: **History:** Derived from Energy Information Administration (EIA), *International Energy Annual 1997*, DOE/EIA-0219(97) (Washington, DC, April 1999). **Projections:** EIA, *Annual Energy Outlook 2000*, DOE/EIA-0383(2000) (Washington, DC, December 1999), Table A2; and World Energy Projection System (2000).

Table E7. World Total Road Use Energy Consumption by Region, Reference Case, 1990-2020
(Million Barrels of Oil per Day)

Region/Country	History			Projections				Average Annual Percent Change, 1997-2020
	1990	1996	1997	2005	2010	2015	2020	
Industrialized Countries								
North America	**9.4**	**10.6**	**10.8**	**13.0**	**14.3**	**15.4**	**16.5**	**1.8**
United States[a]	8.1	9.2	9.4	11.2	12.1	12.9	13.5	1.6
Canada.	0.7	0.7	0.8	0.9	0.9	1.0	1.0	1.3
Mexico	0.6	0.7	0.7	1.0	1.2	1.6	2.0	4.6
Western Europe	**4.6**	**5.1**	**5.2**	**5.8**	**6.1**	**6.2**	**6.3**	**0.8**
United Kingdom	0.7	0.8	0.8	0.9	0.9	1.0	1.0	1.0
France	0.7	0.8	0.8	0.9	1.0	1.0	1.0	0.9
Germany	1.0	1.1	1.1	1.3	1.3	1.3	1.4	0.8
Italy.	0.6	0.7	0.7	0.8	0.8	0.8	0.8	0.5
Netherlands	0.2	0.2	0.2	0.2	0.2	0.2	0.2	0.9
Other Western Europe	1.4	1.6	1.6	1.8	1.8	1.8	1.9	0.7
Industrialized Asia	**1.6**	**1.9**	**1.9**	**2.1**	**2.2**	**2.3**	**2.3**	**0.8**
Japan	1.2	1.4	1.5	1.6	1.7	1.7	1.7	0.6
Australasia	0.4	0.5	0.5	0.5	0.6	0.6	0.6	1.4
Total Industrialized	**15.6**	**17.6**	**18.0**	**21.0**	**22.5**	**23.9**	**25.1**	**1.5**
EE/FSU								
Former Soviet Union	1.5	0.5	0.6	1.0	1.1	1.3	1.4	3.8
Eastern Europe	0.5	0.4	0.5	0.7	0.8	0.9	0.9	2.8
Total EE/FSU	**2.0**	**0.9**	**1.0**	**1.7**	**1.9**	**2.1**	**2.3**	**3.4**
Developing Countries								
Developing Asia	**2.0**	**3.3**	**3.6**	**6.0**	**7.7**	**9.5**	**10.2**	**4.6**
China.	0.5	0.8	0.9	1.9	3.1	4.5	4.9	7.6
India	0.4	0.7	0.7	1.3	1.4	1.5	1.5	3.3
South Korea	0.2	0.4	0.5	0.7	0.8	0.8	0.9	2.6
Other Asia	0.9	1.4	1.5	2.2	2.4	2.7	2.9	2.8
Middle East	**0.8**	**1.2**	**1.2**	**1.4**	**1.6**	**1.9**	**2.2**	**2.6**
Turkey	0.2	0.2	0.2	0.3	0.3	0.4	0.4	2.1
Other Middle East	0.7	1.0	1.0	1.1	1.3	1.5	1.8	2.7
Africa	**0.6**	**0.7**	**0.7**	**1.0**	**1.2**	**1.6**	**1.9**	**4.5**
Central and South America . . .	**1.3**	**1.7**	**1.8**	**2.4**	**3.0**	**3.8**	**4.7**	**4.3**
Brazil.	0.6	0.7	0.7	1.0	1.3	1.6	2.0	4.4
Other Central/South America . .	0.7	0.9	1.0	1.4	1.8	2.2	2.7	4.3
Total Developing	**4.7**	**6.9**	**7.3**	**10.8**	**13.6**	**16.7**	**19.0**	**4.2**
Total World	**22.3**	**25.5**	**26.4**	**33.4**	**38.1**	**42.7**	**46.4**	**2.5**

[a]Includes the 50 States and the District of Columbia. U.S. Territories are included in Australasia.
Notes: EE/FSU = Eastern Europe/Former Soviet Union. Totals may not equal sum of components due to independent rounding.
Sources: **History:** Derived from Energy Information Administration (EIA), *International Energy Annual 1997*, DOE/EIA-0219(97) (Washington, DC, April 1999). **Projections:** EIA, World Energy Projection System (2000).

Table E8. World Total Air Use Energy Consumption by Region, Reference Case, 1990-2020
(Million Barrels of Oil per Day)

Region/Country	History			Projections				Average Annual Percent Change, 1997-2020
	1990	1996	1997	2005	2010	2015	2020	
Industrialized Countries								
North America	**1.7**	**1.8**	**1.8**	**2.3**	**2.7**	**3.1**	**3.5**	**2.9**
United States[a]	1.6	1.7	1.7	2.1	2.5	2.8	3.2	2.8
Canada.	0.1	0.1	0.1	0.1	0.1	0.1	0.1	1.3
Mexico	0.0	0.0	0.1	0.1	0.1	0.2	0.2	6.3
Western Europe	**0.6**	**0.8**	**0.8**	**1.0**	**1.2**	**1.4**	**1.6**	**3.1**
United Kingdom	0.1	0.2	0.2	0.2	0.3	0.3	0.4	3.4
France	0.1	0.1	0.1	0.1	0.2	0.2	0.2	2.8
Germany	0.1	0.1	0.1	0.2	0.2	0.2	0.2	2.8
Italy.	0.0	0.1	0.1	0.1	0.1	0.1	0.1	2.9
Netherlands	0.0	0.1	0.1	0.1	0.1	0.1	0.1	3.2
Other Western Europe	0.2	0.3	0.3	0.4	0.4	0.5	0.6	3.2
Industrialized Asia.	**0.2**	**0.3**	**0.3**	**0.4**	**0.4**	**0.5**	**0.5**	**2.1**
Japan	0.1	0.2	0.2	0.2	0.2	0.3	0.3	1.5
Australasia	0.1	0.1	0.1	0.1	0.2	0.2	0.2	3.2
Total Industrialized	**2.5**	**2.9**	**2.9**	**3.7**	**4.3**	**5.0**	**5.6**	**2.9**
EE/FSU								
Former Soviet Union	0.4	0.2	0.2	0.2	0.3	0.4	0.6	4.3
Eastern Europe	0.0	0.0	0.0	0.1	0.1	0.1	0.1	5.7
Total EE/FSU	**0.5**	**0.3**	**0.3**	**0.3**	**0.4**	**0.5**	**0.7**	**4.5**
Developing Countries								
Developing Asia	**0.3**	**0.5**	**0.5**	**0.7**	**1.0**	**1.3**	**1.7**	**5.4**
China.	0.0	0.1	0.1	0.1	0.2	0.2	0.3	6.0
India	0.0	0.0	0.1	0.1	0.1	0.2	0.3	7.6
South Korea	0.0	0.0	0.1	0.1	0.1	0.2	0.2	6.0
Other Asia	0.2	0.3	0.3	0.4	0.6	0.8	1.0	4.7
Middle East.	**0.1**	**0.1**	**0.1**	**0.2**	**0.2**	**0.3**	**0.4**	**4.2**
Turkey	0.0	0.0	0.0	0.0	0.1	0.1	0.1	5.3
Other Middle East	0.1	0.1	0.1	0.1	0.2	0.2	0.3	3.8
Africa	**0.1**	**0.1**	**0.1**	**0.2**	**0.2**	**0.3**	**0.4**	**5.2**
Central and South America . . .	**0.1**	**0.1**	**0.2**	**0.2**	**0.3**	**0.4**	**0.5**	**5.2**
Brazil.	0.0	0.0	0.0	0.1	0.1	0.1	0.2	5.3
Other Central/South America . .	0.1	0.1	0.1	0.2	0.2	0.3	0.4	5.2
Total Developing	**0.6**	**0.9**	**0.9**	**1.3**	**1.8**	**2.3**	**3.0**	**5.2**
Total World	**3.7**	**4.0**	**4.1**	**5.3**	**6.5**	**7.8**	**9.3**	**3.6**

[a]Includes the 50 States and the District of Columbia. U.S. Territories are included in Australasia.
Notes: EE/FSU = Eastern Europe/Former Soviet Union. Totals may not equal sum of components due to independent rounding.
Sources: **History:** Derived from Energy Information Administration (EIA), *International Energy Annual 1997*, DOE/EIA-0219(97) (Washington, DC, April 1999). **Projections:** EIA, World Energy Projection System (2000).

Table E9. World Total Other Transportation Use Energy Consumption by Region, Reference Case, 1990-2020
(Million Barrels of Oil per Day)

Region/Country	History			Projections				Average Annual Percent Change, 1997-2020
	1990	1996	1997	2005	2010	2015	2020	
Industrialized Countries								
North America	**1.5**	**1.4**	**1.4**	**1.5**	**1.6**	**1.7**	**1.8**	**1.3**
United States[a]	1.3	1.2	1.1	1.2	1.3	1.5	1.6	1.4
Canada.	0.2	0.2	0.2	0.2	0.2	0.2	0.3	1.1
Mexico	0.0	0.0	0.0	0.0	0.0	0.0	0.0	0.8
Western Europe	**1.0**	**1.0**	**1.0**	**1.0**	**1.1**	**1.1**	**1.1**	**0.3**
United Kingdom	0.1	0.1	0.1	0.1	0.1	0.1	0.1	0.3
France	0.1	0.1	0.1	0.1	0.1	0.1	0.1	0.2
Germany	0.1	0.1	0.1	0.1	0.1	0.1	0.1	0.1
Italy.	0.1	0.1	0.1	0.1	0.1	0.1	0.1	0.2
Netherlands	0.2	0.2	0.2	0.3	0.3	0.3	0.3	0.3
Other Western Europe	0.4	0.4	0.4	0.4	0.5	0.5	0.5	0.3
Industrialized Asia	**0.3**	**0.3**	**0.3**	**0.3**	**0.3**	**0.3**	**0.3**	**0.4**
Japan	0.2	0.2	0.2	0.2	0.3	0.3	0.3	0.3
Australasia	0.0	0.1	0.1	0.1	0.1	0.1	0.1	0.7
Total Industrialized	**2.7**	**2.8**	**2.7**	**2.8**	**3.0**	**3.1**	**3.3**	**0.8**
EE/FSU								
Former Soviet Union	0.8	0.5	0.5	0.6	0.6	0.7	0.7	1.3
Eastern Europe	0.1	0.1	0.1	0.1	0.1	0.1	0.1	1.1
Total EE/FSU	**0.8**	**0.6**	**0.6**	**0.6**	**0.7**	**0.7**	**0.8**	**1.3**
Developing Countries								
Developing Asia	**0.9**	**1.0**	**1.1**	**1.2**	**1.2**	**1.3**	**1.4**	**1.2**
China.	0.3	0.4	0.4	0.4	0.4	0.5	0.5	1.2
India	0.1	0.1	0.1	0.1	0.1	0.1	0.1	0.9
South Korea	0.1	0.2	0.2	0.2	0.2	0.2	0.2	1.4
Other Asia	0.4	0.4	0.4	0.5	0.5	0.5	0.6	1.2
Middle East.	**0.2**	**0.3**	**0.3**	**0.3**	**0.3**	**0.4**	**0.4**	**1.0**
Turkey	0.0	0.0	0.0	0.0	0.0	0.0	0.0	0.8
Other Middle East	0.2	0.3	0.3	0.3	0.3	0.4	0.4	1.0
Africa	**0.1**	**0.2**	**0.2**	**0.2**	**0.2**	**0.2**	**0.3**	**1.2**
Central and South America . . .	**0.2**	**0.2**	**0.2**	**0.2**	**0.2**	**0.2**	**0.3**	**0.8**
Brazil.	0.1	0.1	0.1	0.1	0.1	0.1	0.1	0.9
Other Central/South America . .	0.1	0.1	0.1	0.1	0.2	0.2	0.2	0.8
Total Developing	**1.5**	**1.8**	**1.8**	**1.9**	**2.1**	**2.2**	**2.3**	**1.1**
Total World	**5.0**	**5.1**	**5.1**	**5.4**	**5.7**	**6.1**	**6.4**	**1.0**

[a]Includes the 50 States and the District of Columbia. U.S. Territories are included in Australasia.
Notes: EE/FSU = Eastern Europe/Former Soviet Union. Totals may not equal sum of components due to independent rounding.
Sources: **History:** Derived from Energy Information Administration (EIA), *International Energy Annual 1997*, DOE/EIA-0219(97) (Washington, DC, April 1999). **Projections:** EIA, World Energy Projection System (2000).

Table E10. World Per Capita Vehicle Ownership (Motorization) by Region, Reference Case, 1990-2020
(Vehicles per Thousand Population)

Region/Country	History			Projections				Average Annual Percent Change, 1997-2020
	1990	1996	1997	2005	2010	2015	2020	
Industrialized Countries								
North America	601	605	609	630	645	662	682	0.5
United States[a]	765	770	773	787	792	795	797	0.1
Canada.	596	578	589	646	665	678	686	0.7
Mexico	119	139	148	200	253	313	386	4.2
Western Europe	455	501	509	554	569	579	587	0.6
United Kingdom	457	491	501	552	569	580	587	0.7
France	502	532	543	598	617	629	636	0.7
Germany	485	534	543	592	609	619	626	0.6
Italy.	525	584	594	649	667	679	687	0.6
Netherlands	385	414	422	462	476	485	490	0.7
Other Western Europe	384	435	440	471	481	488	492	0.5
Industrialized Asia.	488	556	565	613	630	640	650	0.6
Japan	467	544	554	603	620	631	638	0.6
Australasia	617	624	631	666	678	686	691	0.4
Total Industrialized	520	556	562	598	615	629	643	0.6
EE/FSU								
Former Soviet Union	357	114	121	162	176	184	190	2.0
Eastern Europe	213	191	200	251	269	280	287	1.6
Total EE/FSU	314	136	144	188	203	212	218	1.8
Developing Countries								
Developing Asia	10	17	18	30	38	45	47	4.2
China.	5	9	10	22	36	51	54	7.5
India	5	8	9	16	19	20	21	4.0
South Korea	79	207	228	344	382	407	422	2.7
Other Asia	18	27	29	41	45	47	49	2.3
Middle East.	38	53	54	65	73	81	91	2.3
Turkey	42	72	77	104	113	119	122	2.1
Other Middle East	37	48	48	55	64	72	84	2.5
Africa	24	23	23	28	32	37	43	2.7
Central and South America . . .	68	83	86	109	133	160	194	3.6
Brazil.	81	99	101	131	163	200	247	4.0
Other Central/South America . .	58	71	76	95	113	133	160	3.3
Total Developing	21	28	30	42	51	61	68	3.7
Total World	124	117	119	130	137	142	146	0.9

[a]Includes the 50 States and the District of Columbia. U.S. Territories are included in Australasia.
Notes: EE/FSU = Eastern Europe/Former Soviet Union. Totals may not equal sum of components due to independent rounding.
Sources: **History:** American Automobile Manufacturers Association, *World Motor Vehicle Data* (Detroit, MI, 1997). **Projections:** Energy Information Administration, World Energy Projection System (2000).

World Energy Projection System

The projections of world energy consumption published annually by the Energy Information Administration (EIA) in the *International Energy Outlook (IEO)* are derived from the World Energy Projection System (WEPS). WEPS is an integrated set of personal-computer-based spreadsheets containing data compilations, assumption specifications, descriptive analysis procedures, and projection models. The WEPS accounting framework incorporates projections from independently documented models and assumptions about the future energy intensity of economic activity (ratios of total energy consumption divided by gross domestic product [GDP]) and about the rate of incremental energy requirements met by natural gas, coal, and renewable energy sources (hydroelectricity, geothermal, solar, wind, biomass, and other renewable sources).

WEPS provides projections of total world primary energy consumption, as well as projections of energy consumption by primary energy type (oil, natural gas, coal, nuclear, and hydroelectric and other renewable resources), and projections of net electricity consumption and energy use in the transportation sector. Projections of energy consumed by fuel type are also provided for electricity generation and for transportation. Carbon emissions resulting from fossil fuel use are derived from the energy consumption projections. All projections are computed in 5-year intervals through the year 2020. For both historical series and projection series, WEPS provides analytical computations of energy intensity and energy elasticity (the percentage change in energy consumption per percentage change in GDP).

WEPS projections are provided for regions and selected countries. Projections are made for 14 individual countries, 9 of which—United States, Canada, Mexico, Japan, United Kingdom, France, Germany, Italy, and Netherlands—are part of the designation "industrialized countries." Individual country projections are also made for China, India, South Korea, Turkey, and Brazil, all of which are considered "developing countries." Beyond these individual countries, the rest of the world is divided into regions. Industrialized regions include North America (Canada, Mexico, and the United States), Western Europe (United Kingdom, France, Germany, Italy, Netherlands, and Other Europe), and Pacific (Japan and Australasia, which consists of Australia, New Zealand, and the U.S. Territories). Developing regions include developing Asia (China, India, South Korea, and Other Asia), Middle East (Turkey and Other Middle East), Africa, and Central and South America (Brazil and Other Central and South America). The transitional economies, consisting of the countries in Eastern Europe (EE) and the former Soviet Union (FSU), are considered as a separate country grouping, neither industrialized nor developing. Within the EE/FSU, projections are made separately for nations designated as Annex I and non-Annex I in the Kyoto Climate Change Protocol.

The process of creating the projections begins with the calculation of a reference case total energy consumption projection for each country or region for each 5-year interval in the forecast period. The total energy consumption projection for each forecast year is the product of an assumed GDP growth rate, an assumed energy elasticity, and the total energy consumption for the prior forecast year. For the first year of the forecast, the prior year consumption is based on historical data. Subsequent calculations are based on the energy consumption projections for the preceding years.

Projections of world oil supply are provided to WEPS from EIA's International Energy Module, which is a submodule of the National Energy Modeling System (NEMS). Projections of world nuclear energy consumption are derived from nuclear power electricity generation projections from EIA's International Nuclear Model (INM), PC Version (PC-INM). All U.S. projections are taken from EIA's *Annual Energy Outlook (AEO)*.

A full description of WEPS is provided in a model documentation report: Energy Information Administration, *World Energy Projection System Model Documentation*, DOE/EIA-M050(97) (Washington, DC, September 1997). The report presents a description of each of the spreadsheets associated with WEPS, along with descriptions of the methodologies and assumptions used to produce the projections. The entire publication can be found through the Internet in portable document format (PDF) at: ftp://ftp.eia.doe.gov/pub/pdf/model.docs/m05097.pdf.

The WEPS model will be made available for downloading through the Internet on EIA's home page by May 2000. The package will allow users to replicate the projections that appear in *IEO2000*. It is coded in Excel, version 5.0, and can be executed on any IBM-compatible personal computer in a Windows environment. The package requires about 8 megabytes of hard disk space for complete installation and model execution.